The Evolutionary History of Witch-hunting

Steije Hofhuis

The Evolutionary History of Witch-hunting

A Qualitative Darwinian Approach

palgrave
macmillan

Steije Hofhuis
Berlin Social Science Center
Berlin, Germany

ISBN 978-3-031-66935-4   ISBN 978-3-031-66936-1 (eBook)
https://doi.org/10.1007/978-3-031-66936-1

This Palgrave Macmillan imprint is published by the registered company Springer Nature Switzerland AG.
The registered company address is: Gewerbestrasse 11, 6330 Cham, Switzerland

For Marietje van Winter

Foreword

Why did people in sixteenth and seventeenth-century Europe burn "witches" at the stake, sometimes their own friends and loved ones, even though these supernatural creatures—flying on broomsticks and cavorting with the devil in nocturnal sabbaths—were just figments of their imagination? In our thoroughly secularized age, where the power of religious belief has become a pale shadow of what it used to be, the European witch-hunts seem difficult to comprehend. Still, despite a common misconception, the largest "outbreaks" did not happen in the so-called Dark Ages, but at the cusp of modernity and Enlightenment, with the most deadly peaks happening around the time of Galileo's discoveries of the solar system and Francis Bacon's program for a new science.

The term "outbreak" is chosen deliberately and, as this book will make abundantly clear, accurate in more ways than we would imagine. Superficial analogies between witch trials and outbreaks of diseases—with one town "infecting" the next or with fears "spreading" through pamphlets—have been made for a long time, indeed even by contemporary critics of the witch craze. But for a long time historians assumed, perhaps from their own overly secular perspective, that the witch hunts must have "really" been about something else: oppressing women, subjugating the poor, appropriating other people's possessions, or just settling personal scores. And they had a point: the whole sordid affair just seemed too calculated and clever—large witches sabbaths, confessions under torture, flying on broomsticks over long distances—to be the result of mere chance.

But as Charles Darwin taught us, sometimes we can have designs without a designer. What if the system of witchcraft beliefs and practices in early modern Europe were designed by a "blind watchmaker," an evolutionary process winnowing and selecting cultural variants unbeknownst to the terrified believers?

It is easy to get carried away by such speculations about "brain viruses" and "parasites of the mind," and we should be careful not to pathologize other people's strange beliefs. If we want to demonstrate that some cultural design is the result of blind evolutionary processes, we can't make up some "just so" stories that strike us as plausible. We have to carefully study historical sources, trace the origins of different cultural variants, and show how they evolved and adapted to cultural environments.

This is exactly what the historian Steije Hofhuis accomplishes in this groundbreaking work on the history of the European witch-hunts. Critics of the notion of "selfish memes" or "mind viruses" (originally coined by Richard Dawkins) have long complained that memetic approaches are merely fanciful redescriptions of what we knew all along, that meme enthusiasts have yet to explain something that scholars of culture don't already understand.

That complaint should now be retired. With meticulous care, Hofhuis describes the problem that historians of witchcraft had been wrestling with, and why previous theoretical approaches failed to solve it. In a nutshell, the conundrum is this. On the one hand, the system of witchcraft beliefs and practices seems intricately designed, suggesting some intelligent designer. On the other hand, nobody has succeeded in identifying some plausible human designer(s), some interested party behind the scenes who stood to gain from the witch trials. Strange though it may seem to believe, the people involved in the witch trials *really* believed in witches.

But what if the witch-hunting beliefs and practices were designed not to serve some human interests, asks Hofhuis, but serve *their own* interests, namely their further propagation? No historian seems to have seriously considered the possibility that witchcraft beliefs and trials could evolve "interests" of their own, subverting human interests.

Hofhuis is not the first one to apply Darwinian thinking to culture, but he is the first to successfully demonstrate a historical case of selfish memes subverting the interests of their hosts. In comparison with other work in the field of cultural evolution, which tends to be highly theoretical and quantitative, Hofhuis also demonstrates that you can have a solid understanding of evolutionary theory *and* offer the sort of "thick" descriptions of human motivation and cultural context that are dear to historians.

As you will see, Hofhuis is a masterful storyteller, taking the reader on an engrossing journey with strange plot twists and tragic ironies. He has an excellent understanding of evolutionary theory, but he never allows his theoretical lens to blind him to personal intentions and feelings of the historical actors involved.

One of the most marvellous findings in Hofhuis's research—almost too good to be true—is that some of contemporary critics more or less understood that the witchcraft system was designed for maximal propagation but—not knowing anything about evolution—attributed the design to the most devious agent in their explanatory arsenal: the Devil. Yes, it was that "crafty old weaver" and "subtle and devious spirit" who had cleverly designed the witch hunts to be as virulent and ravaging to human communities as possible. Like the natural theologian William Paley in 1802 writing about the similarity between a camera eye and a watch, these critics *almost* understood what was going on, but erroneously personified the "watchmaker."

In many ways, this is a timely book. With the emergence of QAnon, 9/11 conspiracy thinking, anti-vaccination beliefs and other bizarre belief systems wreaking havoc in our societies, we have become more attuned to the possibility of viruses of the mind, beliefs that spread like diseases while ravaging our societies and also hurting their "hosts" (as in the case of anti-vaccination beliefs). The World Health Organization's recent coinage of the term "infodemic" is one of many examples of this growing awareness. Hopefully Hofhuis's book will inspire many others to explore other possible viruses of the mind, with equal theoretical rigour and attention to detail.

In only one respect, I'm sad to say, the moment of publication is untimely. Daniel Dennett, one of the most important evolutionary thinkers of our age who recently passed away, was an avid (and lonely) proponent of "selfish memes," as in his seminal books *Breaking the Spell* and *From Bacteria to Bach and Back*. It is a pity that he is no longer around for this book to see the light of day, as he was a generous supporter of Steije's work and I'm sure that he would've been thrilled to read it (he also would've loved to write this foreword in my stead). Dan "infected" us with a lot of good ideas, and we'll miss him.

But the "selfish meme" meme itself is anything but dead. With scholars like Steije Hofhuis, it promises to have a long life-span, fruitfully multiplying in many academic quarters.

Ghent University, Gent, Belgium Maarten Boudry

Acknowledgements

The book is based on a dissertation written at Utrecht University between 2016 and 2022. From the moment I developed this line of investigation as a history student at the University of Amsterdam, I encountered significant scepticism. After a long search, I finally obtained a PhD position, which resulted in the dissertation and subsequently in this book. None of that would have been possible if a range of people had not offered wonderful support in many ways—in some cases despite their scepticism about the project. I hope that mentioning and thanking them here will give a glimpse of my profound gratitude to them.

First of all, I would like to thank Jacques Bos and Wyger Velema for their willingness to supervise my rather unusual MA thesis. After graduation, several sociologists from the University of Amsterdam were interested in my approach and most willing to offer help to turn my ideas into a project. I particularly like to thank Nico Wilterink and Jeroen Bruggeman for their continued support and thought-provoking input. My years as a junior-lecturer at the Institute for Interdisciplinary Studies of the University of Amsterdam have further broadened my mind. I wish to thank my wonderful colleagues and the students whose lively discussions have inspired me greatly. I also had the pleasure of learning to know many people interested in the interdisciplinary significance of Darwinian theory, who offered help. I especially wish to express my thanks to Derk Venema, Chris Buskes, and John Neubauer. Outside of The Netherlands, I particularly think of Daniel Dennett, Alex Mesoudi, Johan Braeckman, Vera Hoorens, Jamie Tehrani, Daniel Lord Smail, Peter Richerson, and David Sloan Wilson. Learning to know Maarten Boudry has produced a sparkling intellectual exchange, and I feel honoured that he has been willing to write a foreword for this book.

Finding funding for this project through regular means remained challenging, but fortunately, Emeritus History Professor Marietje van Winter provided an outcome. Always curious about new scientific ideas, she encouraged me to apply for funding at the Professor van Winter Fonds, a private fund that she has created from a family legacy. The fund enabled the project, which is why the book is devoted to Marietje. I then wish to express my gratitude to my supervisor, Joris van Eijnatten, for his unrelenting support and innumerable wise words of advice, as well as the co-supervisors, Bert Theunissen and Ed Jonker. I also wish to thank the various scholars of witchcraft who have been willing to help at various stages of the project, including Julian Goodare, Alison Rowlands, and Robin Briggs. In addition, I think of the organizers of the many conferences of the *Arbeitskreis Interdisziplinäre Hexenforschung*, which helped me better understand German witchcraft scholarship. I wish to thank all my great colleagues in Utrecht, who made working on the project a pleasure, like Jaap de Haan, Lorella Viola, and not the least, Frans Willem Lantink, whose support has been enjoyably multifaceted.

Then there are the many friends who helped me in the process in multiple and wonderful ways, particularly I think of Max Joles, Marjolein Frijling, Sophie van Oers, Jean-Bernard Franken, Pablo M. Lamberti, Edzard van Engelsdorp Gastelaars, and Jeroen Joosse. In addition, I wish to thank the editor, Anthony Sami, and the anonymous reviewer of Palgrave, who helped me turn the dissertation into this book. Furthermore, I want to thank my family. Most particularly, my uncle and aunt Emile Hofhuis and Margriet Mulder, my cousins Irene and Marleen Hofhuis, my sister Maartje Hofhuis, and my brother Jesse Hofhuis. Last but not least, I wish to thank my parents for their unrelenting and multifaceted support. It has been greatly appreciated.

Contents

About the Author

Steije Hofhuis is a cultural historian and currently works as a postdoctoral researcher at the WZB Berlin Social Science Center. He graduated with a research master's degree in history from the University of Amsterdam (cum laude) in 2011. From 2013 to 2016, he was a junior lecturer at the Institute for Interdisciplinary Studies at the UvA. From 2016 to 2022, he wrote his dissertation *Qualitative Darwinism: An evolutionary history of witch-hunting* at the Cultural History group of Utrecht University. From 2021 to 2024, Hofhuis worked as a teacher in cultural history at UU. His specializations include interdisciplinary applications of Darwinian theory of evolution, social and cultural theory, and the history of European witch-hunting. Since July 2024 he has moved to the WZB Berlin Social Science Center, becoming a member of the research area Migration and Diversity. His new field of study concerns the uses of migration history in public debates about migration and integration today.

CHAPTER 1

Introduction

As a young boy, the renowned philosopher Daniel Dennett contemplated inventing a "universal acid."[1] With some of his schoolboy friends, he fantasized about a liquid so corrosive that it would eat through anything it encountered. Keeping the substance in a glass bottle or a stainless steel canister would be useless, for the acid would dissolve its container as readily as a paper bag. What would happen if a liquid like this were invented? Indeed, it would transform everything in its path, or so Dennett and his schoolboy friends thought. But would it eventually destroy everything? And what would it leave behind?

An acid like this has never been invented and remained the topic of Dennett's childhood fantasy. But the tale has become a classic story of science history, as Dennett believes to have come across a scientific idea resembling this universal acid. As he professed in 1996, this scientific idea would penetrate anything it encounters. Attempting to restrict it would lead to nothing: "It eats through just about every traditional concept, and leaves in its wake a revolutionized world-view, with most of the old landmarks still recognizable, but transformed in fundamental ways."[2] The idea that Dennett was speaking of is Darwin's theory of evolution by means of natural selection. The theory's critical insight is that a mindless process of

[1] Dennett, *Darwin's Dangerous Idea*, 63.
[2] Dennett, 63.

S. Hofhuis, *The Evolutionary History of Witch-hunting*,
https://doi.org/10.1007/978-3-031-66936-1_1

selecting and accumulating accidentally adaptive variants can produce adaptations to specific environments. While the process has neither intentions nor goals, its outcomes look like they were intelligently and purposefully designed; it concerns design without designer.[3] This core insight has been widely acknowledged in biology. But in his 1996 book Darwin's Dangerous Idea, Dennett proclaimed that Darwinism reached beyond living nature. To him, the theory was also about the cosmos, the human mind, language, culture, morality, and even the most profound questions about the meaning of existence. The far-reaching implications of the "universal acid" had not yet been thoroughly thought out yet.

Dennett deserves credit for his predictive abilities. Over the past decades, Darwinian approaches have experienced significant growth outside of biology.[4] Evolutionary perspectives on the biological foundations of human behaviour have garnered ever more attention in cognitive science and evolutionary psychology.[5] An approach named gene-culture coevolution has shed new light on many human traits.[6] Dennett also saw much potential for Darwinian views on cultural change: "The open question is not whether there will be a Darwinian theory of culture but what shape such a Darwinian theory will take."[7] Indeed, a new and interdisciplinary field of cultural evolution is emerging, which aims to analyse cultural change through a Darwinian lens, focusing on processes of variation, selection, and transmission, resulting in adaptations.[8] Darwinian tools and methods are entering disciplines like psychology, linguistics, sociology, anthropology, archaeology, and economics. Eventually, or so some proponents assume, the impact of Darwinian theory on the humanities and social sciences will be similar to the effect it once had on biology.[9]

At the same time, much research into cultural phenomena has hardly been affected by the "universal acid." Trained as a qualitative historian, I have seldom encountered colleagues who took much notice of the new

[3] Dennett, *Darwin's Dangerous Idea*; Ayala, "Darwin's Greatest Discovery."

[4] Buskes, *Evolutionair denken*; Wilson, *Evolution for Everyone*; Ridley, *The Evolution of Everything*; Stewart-Williams, *The Ape That Understood the Universe*; Watson, *Cultural Evolution and Its Discontents*; Sauer, *Moral*; Bruggeman, *A Sociology of Humankind*.

[5] Buss, *Evolutionary Psychology*.

[6] Henrich, *The Secret of Our Success*; Laland, *Darwin's Unfinished Symphony*.

[7] Dennett, "Foreword," ix.

[8] Blute, *Darwinian Sociocultural Evolution*; Mesoudi, *Cultural Evolution*, 2011; Wilson, *This View of Life*; Henrich, *The WEIRDest People in the World*.

[9] Hodgson and Knudsen, *Darwin's Conjecture*, x; Mesoudi, *Cultural Evolution*, 2011.

Darwinian developments, let alone apply them to their research. A few historians have called for more integration, but these have largely fallen on deaf ears.[10] But it is not just in qualitative history that the new Darwinians fail to have much impact. Their influence has remained similarly negligible in other disciplines oriented towards qualitative research. Scholars familiar with cultural symbols and specialized in interpreting their complex and multilayered meanings have traditionally been wary of science-oriented approaches.[11] The emphasis in these domains is typically on hermeneutics, that is, the interpretation of texts within their historical context. The "hard" natural sciences models are thought to be too rude to do justice to the intricacies and subtleties of human history and socio-cultural life. Moreover, existing reservations are strengthened by the fact that the field of cultural evolution has taken a clear quantitative direction, paying little attention to qualitative methodologies. The novel evolutionary approaches are thought either to misrepresent and simplify human history and culture or to merely rephrase what social scientists and scholars of the humanities knew all along.[12]

Has the "universal acid" thus reached an impenetrable barrier after all? Or is this verdict premature, and has it not yet been adequately assessed how Darwinian theory may change and revitalize these domains of academic scholarship? This question will take centre stage in this book, exploring the prospects of qualitative Darwinism. The critical question will be: Can Darwinian theory, particularly the idea of Darwinian cultural evolution, renew qualitative historical research? The potential integration of Darwinian theory and qualitative historical scholarship deserves consideration since it has hardly been examined. This is a loss since cultural evolution offers exciting new avenues for historical investigation. Two key ideas will be explored in detail in the chapters that follow.

The first is the idea that culture and society undergo processes of Darwinian cultural evolution, resulting in socio-cultural forms of "design

[10] Ferguson, *The Ascent of Money*; Smail, *On Deep History and the Brain*; Mokyr, *A Culture of Growth*.

[11] Collingwood, *The Idea of History*; Geertz, *The Interpretation of Cultures*; Kleinberg, "Just the Facts"; Howell and Prevenier, *From Reliable Sources*, 94.

[12] Kuper, "If Memes Are the Answer, What Is the Question?"; Ingold, "The Trouble with 'Evolutionary Biology'"; Goodheart, *Darwinian Misadventures in the Humanities*; Fracchia and Lewontin, "Does Culture Evolve?"; Wilterdink, "Te veel darwinisme, te weinig sociologie"; Kronfeldner, *Darwinian Creativity and Memetics*.

without designer."[13] Even in the cultural evolutionary field, it is widely accepted that human intentions play an essential role in how socio-cultural functions, or adaptations, come about. Nevertheless, human intentional design is thought to be an insufficient explanation for all cultural functions. In many cases, human actors do not recognize how and why specific cultural variants spread or do not intend them to do so. Similar to genetic mutations, some cultural variants may be accidentally adapted to their environment. As soon as such chance hits are then cumulatively preserved, they may gradually evolve into hidden cultural adaptations.

The second key idea concerns what cultural adaptations are built *for*. Who or what benefits from them? Scholars of socio-cultural disciplines are traditionally very interested in this question. Do socio-cultural phenomena serve those in power, the needs of individuals, or perhaps social systems? Drawing inspiration from selfish gene theory in evolutionary biology, authors like Daniel Dennett, Richard Dawkins, and Susan Blackmore have pointed to possible beneficiaries thus far overlooked: the socio-cultural phenomena themselves.[14] They argue that cultural items may be adapted to serve no ends other than their own "selfish" reproduction. Cultural phenomena like harmful viruses or other pathogens can evolve at the expense of their human hosts.

Qualitative historical investigation is indispensable in the examination of these hypotheses. If the latter are to be proven correct, the historical record should offer examples of cultural phenomena with adaptive features that evolved outside of people's intentions and recognition. To examine the issue empirically, we need to compare the adaptive or functional traits of cultural items to how actors themselves view their functionality. Studying what historical actors did or did not understand requires the tools qualitative hermeneutical historians are traditionally familiar with. We need to comprehend people's intentions, recreate their thoughts in our minds, and interpret the rich symbolic worlds they were living in.

The claim that cultural phenomena can spread "selfishly" while harming their hosts' interests also raises questions about whether the historical record provides convincing examples. The Darwinian proposal that cultural phenomena can resemble pathogens is highly contested even in

[13] Wilson, *Darwin's Cathedral*, 2002; Dennett, *Breaking the Spell*, Henrich, *The Secret of Our Success*.

[14] Blackmore, *The Meme Machine*; Dennett, *From Bacteria to Bach and Back*; Dennett, *Breaking the Spell*; Dawkins, *The Selfish Gene*, 189–201.

cultural evolution, primarily for theoretical reasons.[15] But why not simply examine it? The hypothesis contains many components that demand qualitative answers. How are we to understand and define "benefits" or "harm," and in what way should we relate that to the understanding of the actors involved?

To concretely address such theoretical and empirical issues, this book will delve into one case study: early modern European witch persecutions. These notorious trials mainly occurred between 1430 and 1750 and constituted a complex phenomenon rife with interpretative difficulties. Since the rich symbolic worldview underlying the early-modern hunts may appear alien today, understanding it requires applying the tools that qualitative history has to offer. Witchcraft scholarship traditionally combines empirical precision with theoretical sophistication. Yet, despite lavish academic attention, specific dimensions of these persecutions remain mysterious. One is that beliefs and practices regarding witchcraft, as they developed from the fifteenth to the early seventeenth centuries, were remarkably well adapted to creating ever-larger witch trials. We might think of the idea of the witches' sabbath, the witches' nightly flight, the idea of child witches, or the practice of using torture to uncover witches' hidden identities. Past historians thus often assumed that witchcraft concepts and trials were intelligently designed for ulterior purposes. They pointed to such things as financial gain, the strengthening of clerical power, or the oppression of women.[16] More recently, however, many experts have concluded that the people responsible for the persecutions were genuinely afraid of witches, while the trials occurred in a somewhat unplanned and haphazard manner.[17] Moreover, the effects of persecution often appear to have been dysfunctional. But if this is so, how did the witch-hunting phenomenon adapt remarkably well to its environment, and why did it recur so often over such a long time? In what follows, I will explore the Darwinian possibility that the witch-hunt system itself was a culturally evolved "design without designer" that spread in a virus-like manner to the detriment of its human hosts.

[15] Lewens, *Cultural Evolution*, 2015.

[16] Hansen, *Zauberwahn, Inquisition und Hexenproceß im Mittelalter*; Muchembled, *Sorcières*; Barstow, *Witchcraze*; Currie, "Crimes without Criminals"; Heinsohn and Steiger, *Die Vernichtung der weisen Frauen*.

[17] Dillinger, *Hexen und Magie*; Behringer, *Hexenverfolgung in Bayern*; Goodare, *The European Witch-Hunt*; Briggs, *Witches & Neighbours*.

By integrating Darwinian insights with qualitative historical scholarship, this book draws heavily on different disciplines, above all, history and biology. Interdisciplinary ventures such as this are known to come with risks.[18] Researchers from various fields often lack understanding of each other's concepts, use terms differently, or work within paradigms that can be mutually exclusive. Moreover, the boundaries between disciplines are often vehemently defended as soon as the respective practitioners feel they are under threat. Such responses certainly apply to efforts to take Darwinism beyond biology. But at the same time, interdisciplinary endeavours are worth the effort. "We are not students of some subject matter, but students of problems. And problems may cut right across the boundaries of any subject matter or discipline," Karl Popper maintained.[19] Disciplines typically have their methodological precision and investigatory depths, but blind spots are inevitable. A look beyond disciplinary boundaries may thus offer crucial inspiration. As many interdisciplinary scholars have stressed, we should not consider interdisciplinary research a one-way street. When disciplines take stock of each other's understandings and find common ground for greater integration, they all reap the benefits. This is precisely what this book aims at; qualitative history will gain new insights, as will cultural evolution.

This unusual, interdisciplinary objective has some consequences for the book. Historians have often been wary of theory; the illustrious Dutch historian Johan Huizinga once cautioned his colleagues not to use more than a teaspoon of it.[20] Here, however, theoretical considerations are crucial. A theoretical groundwork in the first three out of seven chapters is necessary to subsequently apply Darwinism to the witch-hunt case study comprehensibly. Many terms central to the analysis, like culture, adaptation, or function, can mean different things to different people. So, the definitions must be clear. Moreover, while this is primarily a historical book—written by a historian, for historians—it also has an interdisciplinary audience in mind, mainly scholars from the field of cultural evolution. Consequently, the text may sometimes dwell on things that will seem overly obvious to historians, like the history of the profession, the need for qualitative research, or the basics of the early modern period. On the other

[18] Repko, *Interdisciplinary Research*.
[19] Quoted in: Rutting et al., *An Introduction to Interdisciplinary Research*, 13.
[20] Otterspeer, *Orde en trouw*, 167.

hand, since many historians are mainly unaware of Darwinian insights, some necessary elaborations on that topic will appear superfluous to experts in evolution. Hopefully, the reading will not require too much perseverance on either side of the disciplinary divide.

Chapter 2, "Foundations," dissects some of the first difficulties any Darwinian approach to history will find on its path. First, I grapple with the long-standing division between history and "hard" science. What grounds do historians have to have been traditionally so dismissive towards approaches that draw inspiration from the natural sciences? While we will see that many past concerns may have been legitimate, current insights from the natural sciences, especially evolutionary biology, are no longer as unfitting for history. In the same chapter, I will address how Darwinian theory enters the human and socio-cultural disciplines. I shall review how scientists have studied our evolved human nature and processes of gene-culture coevolution. However, since this book's key topic is cultural evolution as a semi-autonomous process, I will address the grounds on which culture can be considered a semi-independent phenomenon that deserves to be studied in its own right.

Chapter 3 then delves into a concept of essential importance here: functions. The observation that cultural phenomena can be functional or adaptive is a precondition for the key hypotheses mentioned above. What are functions? How do they come into being? And what are they for? Scholars of cultural evolution have often been accused of ignoring established insights and operating as if they were the first to study the socio-cultural world.[21] This objection is partly justified, so this chapter thoroughly engages with the existing insights. How have social scientists and scholars of the humanities analysed the functionality of socio-cultural phenomena so far? Questions regarding functions, especially the relation between functions and human meaning-making, have been addressed most explicitly by sociologists and anthropologists. Their theories will thus receive the lion's share of attention. Established views on the origin and purposes of socio-cultural functions will be compared with how biologists have dealt with similar questions regarding functions in living nature.

These comparisons lay the groundwork for Chap. 4, which introduces the theory of Darwinian cultural evolution. It will be argued that Darwinian evolutionary biology can offer breakthroughs in long-standing,

[21] Ingold, "The Trouble with 'Evolutionary Biology'"; Benton, "Social Causes and Natural Relations."

unresolved issues. Many older approaches, like Durkheimian functionalism or Marxism, have already assumed that socio-cultural phenomena possess unintended and unrecognized functions. Cultural Darwinian theory can now explain how such functions arise. Moreover, the idea of cultural phenomena spreading "selfishly" opens up a new investigatory dimension. Notably, the proposed explanatory models distinguish between the unreal effects of cultural phenomena, which are perceived as real by involved actors, and their real effects, which scientists perceive as such. This distinction, however, is bound to raise eyebrows among socio-cultural scholars today with a relativist orientation. An intermediate section in Chap. 3 will address these objections.

Subsequent chapters explore the witch-hunt case study. Chapter 5 introduces witch-hunting itself. Chapter 6 addresses the innumerable approaches that have been taken to make sense of or explain witchcraft persecutions. The chapter culminates in the proposal of witch-hunting as an evolved "design without designer" that spread "contagiously." This scenario assumes that cultural variants that *accidentally* created larger witch-hunts cumulatively survived because of that enhancing effect. Chapter 7 examines how this purported selection process may have operated concretely, mainly by examining interactions between beliefs and trials in Germany from 1560 to 1630. This was the time and place where witch-hunting experienced its most significant surge. Lastly, Chap. 8 reflects on applying cultural evolution to witch-hunting. It begins with the most challenging alternative model, developed by various German historians. The latter postulates that witchcraft accusations and trials were instrumentalized for multifunctional ulterior purposes. Subsequent sections scrutinize a range of primary sources to see which of the interpretations of those sources provide the best clues. I then consider the comparison between witch-hunts and contagious disease, and the book ends with the question of why the persecutions eventually ceased.

Before we start, some words on a thorny issue are in order. That applications of Darwinism outside biology so often face scepticism is partly the consequence of important moral and political considerations. Often lumped together under the label "social Darwinism," earlier attempts at extending Darwinian insights to humans and their societies have been associated with horrific ideological derailments.[22] Darwin spoke of a

[22] Hawkins, *Social Darwinism in European and American Thought 1860–1945*; Browne, "Charles Darwin and ideology."

"struggle for existence" and a "war of nature," interpreting competition and death in a relatively positive light.[23] Certain scholars who oppose applying the theory to the socio-cultural domain argue that Darwin's ideas have inspired attempts to organize human societies along harsh ideological lines.[24] Darwinian theory has been linked to the eugenic movement, starvation of the poor, ruthless capitalism, colonialism, Nazism, and imaginations of racial superiority. Conversely, some proponents of broader applications of Darwinian theory play down harmful ideological consequences.[25] In Kevin Laland's and Gillian Brown's view, past ideologies have used "crude distortions of Darwin's theory, which derive more from the work of other 19th-century intellectuals such as Jean Lamarck and Herbert Spencer, although it is Darwin's name that is often unfairly linked to these views."[26] Indeed, there are good arguments for this line of defence.[27] Darwin had little to say about politics, but one political topic that greatly motivated him was the abolition of slavery.[28] The ideologies mentioned also drew inspiration from countless other sources. Darwinism has guided all sorts of persuasions, not all callous or evil. Importantly, key Nazis like Hitler or Himmler did not explicitly refer to Darwin; Himmler's worldview, for instance, was highly spiritualist.[29]

At the same time, however, this defensive response is a little too facile, as there are links between Darwin's work and crude political programmes. Darwin's personal friends, like his nephew Francis Galton and his colleague Ernst Haeckel, used elements of his theory to legitimize political proposals that aimed to remove the weak from society.[30] Leading German Darwinists, like again Haeckel, or Friedrich Ratzel, also popularized terms such as "Lebenswerth" (life worth) and "Lebensraum" (living space) that would later influence Nazi thought.[31] In his role as a Darwin enthusiast, Daniel Dennett is therefore rightly somewhat more self-critical: "It is important to recognize that Darwinism has always had an unfortunate

[23] Darwin, *The Origin of Species*, 114, 459.
[24] Weikart, *Hitler's Ethic*; Pichot, *La société pure*; Saini, *Superior*.
[25] Wilson, *This View of Life*; Richards, *Was Hitler a Darwinian?*
[26] Laland and Brown, *Sense and Nonsense*, 19.
[27] Theunissen, "Sociaal-darwinisme"; Allen Orr, "Darwin and Darwinism."
[28] Desmond and Moore, *Darwin's Sacred Cause*.
[29] Richards, *Was Hitler a Darwinian?*, 192–243; Kurlander, *Hitler's Monsters*.
[30] Hawkins, *Social Darwinism in European and American Thought 1860–1945*, 216–48.
[31] Haeckel, *Die Welträtsel*, 442–72; Ratzel, *Der Lebensraum*.

power to attract the most unwelcome enthusiasts – demagogues and psychopaths and misanthropes and other abusers of Darwin's dangerous idea."[32]

How, then, to deal with this tainted history? Let us first remind ourselves that we should not succumb to the moralistic fallacy, which holds that normative claims can justify factual claims—something is supposed to be a particular way because it morally ought to be that way.[33] Whether something is true or false does not depend on what we like to be confirmed on moral grounds. Still, academic scholars are responsible for treating delicate topics with care and being aware of the moral hazards that their scientific ideas might bring. At various points, I will thus reflect on the possibly harmful ideological ramifications of current Darwinian ideas. Reassuringly, the reader will find out that this book mainly elaborates on ideas that differ significantly from the old social Darwinism. The ideas even carry the promise of enhancing human welfare.

Let us now move to an argument that aims to expand our comprehension of socio-cultural processes. Hopefully, it will become apparent that the approach advanced here is neither about reducing cultural complexity to some lower-level explanation nor making qualitative history subservient to "hard" science. On the contrary, the aim is to give qualitative methodologies a new relevance and to make our understanding of humanity's meaningful and symbolic worlds more multifaceted and sophisticated. In his recent book with the appealing title *Darwin's Unfinished Symphony*, Kevin Laland stated that: "Far from destroying culture, our understanding of the underlying science feeds back to make the historical analysis richer and less mysterious."[34] I could not agree more.

[32] Dennett, *Darwin's Dangerous Idea*, 264.
[33] Stewart-Williams, *The Ape That Understood the Universe*, 164, 284.
[34] Laland, *Darwin's Unfinished Symphony*, 320.

CHAPTER 2

Foundations

2.1 History and the "Hard" Natural Sciences

In a speech at the University of Strasbourg on 1 May 1894, the German philosopher Wilhelm Windelband presented a distinction that he believed marked the essential difference between history and the natural sciences.[1] At the time, Windelband was not the only one seeking a philosophical foundation undergirding the separation of scientific domains. His colleague and compatriot Wilhelm Dilthey had already introduced a division between the so-called *Geisteswissenschaften*, the sciences of the spirit or the mind, and *Naturwissenschaften*, the sciences of the natural world. Windelband found this division unsatisfying, as some disciplines that studied the human mind, like psychology, used methods more akin to the natural sciences. His proposal was thus not to distinguish the kinds of objects studied but, rather, in the method of studying them. While the natural sciences searched for general laws, history focused on individual and unique events. Windelband described it as a distinction between nomothetic and idiographic inquiry, and his conceptual pairing would become a lasting success.[2]

[1] Windelband, *Geschichte Und Naturwissenschaft*; Jaeger and Rüsen, *Geschichte des Historismus*, 146–56.

[2] Kleinberg, "Just the Facts"; Vries, "Changing the Narrative," 322.

S. Hofhuis, *The Evolutionary History of Witch-hunting*,
https://doi.org/10.1007/978-3-031-66936-1_2

During the lecture, Windelband painted a cold picture of the natural sciences. It was a world of abstractions, timeless immutability, mathematical formulations, and deterministic necessity. Natural scientists had no interest in anything transient, and a single object was only relevant to them insofar as it served the formulation of a general law. On the other hand, historical scholars cared about "the loving expression of the particular," aiming to evoke rich images of humankind in all its peculiar forms and individual liveliness.[3] The German philosopher painted a warm picture of how historians studied people's past languages, faiths, values, poetry, thought, and struggles for power and freedom. The gap between the two domains of knowledge was deep and irreconcilable. His philosophical colleague Heinrich Rickert soon elaborated upon the distinction, maintaining that historical scholars neither could nor wanted to generalize along the lines of the natural sciences.[4]

The two philosophers were defending one of the most significant intellectual achievements of the nineteenth century, which often goes under the name of historicism.[5] This academic movement mostly took shape in Germany and brought historical scholarship to unprecedented professionalization and sophistication. The historicists were not the first to study the past, but what made them different was that they no longer understood historical people, institutions, or events against a somewhat ahistorical background. Instead, they tried to comprehend how all human creations undergo historical change. A desire to distinguish oneself from the natural sciences was not there from the beginning; some of the early exponents of the movement even drew inspiration from the organic, irregular development of plants or animals.[6] Nations, religions, languages, or law systems were thought to have no immutable essences but to be fundamentally malleable and always connected to a particular historical context. The best-known representative of the movement was the Berlin professor of history Leopold von Ranke.[7] In Ranke's view, history ranked amongst the sciences, as its first demand was the "pure love of truth," and its final aim

[3] Windelband, *Geschichte Und Naturwissenschaft*, 16.

[4] Rickert, *Kulturwissenschaft und Naturwissenschaft*, 54.

[5] Jaeger and Rüsen, *Geschichte des Historismus*; Collingwood, *The Idea of History*; Beiser, *The German Historicist Tradition*.

[6] Beiser, *The German Historicist Tradition*, 105, 136, 151, 250.

[7] Ranke, *The Theory and Practice of History*; Jaeger and Rüsen, *Geschichte des Historismus*, 81–86, 92–95; Iggers, *Historiography*, 1–3, 23–27.

was "to show how, essentially, things happened."[8] Yet, it was simultaneously a distinct form of science, which turned sympathetically to the particular. Historical epochs were to be understood within their own terms. Through the critical and meticulous examination of historical sources, preferably in archives, Ranke aimed to reconstruct individual human beings' concrete intentions and thoughts. History was to be told as a narrative, simultaneously making it a science and an art.

However, by the late nineteenth century, philosophers like Dilthey, Windelband, and Rickert felt a need to defend this tradition. In the nineteenth century they had also witnessed a continuing ascent of natural science and some scholars wanted to extend its successes into the domain of human society.[9] Take Karl Marx, who proclaimed to have uncovered history's ultimate laws of how material conditions and class struggle determined the inevitable collapse of capitalism and the establishment of socialism.[10] In France, Auguste Comte had created a positivist science of society, sociology, which aimed to find the general causal explanations for social processes. Emile Durkheim soon followed in Comte's footsteps, making derogatory comments about the work of their historical colleagues. Historians only focused on "series of accidents."[11] Some historians decided to jump on the bandwagon of more scientific history.[12] Around 1890, the situation came to a head when a history professor in Germany broke the historicist consensus in his country.[13] Karl Lamprecht aimed to unveil the social and material collective forces and "biogenetic fundamental laws" that underlay the allegedly superficial events that his colleagues were studying.[14] The response of the German historical establishment was vitriolic, triggering one of the most bitter conflicts in the profession's history. This so-called *Methodenstreit*, "the methods dispute," provided the context in which Windelband and Dilthey came to the historians' defence. Their dichotomy between disciplines was one of many attempts to restrict history's terrain. Another famous distinction assumed

[8] Ranke, *The Theory and Practice of History*, 39, xix, xx.

[9] Bentley, *Modern Historiography*, 43–52.

[10] Wiersing, *Geschichte des historischen Denkens*, 395–413.

[11] Durkheim, *Cours de science sociale. Leçon d'ouverture*, 6.; Burke, *History and Social Theory*, 6–9.

[12] Tollebeek, "Het Duitse debat," 27.

[13] Jaeger and Rüsen, *Geschichte des Historismus*, 141–46; Wiersing, *Geschichte des historischen Denkens*, 459–79.

[14] Jaeger and Rüsen, *Geschichte des Historismus*, 145.

that the natural sciences aimed at *erklären* (explaining), meaning the search for causal connections between facts. In contrast, historians focused on *verstehen* (understanding), referring to interpreting people's meaningful intentions within a historical context.

The outcome of the *Methodenstreit* was that the historical establishment in Germany largely kept their discipline clean from nomothetic approaches. However, social sciences, like economics, psychology, and sociology, established themselves at universities, sometimes combining a "hard" scientific orientation with an interest in history. Beyond Germany, history departments also offered room for new approaches.[15] The most influential movement of the twentieth century, the Annales school, took shape in France and aimed to move beyond Rankean history.[16] One of its leading figures, Fernand Braudel, somewhat disdainfully described his nineteenth-century colleague's work as "l'histoire evenementielle" (the history of events). In his view, it only focused on political oscillations and superficial excitements. Instead, Braudel explored the "abysmal depths" of history, like the immobile structures of geographical and climatic constraints, the slow rhythms of economic conjunctures, or the regularities of social and mental processes.[17] He aimed to uncover the interconnections of *all* human activities, searching for a constructive rapprochement to the social sciences. More twentieth-century schools aimed to model history more closely after the "hard" scientific model.[18] Marxist history gained widespread traction with its claim that economic conditions determined social, political, and cultural developments. By the 1960s, some nomothetically oriented historians even managed to breach the walls of historicist dominance in Germany, like Ulrich Wehler, who called for a "historical social science" that studied the transformation and modernization of social structures.[19] There was a new thirst for quantitative methods, and the new opportunities afforded by computers attracted lively interest.[20] Overall, the era was characterized by a widespread confidence that major problems of historical explanation would soon be resolved.

[15] Iggers, *Historiography*, 33–47; Bentley, *Modern Historiography*, 81–101.
[16] Burke, *The French Historical Revolution*; Braudel, *Ecrits sur l'histoire*.
[17] Braudel, *Ecrits sur l'histoire*, 22–4.
[18] Iggers, *Historiography*, 3–5, 65–96; Vries, "Changing the Narrative," 313.
[19] Jaeger and Rüsen, *Geschichte des Historismus*, 181–85.
[20] Iggers, *Historiography*, 44.

But the nomothetic and the idiographic relationship resembles the tides; just as the structural models reached an unprecedented peak in the 1970s, a counter-movement was already in the making.[21] In 1979, the historian Lawrence Stone captured a new mood when publishing a landmark article named "The Revival of Narrative." Stone had been working on quantitative social history for quite some time, but by now, he expressed a shared feeling that the grand theories had failed to deliver. Many historians were fed up with what Stone called "turgid and excruciatingly dull tomes full of tables of figures, abstruse algebraic equations, and percentages given to two decimal places."[22] What followed was "a return of the event," "a return of the actor," and a new focus on the past's discontinuities, contingency, diversity, and otherness. No longer the material conditions mattered, but how people experienced them. The model of scientific history was dismissed as naïve or even as a myth. An apparent problem of the generalizing theoretical models was that they were either too vague or false on many details.[23] As soon as general theories made information-rich predictions about specific processes, it always turned out that historical reality was more complex and that there were too many exceptions for a general rule to work.

The revival of narrative did not imply a return to Rankean political and diplomatic history. Most of all, cultural history won the day, with its interest in interpreting language, rituals, and symbolic meaning.[24] New inspiration was drawn from highly idiographic and qualitative forms of cultural anthropology, such as Clifford Geertz's "thick description" methodology.[25] In contrast to "thin description," which only describes human behaviour, "thick description" aims to interpret the complex layers of meaning of human behaviour within its socio-cultural contexts. Some historians rejected science even further and relativized the idea of historians studying history as it happened.[26] Take the American historian Hayden White, who went to lengths to compare the works of historians to the writings of novelists. He maintained that there are no theoretical grounds for judging one narrative as more realistic than the others.[27] This relativist move was

[21] Vries, "Changing the Narrative," 314–22; Iggers, *Historiography*, 10–16, 97–140.
[22] Stone, "The Revival of Narrative," 12.
[23] Berlin, "History and Theory."
[24] Stone, "The Revival of Narrative," 17; Burke, *What Is Cultural History?*, 32–51.
[25] Geertz, *The Interpretation of Cultures.*
[26] Iggers, *Historiography*, 10–14; Thompson, *Postmodernism and History.*
[27] White, *Metahistory*, xii.

too extreme for many other historians, who remain committed to the idea that their stories must be based on factual reality.[28] But the move illustrates how deep the schism between "history" and "science" had become.

But, as I said, the changing relationship between the nomothetic and the idiographic resembles the tides, and there are indications that the tide is turning again. The economic historian Jan de Vries observes how historians "are emerging from the comfortable refuge that their predecessors occupied nearly forty years ago." He sees them move away from narrative, micro, and subjectivist histories towards a "coherent, causal explanation of societal change."[29] *History Manifesto* by David Armitage and Jo Guldi calls for a renewed focus on big data and long-term processes, referring positively to Fernand Braudel.[30] On the other hand, the historian Ethan Kleinberg criticizes the new turn towards the sciences, warning his colleagues not to subordinate or sublimate "the practice of history to a fantasy of 'scientific method.'"[31] After all, historical inquiry has rendered general laws problematic, which is why historians should stick to their idiographic and interpretative methods. Otherwise, they risk erasing themselves in unconditional surrender.

The gap between the two forms of history may seem unbridgeable. Considering the depth and age of the divide, the universal acid of Darwinism is surely confronted with a formidable barrier. Yet, there might still be an opening. It does not lie in finally making one model dominant over the other but, instead, in a more critical look at this old distinction between the nomothetic and the idiographic. Because interestingly, through all the skirmishes, this division itself has mainly remained uncontested. Scholars disagree about the methodological course that history should take but not necessarily about the underlying landscape of possibilities. It almost goes without saying that on one side of the spectrum, we find the model of the "hard" natural sciences, with its universal laws, macro perspectives, structures, determinism, quantification, and abstraction. Conversely, we find the "soft" hermeneutic model of qualitative interpretation, meaning, events, intentionality, contingency, irregularity, and complexity. Within this paradigm, it also makes sense to put the social

[28] Evans, *In Defense of History*; Appleby, Hunt, and Jacob, *Telling the Truth about History*.
[29] Vries, "Changing the Narrative," 330, 32, 33.
[30] Guldi and Armitage, *The History Manifesto*.
[31] Kleinberg, "Just the Facts," 99.

sciences somewhere in between. The historical sociologist Max Weber once saw sociology as a mixed discipline that combined the methods of *verstehen* and *erklären*.[32]

But do these dichotomies from the nineteenth century still make sense in 2024? Should a further integration of history and the sciences inevitably depend on models that turn history into something more abstract, inflexible, and deterministic? Fortunately, things do not have to be like that, as the natural sciences have developed a much "softer" side. Historians have disregarded this side for far too long, so we will now explore what it looks like.

2.2 History and the "Soft" Natural Sciences

When Windelband and Rickert created their distinction in the late nineteenth century, they admitted that it was somewhat outdated.[33] Geologists like James Hutton and Charles Lyell had already shown long before that Earth's surface undergoes slow and continuing processes of upheaval, erosion, and sedimentation, or, in other words, change. Since 1859, Darwin's biological theory of descent by modification has also made science's view of living nature more malleable. Darwin assumed that all the branches of life on Earth took their own divergent and unique course in an ongoing struggle for survival. This caused Rickert to complain that biology's new hybridity "caused some confusion" to the dichotomy between the idiographic and the nomothetic.[34] Still, the two philosophers saved their distinction by admitting that some of the natural sciences used idiographic methods, too. For the rest, they focused on scientific theories that fitted more nicely into their model, such as the fixed laws of planetary orbits, embryological processes, or the mechanics of electrons. Windelband spoke of a "world of atoms, colourless and soundless, without all the qualities of sensory experience."[35]

But not long after Windelband made this statement, the eternal world of atoms also began to show some cracks.[36] In 1896, the French physicist Henri Becquerel accidentally discovered that uranium salts in his cabinet

[32] Weber, *Wirtschaft Und Gesellschaft*.

[33] Windelband, *Geschichte Und Naturwissenschaft*, 12–13; Rickert, *Kulturwissenschaft und Naturwissenschaft*, 7, 107–11.

[34] Rickert, *Kulturwissenschaft und Naturwissenschaft*, 7.

[35] Windelband, *Geschichte Und Naturwissenschaft*, 18.

[36] Bauer, *The Story of Science*, 136–39.

emitted a mysterious form of radiation. Shortly after, Marie and Pierre Curie saw the same energy coming from thorium and coined it radioactivity. Early twentieth-century science soon found out that this radiation was caused by the disintegration of atoms, implying that atoms were not as "colourless and soundless" as Windelband had assumed but were subject to change. The destabilization of the natural world was taken even further when Albert Einstein put time and space on more shaky grounds.[37] Newtonian physics had always assumed an immutable universe in which motion and time could be measured according to some fixed and unchanging point in a fixed and permanent space. Yet Einstein's theories of relativity showed that time and space are not absolute but relative to speed. In the 1920s and 1930s, the new science of quantum mechanics also found that subatomic particles do not obey the predictable laws of Newtonian mechanics.[38] These particles could be simultaneously at different places, acquiring a definite location only when interfered with by measuring instruments. Without interference, what was "out there" was undetermined and could only be talked about in terms of probabilities.

There was a prominent historian who noticed it all.[39] Marc Bloch observed that the natural sciences were moving in the direction of historians. In his book *The Historian's Craft,* Bloch stated that historians no longer had anything to be ashamed of when placing themselves amongst the sciences. Due to his early death as a French resistance fighter against the Nazis, Bloch would not experience how these developments would continue even further. Over the following decades, Big Bang cosmology won the day with its claims that the universe was not a fixed and ahistorical place but that it had a beginning, the Big Bang, and that all the atoms, molecules, stars, and galaxies had appeared at some point in time and have been subject to change since that point.[40] In the 1950s and 1960s, conclusive evidence appeared for Alfred Wegener's hypothesis from 1912, namely that the fixed and solid continents were continually moving.[41]

Again, there was an influential historian who noticed these developments. In his short book, *What is History*? (1961), Edward H. Carr proclaimed that the "historian has some excuse for feeling himself more at

[37] Bauer, 215–24.
[38] Bauer, 225–37.
[39] Bloch, *Apologie pour l'histoire ou Métier d'historien*, xv–xvii, 5–23.
[40] Bauer, *The Story of Science*, 238–51.
[41] Bauer, 141–48.

home in the world of science today."[42] The new developments made Carr realize that the old distinction between the generalizing natural sciences and the singularizing historical sciences was becoming increasingly obsolete. He remarked that many natural scientists, such as palaeontologists or geologists, could not test their hypotheses in laboratories but relied on traces of past processes that were often ambiguous and contradictory.[43] In the natural world, "no two geological formations, no two animals of the same species, and no two atoms are identical."[44] However, the uniqueness of every object did not prevent natural scientists from searching for more general patterns and principles.

After Carr made his remarks, the softening of the natural sciences would be taken even further. An essential development of the latter part of the twentieth century was the rise of so-called chaos theory and its importance to contingent events.[45] In the late 1960s, the American meteorologist Edward Norton Lorenz made computer simulations of weather conditions and found that minor perturbations could have enormous consequences. He eloquently summarized his argument by stating that flapping a butterfly's wings in Brazil could set off a tornado in Texas.[46] Chaos theory, of which Lorenz was a critical exponent, highlights how even tiny events could generate endless cascades of changing conditions, resulting in drastically different outcomes. Closely related is complex systems theory. It describes how systems in the natural world can consist of innumerable components that interact with each other in infinitely intricate ways, disallowing predictions in a precise Newtonian sense.[47]

Evolutionary biology is one domain in which this trend towards increased complexity, contingency, and irregularity has become apparent.[48] Darwin himself was already much of a complexity thinker who stressed the irregular and undirected aspects of evolution. He emphasized that species have no fixed essences but continually adapt to particular environments. For a long time, such assumptions were not mainstream

[42] Carr, *What Is History?*, 51.
[43] Carr, 41–44.
[44] Carr, 57.
[45] Gleick, *Chaos Making a New Science*; Bauer, *The Story of Science*, 252–56.
[46] Erdi, *Complexity Explained*, 16.
[47] Holland, *Complexity*.
[48] Zimmer, *Evolution*; Gould, *Wonderful Life*; Ward and Kirschvink, *A New History of Life*; Gould, *Full House*; Rose, *Lifelines*.

amongst his colleagues.[49] During much of the twentieth century, the ascent of complex life was often seen as a rather ladder-like phenomenon. It started with an "age of invertebrates," followed by an "age of fishes," then ages of amphibians, reptiles, and birds, and finally, leading to our own present "age of mammals." Even though they are not entirely incorrect, such notions have begun to feel increasingly uncomfortable to biologists. The trajectories of life on Earth turned out to be messier.[50] Newly discovered mechanisms of symbiosis and lateral gene transfer also demonstrated that various evolutionary branches had merged again, making the metaphor of a splitting "tree of life" problematic. Some biologists thus suggest we might speak of a "bush," "mangrove," or cluttered "net" of life.[51]

Evolutionary biology also learned to appreciate the significance of events. In the late twentieth century, much evidence appeared to suggest that the extinction of the dinosaurs was caused by an asteroid impact around sixty-six million years ago.[52] This chance event had dramatic consequences, creating opportunities for mammal species to develop into much more significant forms. In a thought experiment that he called "rewinding the tape of life," the palaeontologist Stephen Jay Gould also contemplated the impact of tiny events: "alter any event, ever so slightly and without apparent importance at the time, and evolution cascades into a radically different channel."[53] Colleagues like Richard Dawkins and Edward O. Wilson suggest that rewinding the tape of life would still produce remarkably similar patterns, and they refer to separate evolutionary processes on different continents that often led to eerily similar species. [54] The positions in this debate are nuanced and far removed from the straightforward determinism of a "nomothetic" model.

Again, some historical scholars took notice. The American world historian William McNeill speaks of a "grand convergence" of history and science, in which history is no longer the "hopelessly inexact laggard" it once was.[55] In France, Paul Ricoeur 1992 argued for a "return of the event" but

[49] Bowler, *Evolution*.

[50] Ward and Kirschvink, *A New History of Life*.

[51] Gould, *Full House*, 61–69; Zimmer, *Evolution*, 119–35; Doolittle, "Uprooting the Tree of Life"; Kunin et al., "The Net of Life."

[52] Dawkins, *The Ancestor's Tale*, 597–629.

[53] Gould, *Wonderful Life*, 51, 283.

[54] Wilson, *The Diversity of Life*, 120–28; Dawkins, *The Ancestor's Tale*, 230–37.

[55] McNeill, "History and the Scientific Worldview," 1.

stressed that it does not necessarily put history further away from the natural sciences. It may bring them closer together.[56] The Scottish historian Niall Ferguson draws inspiration from Stephen Jay Gould and chaos theory, arguing for what he calls "chaostory," which reconciles causation and contingency.[57] The historian John Lewis Gaddis maintains that the social sciences are no longer the bridge between "hard" and "soft" methodologies they used to be.[58] Many economists, for instance, still strive for abstract and parsimonious models in which rational agents interact in systems that move towards equilibrium. Natural scientists now often level criticisms against such models that historians have expressed for ages.[59]

Critics of an alleged "grand convergence" might retort that the fixed law of gravity, Newton's three laws of motion, or electromagnetic forces laws are still generally applicable. But even here, the situation might be more nuanced. It is still a speculative theory, but many cosmologists now assume that our universe might be a part of a multiverse in which various universes have specific physical laws and constants.[60] Step by step, many old distinctions between science and history have thus evaporated, making a further integration of the sciences more palpable. Illustratively, the historian David Christian and the anthropologist Fred Spier came to the beautiful idea of turning the history of physical nature, living nature, and human history into one continuing narrative. They call it *Big History*, which tells the story of everything from the Big Bang until now.[61]

One might expect that this "softening" of the natural sciences has pleased and fascinated innumerable historians eagerly contemplating the consequences for their work. But surprisingly, over the past decades, the stereotypes from the days of Windelband have proved remarkably resilient.[62] The philosopher of history, Frank Ankersmit, calls any notion of a unity of the sciences "a dumb dogma," as history will never live up to science's generalizing and exact standards.[63] The theorist of history, Chris

[56] Ricœur, "Le retour de l'Événement."

[57] Ferguson, "Virtual History," 79.

[58] Gaddis, *The Landscape of History*, 71.

[59] Beinhocker, *The Origin of Wealth*, 46–75; Wilson, *Consilience*, 218.

[60] Gribbin, *In Search of the Multiverse.*

[61] Christian, *Maps of Time*; Spier, *Big History and the Future of Humanity*; Christian, *Origin Story.*

[62] See, for instance: Martin, "The Essential Difference between History and Science"; Howell and Prevenier, *From Reliable Sources*; Dussen, *Filosofie van de geschiedenis.*

[63] Ankersmit, "Representatie Als Cognitief Instrument," 261.

Lorenz, states that the natural sciences typically deal with closed systems, making them fundamentally different from history, which always deals with open and undetermined systems.[64] Or, in the words of Hayden White:

> To be sure, history is the last of the disciplines of the human sciences to presume that society is radically other than the rest of the natural world, that it is a product of human work, labor, and creativity, and that the understanding of any of its processes must always be directed at the search for its origins, its relations to its time and space and socially specific contexts, and the employment of its transformations over time.[65]

Hayden White blurred the distinction between fact and fiction, which may help explain his somewhat fictional account of the natural sciences. But even crucial historians who defend history against relativism fail to mention how history and science converged.[66]

This raises the question of why the softening of the natural sciences has failed to penetrate mainstream historical consciousness. Perhaps the bogeyman of the "hard" sciences is too convenient an enemy to say goodbye to. Or maybe historians fear the shifting borders and destabilization that a new integration of insights might trigger. Whatever the causes, it is a loss. Past idiographic historians were correct that nomothetic models provide an overly tight straitjacket that will never fit history's irregular shapes. But now that the natural sciences have become softer, we might give "scientific history" a new chance. Not only on grand macroscopic levels, but also on small and more idiographic scales.

In this context, we will begin exploring Darwinian theory's methodological potential. Niall Ferguson already drew comparisons between the evolution of species and the evolution of financial markets.[67] Daniel Lord Smail keenly observes how Darwinism dovetails with historians' traditional scepticism of essentialist and ahistorical categories.[68] The economic historian Joël Mokyr maintains that Darwinian theory can help us find a middle ground between "the extremes of a materialist analysis that regards historical outcomes as inexorable and foreordained and a nihilist approach

[64] Lorenz, *De constructie van het verleden*, 240.

[65] White, "Afterword," 318.

[66] Appleby, Hunt, and Jacob, *Telling the Truth about History*; Evans, *In Defense of History*; Hunt, *History: Why It Matters*.

[67] Ferguson, *The Ascent of Money*, 370–408.

[68] Smail, *On Deep History and the Brain*, 124–25.

that sees nothing but randomness everywhere."[69] But how can the universal acid of Darwinism be used in concrete historical research? We will first look at its most well-known application: the idea of an evolved human nature.

2.3 Human Nature

In 1978, the normally relatively calm annual meeting of the *American Association for the Advancement of Science* was suddenly disrupted.[70] The seminar was devoted to a new scientific field, sociobiology, and the audience was preparing for a lecture by the famous sociobiologist Edward O. Wilson. This professor in entomology from Harvard University had shifted his attention from the social life of insects to that of humans and was about to present his research into the biological basis of human social behaviour. But when Wilson started his talk, a group of activists from the *Committee Against Racism* rushed forward, carrying placards with crossed-out swastikas, and they shouted, "Racist Wilson, you can't hide, we charge you with genocide!" Two of them poured a jug of water over his neck, yelling triumphantly: "Wilson, you are all wet!"[71] After much commotion, the activists left again, and the lecture was continued. But when Wilson finished his talk, he was still dripping water.

The incident marks the atmosphere of Darwinism's comeback in the human sciences. In the late nineteenth century, it had been quite common to understand human behaviour and societies in Darwinian biological terms. Still, over the twentieth century, such approaches were increasingly resisted.[72] The academic debate over "nature versus nurture" already came to a head in the nineteenth century, and after the horrors of the Nazi regime, the nature perspective on human behaviour and human differences became largely taboo. A broad consensus was that socio-cultural factors, not biological predispositions, primarily determine human conduct. However, by the late 1970s, this model was under pressure. Ethologists and evolutionary biologists have made much progress in studying animal behaviour. They now claimed that not only the physical traits of animals but also many of their specific behavioural patterns had a

[69] Mokyr, *A Culture of Growth*, 32.

[70] Wilson, *Naturalist*, 247–50; Pinker, *The Blank Slate*, 110–11.

[71] Wilson, *Naturalist*, 247–50.

[72] Pinker, *The Blank Slate*, 1–136; Fracchia and Lewontin, "Does Culture Evolve?," 54–55.

genetic evolutionary basis. An example was the mechanism of imprinting, whereby species such as geese or ducks become attached to the first large object they see after birth, usually their mother.[73]

In animal sociobiology, it was assumed that behaviour was shaped by natural selection and eventually served the reproduction of the organisms' genes. It resulted in adaptations like parental care, altruism between siblings, or specific mating patterns. Wilson's essential claim from the mid-1970s was that the same applies to human behaviour.[74] During our long evolutionary past, natural selection has endowed our species with hereditary behavioural mechanisms that enhance the reproduction of our genes. Examples include aggression, dominance, ethics, tribalism, altruism, or sex differences. In addition, Wilson maintained that such universal biological propensities continue to shape behaviour today. Wilson did not deny the impact of socio-cultural factors but believed their scope was limited: "The genes hold culture on a leash."[75]

Wilson's proposals triggered a new and heated round in the old "nature versus nurture" debate.[76] The Darwinian take on humans encountered fierce resistance from a broad alliance of academics, ranging from social scientists and scholars of the humanities to biologists.[77] Wilson was accused of biological imperialism, genetic determinism, and reductionism. His approach was allegedly built upon speculative reconstructions of prehistoric selection pressures and equally speculative claims about the links between genes and behaviour. His Harvard colleague Stephan Jay Gould described Wilson's accounts as "Just so stories"—explanations that may sound plausible but are not supported by rigorous evidence.[78] By claiming that aggression or sex differences had biological roots, Wilson was not just explaining them; he legitimized them. In their turn, Wilson and his supporters accused the critics of an ideological attack on the spirit of free inquiry.[79] The nurture camp purportedly clung to an unscientific and ideologically motivated belief in the all-powerful force of learning and culture. While resistance against sociobiology limited much of its academic

[73] Buss, *Evolutionary Psychology*, 10–17, 28–31.

[74] Wilson, *Sociobiology*; Segerstråle, *Defenders of the Truth*.

[75] Wilson, *On Human Nature*, 167.

[76] Segerstråle, *Defenders of the Truth*.

[77] Lewontin, Rose, and Kamin, *Not in Our Genes*; Schreier et al., "Against 'Sociobiology.'"

[78] Gould, "Sociobiology: The Art of Storytelling"; Lewontin, Rose, and Kamin, *Not in Our Genes*, 258–64.

[79] Wilson, "For Sociobiology"; Segerstråle, *Defenders of the Truth*.

outreach, the genie was out of the bottle. From the 1970s onwards, the study of the biological causes of human behaviour developed into an impressive interdisciplinary research programme.[80]

Sociobiology is heard less often because the research programme mostly lives on under "evolutionary psychology."[81] Its groundwork has been laid by John Tooby and Leda Cosmides, who focus not so much on what genes make humans do but, instead, on what our brains are ***designed*** to do. They write: "Evolutionary processes are the 'architect' that assembled, detail by detail, our evolved psychological and physiological architecture."[82] Since biological evolution is a slow process, they assume that our brains and bodies are not adapted to our current environments but to the environment where our species mostly evolved, the African savannah. We still house a "stone age mind," largely built for survival within our former "Environment of Evolutionary Adaptedness."[83] Like sociobiology, evolutionary psychology takes a keen interest in sex differences, such as mating preferences.[84] The evolutionary psychologist David Buss presented cross-cultural data that supposedly show that men have stronger preferences for quickly mating with young, healthy-looking partners. In contrast, women are more selective and eager to look for partners with resources. Buss relates this to the fact that men have nearly limitless sperm cells and can look for more mates when fertilization is complete. Women, on the other hand, have limited numbers of egg cells and need to make more of an investment—think of gestation and breastfeeding. This would explain why women, on average, are instinctively more careful when choosing a partner and look for optimal genes and prospects of continued support.

Sociobiology and evolutionary psychology proponents have expressed high expectations of putting the human sciences on this new Darwinian footing.[85] Tooby and Cosmides call for a "vertical integration" and a "unity of science" that will make the various disciplines of the social sciences and the humanities more consistent with the biological level and

[80] Schmitt and Pilcher, "Evaluating Evidence of Psychological Adaptation"; Buss, *Evolutionary Psychology*.

[81] Pinker, *The Blank Slate*; Buss, *Evolutionary Psychology*.

[82] Tooby and Cosmides, "The Psychological Foundations of Culture," 50.

[83] Cosmides and Tooby, "Evolutionary Psychology: A Primer"; Buss, *Evolutionary Psychology*, 88–90.

[84] Buss, *Evolutionary Psychology*, 105–204, 359–92.

[85] Pinker, *The Blank Slate*; Slingerland, *What Science Offers the Humanities*; Barkow, "Missing the Revolution"; Wilson, *Consilience*.

with each other.[86] A key hindrance is what they call the "Standard Social Science Model," which dominates many socio-cultural disciplines. It treats the mind as an empty vessel that can be filled with socio-cultural processes in almost any direction. Of course, there is much cultural variation, but Tooby and Cosmides maintain that human behavioural flexibility is only possible because humans possess a plethora of highly specialized mental modules built for the purpose.[87] Recognizing those modules, or so they promise, will revitalize the human and socio-cultural disciplines.

Fortunately, the fury of the late 1970s debates calmed down over subsequent decades as more scholars became open-minded about interactions between nature and nurture factors. Newly found mechanisms like neuroplasticity and epigenetics further indicate how biological and environmental explanations complement each other.[88] Still, resistance against the biological take on human behaviour remains paramount and seems to have gained new strength in recent years.[89] In many ways, the criticisms of evolutionary psychology reiterate earlier criticisms of sociobiology. This resistance can sometimes take on radical forms, for instance, when gender theorists maintain that categories like "men" and "women" are almost entirely socially constructed.[90] Evolutionary psychologists rightly complain about gross misrepresentations of their field and disregarding their vast and fruitful body of interdisciplinary research.[91] In general, it looks implausible indeed that natural selection would have shaped our bodies in the minutest details but left our psychological and behavioural makeup almost entirely unaffected.[92]

This is not to say that evolutionary psychology is without problems. This becomes apparent in Buss's claim about women being more eager to mate with financially secure partners.[93] Buss offers cross-cultural data on the widespread occurrence of this preference, like those collected through

[86] Cosmides, Tooby, and Barkow, "Introduction," 13; Tooby and Cosmides, "The Psychological Foundations of Culture," 14.

[87] Tooby and Cosmides, "The Psychological Foundations of Culture," 38–39, 93.

[88] Rose, "The Human Sciences in a Biological Age"; Ridley, *Nature via Nurture*.

[89] Prinz, *Beyond Human Nature*; Rippon, *Gender and Our Brains*.

[90] Butler, *Undoing Gender*.

[91] Flock, "What Good Is Evolutionary Psychology?"; Perry and Mace, "The Lack of Acceptance."

[92] Scheidel, "Evolutionary Psychology and the Historian," 1565.

[93] Smail, *On Deep History and the Brain*, 138–49; Wilterdink, "Darwinisme in de Sociale Wetenschappen," 448–49.

psychological questionnaires, but such findings do not yet prove a biological origin. A complementary, if not alternative, explanation might consist of a mixture of economic disparities, cultural learning, and stereotyping. Moreover, it is questionable whether ancestral women in the African savannah were as economically dependent on men as Buss's model suggests. Evolutionary psychology often faces such problems. Nature and nurture factors can lead to similar results, making it hard to determine what exactly causes what. The nature camp thereby finds itself in a disadvantaged position. Cultural causes can often be proven convincingly when a behavioural pattern changes within one or a few generations, while biological causes are more difficult to substantiate. Even when a behavioural pattern exists across all human societies, it still offers no conclusive proof for an innate module. The universality may result from an early introduction and common environmental pressures.[94]

To what extent can evolutionary psychology renew and revitalize historical research? Long before Darwin, the notion of a universal human nature attracted attention among historians. Take the eighteenth-century historian and philosopher David Hume: "Mankind is so much the same, in all times and places, that history informs us of nothing new or strange in this particular. Its chief use is only to discover the constant and universal principles of human nature."[95] Yet, one of the breakthroughs of historicism was that it problematized such notions.[96] The movement clarified how deeply particular historical contexts shape human thought and behaviour. What primarily came to interest historians is change and difference, which may help explain why evolutionary psychology failed to gain much traction.

This lack of attention is a loss, though. Evolutionary psychology tries to map biological universals, and historians possess invaluable information on the range of behavioural diversity.[97] Moreover, historical scholars often prioritize comprehending historical people by re-enacting past human thoughts. But such abilities result from the fact that humans always have

[94] Lewontin, Rose, and Kamin, *Not in Our Genes*, 255–58; Stewart-Williams, *The Ape That Understood the Universe*, 190.

[95] Brown, *Hierarchy, History and Human Nature*; Hume, *An Enquiry Concerning Human Understanding*, 78.

[96] Iggers, *Historiography*, 28–29; Collingwood, *The Idea of History*, 205–31.

[97] Barkow, "Missing the Revolution," 41; Harper, "The Sentimental Family," 1549; Brown, "Human Nature and History."

so many traits in common.[98] Without a broad intercultural range of emotional and cognitive repertoires, the whole method of *verstehen* would have been impossible in the first place.[99] In addition, the approach can produce new empirical questions. Martin Daly and Margo Wilson, for example, looked into historical data on violence and homicide and discovered that stepparents are significantly more likely to kill or harm their children than biological parents—precisely what a Darwinian perspective leads us to expect.[100] Donald Brown also reflected on how inborn traits helped to shape historical events, like tendencies towards in- and out-group biases that trigger violence between groups. He referred to the Balkans as a case in point.[101]

But in the latter example, a stark limitation of this evolutionary model comes to the surface. Let us take the Yugoslav wars of the 1990s as an illustration.[102] Donald Brown may be right that biological propensities for in-group and out-group biases were activated during these events and that we are unlikely to find historical societies where such biases were absent. Yet, when a historian wants to understand how and why the Yugoslav wars occurred at that particular time and place, an appeal to universal mental modules remains rather hollow. What exactly would they explain? It seems more fruitful to consider things like the death of President Tito, the worldwide decline of socialism, the resurgence of religion and nationalism in the region, some political decisions of the Serbian leader Slobodan Milošević, and the inability of external powers to intervene. In other words, this is just history, as we already know it. The example illustrates that a nomothetic human nature cannot fully explain idiographic and highly complex historical events.

But do evolutionary psychology's limitations also limit evolutionary theory's explanatory powers per se? When arguing for Darwinism's broad applicability, sociobiologists and evolutionary psychologists stress the importance of biological factors while downplaying the cultural ones.[103] When opponents of Darwinism, on the other hand, try to curtail the

[98] Tooby and Cosmides, "The Psychological Foundations of Culture," 92; Slingerland, *What Science Offers the Humanities*.

[99] Brown, "Human Nature and History"; Scheidel, "Evolutionary Psychology and the Historian."

[100] Daly and Wilson, *Homicide*.

[101] Brown, "Human Nature and History," 142–43.

[102] Berry, "The Laws of History," 164.

[103] See, for instance: Wilson, *On Human Nature*, 78; Barkow, *Darwin, Sex, and Status*, 142.

theory's scope, they stress culture's diversity, power, and prevalence. But such moves and counter-moves are increasingly out of date. Many evolutionary theorists recognize that culture must be taken very seriously. We shall now see how culture *itself* is becoming a subject of Darwinian attention.

2.4 Gene-Culture Coevolution

An evolutionary biologist who tries to look at *Homo sapiens* as objectively as possible will encounter mysterious features.[104] Some oddities already become apparent at the moment we are born.[105] For many animal species, giving birth is a relatively smooth operation. Chimpanzees, for instance, can do it independently and are finished within two hours on average. The risks are also relatively marginal. Human females, in contrast, need nine hours on average and are crucially dependent on help from others in most cases. The hazards are also significant, as the death of the mother and the child was a severe risk before the rise of modern medicine. After birth, new problems arise. Many animals can quickly function independently—reptiles can care for themselves from day one, and a giraffe cub already walks around within a few hours. Conversely, human babies cannot perform even the simplest tasks for months and will remain almost entirely dependent on help from others for many years.[106] Later in life, biological peculiarities continue to accumulate. People do not develop into physically strong beings like our close relatives, the chimpanzees and gorillas.[107] Human weaknesses also extend to our food processing capabilities because our mouths are small and our teeth feeble. These deficiencies are not compensated in the digestive tract, as our colons and stomachs are short. This is surprising because humans need much energy for the nutrition of an energy-absorbing, disproportionally large head. From a purely biological point of view, we thus seem unlikely candidates for evolutionary success.

But we did become an evolutionary success, and one without parallel. Humans have evolved into an ecologically dominant species that spread over all the continents and uses a hundred times more biomass than any

[104] Henrich, *The Secret of Our Success*, 1–7.
[105] Rosenberg and Trevathan, *Costly and Cute Helpless Infants and Human Evolution*.
[106] Henrich, *The Secret of Our Success*, 61–65.
[107] Henrich, 69–71; Pagel, *Wired for Culture*, 233.

species before us.[108] Any scientific model that attempts to explain this from our biotic traits alone will remain painfully incomplete. Over the past decades, our cultural skills have thus become a focal point of evolutionary research. Innovative perspectives are shedding new light on humanity's reproductive success and the aforementioned mysterious traits.[109]

What exactly is culture? The social sciences and the humanities have been thinking about this for a long time.[110] Some people interpret culture in a relatively narrow and normative way, like the nineteenth-century poet Matthew Arnold, who famously called culture "The best which has been thought and said in the world."[111] Anthropological approaches are broader, such as Edward Tylor's well-known designation from 1871, framing culture as "that complex whole which includes knowledge, belief, art, morals, law, custom, and any other capabilities and habits acquired by man as a member of society."[112] Notably, current evolutionary definitions often come close to Tylor.[113] Peter Richerson and Robert Boyd describe culture as: "Information capable of affecting individuals' behaviours that they acquire from other members of their species through teaching, imitation, and other forms of social transmission."[114] In this book, too, culture is seen as all information that humans acquire beyond their biological heredity.

One of the critical insights of current evolutionary research is that culture, in this broad, inclusive sense, is not uniquely human. It can also be found to varying degrees in animals.[115] Recent research has produced an avalanche of studies demonstrating that some behavioural patterns are likely products of social copying. Apes can imitate each other's techniques for using stones to crack nuts or sticks to catch insects, and some whale and bird species copy each other's songs. Yet, when considering such cases, it remains clear that human culture differs.[116] Animal traditions do

[108] Stewart-Williams, *The Ape That Understood the Universe*, 231–43; Henrich, *The Secret of Our Success*, 8.

[109] Richerson and Boyd, *Not By Genes Alone*; Henrich, *The Secret of Our Success*; Laland, *Darwin's Unfinished Symphony*.

[110] Kroeber and Kluckhohn, *Culture*.

[111] Arnold, *Culture and Anarchy*, 6.

[112] Tylor, *Primitive Culture 1. 1*, 1.

[113] Henrich, *The Secret of Our Success*, 3; Mesoudi, *Cultural Evolution*, 2011, 2,3.

[114] Richerson and Boyd, *Not By Genes Alone*, 5.

[115] Jablonka, Lamb, and Zeligowski, *Evolution in Four Dimensions*, 153–88; Laland, *Darwin's Unfinished Symphony*, 31–50.

[116] Pagel, *Wired for Culture*; Richerson and Boyd, *Not By Genes Alone*, 103–11.

not even come close to writing novels, holding religious festivals, or sending satellites to outer space. Human culture is more robust, cumulative, and complex, and evolutionary researchers try to figure out what makes the difference. Experts today thereby mainly focus on our capacity for composite symbolic communication.[117] Some animal species have signals for communication, such as primates who use alarm calls to warn each other of snakes. But animals cannot communicate about snakes when there are no snakes around, let alone integrate the concept of "snake" into complex stories. Through our proficiency in using open-ended language with syntax, humans transfer information about things that are elsewhere, that happened in the past, or that will occur in an envisioned future. We can create a parallel world of the imagination, significantly expanding our capacities for transmitting and accumulating cultural information.

Evolutionary theorists are working on scenarios for how it could have happened and often rely on a critical concept from evolutionary biology: coevolution. This term concerns processes in which species reciprocally adapt to each other's advances.[118] Lions, for instance, are well adapted to run after gazelles, but gazelles, in turn, are well adapted to stay out of the lion's claws. Coevolution can also lead to harmonious coevolutionary cooperation, like flower species that provide bees with nectar and bees that carry the pollen that fertilizes the flowers. As they became entangled in a coevolutionary dance, bees and flowers adapted to each other in shapes, smells, and colours. This opened up new evolutionary vistas for both of them.

Many current evolutionary theorists propose that the human species has entered a coevolutionary dance with a product of its own making: its culture. According to this scenario, it all started modestly with ancestors who used simple forms of cultural transmission comparable to animal culture today. However, incremental changes, such as the culturally acquired use of slightly more complex tools, gradually changed our hominid ancestors' environment. When some individuals became better at using those tools, it created competitive advantages and new selection pressures on the mastery of culture. This stimulated even more complex forms of culture,

[117] Laland, *Darwin's Unfinished Symphony*, 9–23, 175–207; Jablonka, Lamb, and Zeligowski, *Evolution in Four Dimensions*, 189–228.

[118] Durham, *Coevolution*; Richerson and Boyd, *Not By Genes Alone*, 191–236; Henrich, *The Secret of Our Success*; Lumsden and Wilson, *Genes, Mind, and Culture*; Ridley, *Evolution*, 613–42.

triggering new selection pressures and vice versa. Ancestral humans underwent an escalating and self-enforcing process of gene-culture interactions. The current archaeological record indicates that the first stone tools appeared around 2.6 million years ago, and from about 1.8 million years ago, the pace of change seems to have intensified.[119]

From this perspective, some of the mysterious biological traits we started with begin making more sense. One of the most important cultural innovations was the control of fire, which probably dates from around 1 to 1.5 million years ago. A critical effect of domesticated fire was that its heat could soften food, making it easier to digest. A significant part of the digestive work done by large teeth and long gastrointestinal tracts could thus be transferred to the cooking process. This created the evolutionary opportunity of letting our teeth and tracts become smaller.[120] The development of cultural assets such as spears, bows, and arrows made success in hunting and warfare less dependent on muscular strength and more on cultural skills, causing evolution to dispose us of much of our physical prowess.[121]

One organ proved to be an exception to this trend of energy cutbacks. The explosion of our culture was both the cause and the effect of an expansion of our heads.[122] The size of the human brain has nearly quadrupled over the last three million years as it became ever more critical for the acquisition and preservation of cultural information. Some scholars argue that its growth happened so quickly that the rest of our primate body hardly managed to keep up. The size of the brain pushed the boundaries of the birth canal, which may help explain why human childbirth is so difficult.[123] The coevolutionary perspective might also shed new light on our helplessness during the formative years. Animal species acquire most of their information through genetic transmission, which implies that the transfer of information is mainly finished when they are born. By contrast, humans obtain much of their essential information *after* birth. Most of the inner connections of the human brain still need to take shape, allowing us to get the cultural information necessary for survival. The unusual long-term dependency on adults might thus be a by-product of

[119] Henrich, *The Secret of Our Success*, 280–95.

[120] Henrich, 65–69; Wrangham, *Catching Fire: How Cooking Made Us Human.*

[121] Henrich, 57–71.

[122] Henrich, 61–65; Laland, *Darwin's Unfinished Symphony*, 26, 57, 235.

[123] The hypothesis is contested. See: Rosenberg and Trevathan, *Costly and Cute Helpless Infants and Human Evolution.*

our elaborate cultural learning skills.[124] Humankind came to inhabit a world of growing socio-cultural complexity, and we physically adapted to it. We now even depend on our culture for our survival.[125] The newly acquired capability of cultural transmission allowed humans to adapt to new ecosystems more quickly than would have been possible through genetic evolution alone. It sparked a spectacular spread over the globe.

However, this book is about the implications of Darwinian theory for history, and these coevolutionary interactions all happened in a distant prehistoric past. Biological adaptive change is often considered a slow process that requires much longer time spans than historical periods.[126] Yet, the effects of gene-culture coevolution may extend into the historical period. There are indications that the coevolutionary dance between nature and culture continued or accelerated in the last millennia.[127] An example is the ability of some humans to digest milk throughout adulthood. All humans can break down lactose milk sugars as infants since they need to drink their mother's breast milk. But amongst various populations, this trait extends into adulthood, and there is compelling evidence that it evolved as a response to the culturally acquired practice of cattle farming; the cultural innovation stimulated new selection pressures on the ability to digest animal milk.[128]

Some authors argue that gene-culture coevolution also affected cognitive and behavioural traits over the past millennia or centuries.[129] However, at this point, the approach suddenly takes a tricky and thorny twist.[130] Coevolution is alleged to have resulted in noteworthy genetic differences between human populations. The "Ashkenazi Intelligence Hypothesis" is a well-known and incendiary case study.[131] Its proponents point out that Ashkenazi Jews, on average, score exceptionally high on IQ tests and are significantly overrepresented amongst Nobel prize winners and chess

[124] Henrich, *The Secret of Our Success*, 61–65.

[125] Henrich, 22–33.

[126] Cochran and Harpending, *The 10,000 Year Explosion*, 1–24.

[127] Akey, "Constructing Genomic Maps of Positive Selection in Humans"; Laland, *Darwin's Unfinished Symphony*, 209–33.

[128] Durham, *Coevolution*, 226–85; Henrich, *The Secret of Our Success*, 88–92.

[129] Clark, *A Farewell to Alms*; Cochran and Harpending, *The 10,000 Year Explosion*; Wade, *A Troublesome Inheritance*.

[130] Henrich, *The Secret of Our Success*, 94–96.

[131] Cochran and Harpending, *The 10,000 Year Explosion*, 187–224; Wade, *A Troublesome Inheritance*, 197–213.

champions. As the argument goes, the Ashkenazi Jews evolved biological propensities for high intelligence in response to their millennium-long European sojourn in a niche of cognitively demanding jobs. Genetic research indicates that this group remained relatively isolated from around 900 to 1800, and more so than other ethnic groups, the Ashkenazi worked in professions like trade and money-lending. Allegedly, survival in this cultural niche strongly depended on innate propensities for high intelligence, making genes for brainpower spread throughout the population.

In his widely read book *A Farewell to Alms*, the historian Gregory Clark developed a similar scenario for the unusual advances of the English during the Industrial Revolution.[132] Clark looked at medieval and early modern datasets on the reproductive success of various groups within English society, maintaining that it demonstrates a "survival of the richest"; the wealthy had more surviving offspring than the poor.[133] In addition, Clark argues that being rich was often related to biological dispositions for thrift, patience, negotiation, prudence, and hard work. Hence, "middle-class culture spread throughout the society through biological mechanisms."[134] Supposedly, these processes were stronger in England than elsewhere, providing us with one of the explanations for why the Industrial Revolution happened in England.

In this form, the Darwinian theory begins to look more idiographic and more relevant for history. A problem, though, is that the evidence is mainly speculative.[135] What stands out are sentences about genetic differences that "may play a part," "may well have operated," but have "yet to be identified," and "if valid," "perhaps" or "maybe" explain particular differential outcomes.[136] The claims are also thoroughly contested by colleague experts.[137] The evidence for these theories is not by a long stretch strong enough for historians to rely on. An additional problem is that these authors endow particular ethnic groups with traits usually interpreted positively, like intelligence and hard work. But this implies that other groups are less endowed with such qualities. According to Cochran

[132] Clark, *A Farewell to Alms.*

[133] Clark, 113.

[134] Clark, 259.

[135] Pinker, *The Better Angels of Our Nature*, 619–21.

[136] Cochran and Harpending, *The 10,000 Year Explosion*, 112,117, 220; Wade, *A Troublesome Inheritance*, 9,12,64,181; Clark, *A Farewell to Alms*, 11.

[137] Bowles, "Genetically Capitalist?"; Hunley, "The 10,000 Year Explosion"; Allen Orr, "A Troublesome Inheritance"; Reich, *Who We Are and How We Got Here*, 247–74.

and Harpending, agricultural societies selected for hard work, while in hunter-gatherer societies, "being lazy made biological sense."[138] Purportedly, this biological heritage still bears upon their descendants today.[139] Clark states that foragers may be genetically "more violent," and Nicolas Wade postulates that sub-Saharan Africans are biologically adapted to their tribal socio-cultural environment.[140] It may even be one of the reasons why African countries suffer from socio-economic problems. Wade refers to "variations in their nature, such as time preference, work ethic and propensity to violence."[141]

What should we make of this? This is an illustration of the dilemma of the moralistic fallacy. Whether something is scientifically valid or false is independent of our prescriptions for moral decency.[142] It is also difficult to rule out entirely that human populations underwent some behavioural and cognitive changes since they diverged. More substantial evidence may appear shortly.[143] But scientists must still handle delicate topics with care. The social activists who once disturbed Edward O. Wilson's lecture in 1978 feared that a resurgence of Darwinism in the human sciences would stimulate racism, and the publications mentioned show that this concern is not necessarily unwarranted. Claims about biological differences between ethnic groups have an excessive potential for arousing racist stereotypes and political abuse. It makes the speculative claims of these coevolutionary enthusiasts worrisome and deplorable. Then, there is a further scientific problem. The authors repeatedly suggest that we *need* biological factors to explain puzzling phenomena like varying IQ scores, or the Industrial Revolution. However, the non-biological causes could be multiple. These gene-culture enthusiasts regularly pay lip service to socio-cultural factors but do not seem to grasp their full explanatory potential.[144] Hence, it is high time to discuss those socio-cultural factors and begin exploring the complex dynamics of the socio-cultural world.

138 Cochran and Harpending, *The 10,000 Year Explosion*, 115.
139 Cochran and Harpending, 85–128.
140 Clark, *A Farewell to Alms*, 121; Wade, *A Troublesome Inheritance*, 181–88.
141 Wade, *A Troublesome Inheritance*, 182.
142 Pinker, *The Better Angels of Our Nature*, 614.
143 Reich, *Who We Are and How We Got Here*, 247–73.
144 Wade, *A Troublesome Inheritance*, 66; Clark, *A Farewell to Alms*, 14.

2.5 Culture as a Semi-Autonomous Force

In the previous section, we encountered some traits of humans' physical constitution that appeared mysterious from a purely biological point of view. But things get even stranger when we look at some of our behavioural patterns. The ultimate goal of an organism's existence in the Darwinian world is reproducing its hereditary material through procreation. Admittedly, some animals, like sterile worker ants and worker bees, lost their breeding capacities. But this is compensated by the enhanced reproductive success of close genetic kin, like the queen ant or queen bee.[145] However, the historical record of humans teems with examples of societies that fostered traditions of celibacy where no direct genetic benefit seems involved.[146] Think of the vestal virgins of Ancient Rome, the monks of Buddhism, or the clergy of the Roman Catholic church. Some communities even banned the generation of offspring altogether. The American sect known as the Shakers required its members to remain sexually continent, marking off the purity of believers in an impure world.[147]

"Natural selection will never produce in a being anything injurious to itself," wrote Darwin, as organisms that cause pain or injury to themselves will be removed from the population.[148] Human history, however, offers countless examples of people tattooing, flagellating, scarring, bloodletting, or piercing themselves, often at the expense of their health.[149] Such practices are even encoded in carefully transmitted cultural traditions, like Christian hermits depriving themselves of food, Shia Muslims flagellating themselves to commemorate the lost battle of Karbala, or Buddhists who set themselves on fire to symbolize their willingness to abandon the body.[150] Particularly notorious are the traditional funerals of Australian Aboriginals. Participants wounded themselves with fiery sticks, knives, or tomahawks, often until their heads and bodies were burned or dripping with blood, with fatal injuries being a risk.[151] Death is sought even more eagerly in traditions of martyrdom, in which people are encouraged to kill themselves or let themselves be killed. Think of the Japanese

[145] Kay, Lehmann, and Keller, "Kin Selection and Altruism."
[146] Sobo and Bell, *Celibacy, Culture, and Society.*
[147] Sobo and Bell, 115.
[148] Darwin, *The Origin of Species*, 229.
[149] Stein and Stein, *The Anthropology of Religion, Magic, and Witchcraft.*, 92–94, 104–8.
[150] Wilson, "Starvation and Self-Mutilation in Religious Traditions."
[151] Stein and Stein, *The Anthropology of Religion, Magic, and Witchcraft*, 177–78.

suicide practice of "seppuku" through disembowelment, the willingness of soldiers to fight for their national armies, or Islamist terrorists blowing themselves up in airplanes, subways, or buses.[152] Underlying genetic explanations will not do in these cases, as the moral entities that these "martyrs" are doing it for, like their honour, nation, or religion, often far extend genetic relatives or do not overlap with genetic kin at all, as can be seen in cases of Islamist terrorists who converted to Islam only individually.

In other words, an evolutionary biologist confronted with *Homo sapiens* is almost like a physicist confronted with a tree full of upward-falling apples. The examples seem to fly in the face of evolutionary theory, so we must, and will, grapple with them here. Evolutionary psychology and gene-culture coevolution deserve credit for elucidating many human traits, but as these examples indicate, human behaviour can be overtaken by forces beyond the genetic framework. In the decades around 1900, when new disciplines like sociology and cultural anthropology were institutionalized, the insufficiency of psycho-biological models was also eagerly brought forward to mark off the new disciplines' distinctive territory.

Presenting himself as an example, the sociologist Emile Durkheim stated that as an individual, he could decide not to speak French anymore, cease using the *Franc*, decline to honour his contracts, and stop fulfilling his obligations as a husband or a citizen.[153] However, in all of these cases, his attempts would fail miserably. The present institutions would quickly compel him to obey, and an optional struggle would likely be lost. As Durkheim argued, from the moment they are born, every human being is confronted with "social facts" of their particular society that they did not themselves create, that they will subsequently acquire and transmit, and which will often persist long after they have passed away. Social facts have an external and coercive character, implying that traits of individuals, such as biological instincts or psychological processes, can never fully explain them. This observation was Durkheim's foundation of sociology as an autonomous science. Durkheim pointed out that social facts hardly overlap with racial categories and change more quickly than biological explanations can account for.[154] He, for instance, referred to the history of art styles and the rapid switch around 1800 from a rationalistic to a romantic

[152] Carrasco, "Sacrifice/Human Sacrifice in Religious Traditions"; Cook, "Martyrdom in Islam."

[153] Durkheim, *Les Règles de La Méthode Sociologique*, 3–14.

[154] Durkheim, 106–9.

mindset. Society had to be understood as a reality *sui generis*, in which social facts are to be explained from other social facts.[155] The Aboriginal funerals were amongst Durkheim's favourite examples, as their harmful practices seem incomprehensible from a biological point of view. It was society imposing the behaviour on its members.[156]

The German-American ethnologist Franz Boas played a similar role in cultural anthropology.[157] In his time, scientists still widely thought that cultural differences reflected biological differences, and some races were supposed to be more advanced than others. Boas made it his life's work to disconnect the two categories. He stressed how the diversity and changeability of cultural forms hardly match racial types.[158] In 1878, a German colleague had explained Japan's inertia as a result of biological dispositions, but as Boas commented in 1911: "How weak appear his conclusions, in view of the modern economic, political and scientific development of Japan which has adopted to the fullest extent all the best and worst traits of western civilization."[159] Further proof of culture's power was found in his newly adopted homeland, the USA. Migrant communities like the Irish, Italians, or Jews had markedly different cultural heritages, but the English that new generations of migrants spoke, the kind of gestures that they used, or the prevalence of crime rates they had changed so quickly that it made biological explanations evaporate.[160]

Boas would have a lasting impact on American cultural anthropology, giving rise to the Boasian school of anthropology.[161] His many pupils turned the diversity, changeability, and autonomy of culture into the trademark of their profession.[162] The concept of "culture" was preferred over that of Durkheim's "social facts," as it was argued that it is not so much our sociality that sets us apart—many animals also live socially—but the transmission of complex cultural forms.[163] Referring to the biological

[155] Durkheim, 9, 109.

[156] Durkheim, *Les formes élémentaires*, 556–92.

[157] Harris, *The Rise of Anthropological Theory*, 250–89; Moore, *Visions of Culture*, 30–41.

[158] Boas, *The Mind of Primitive Man*.

[159] Boas, 139.

[160] Boas, 30–75.

[161] Harris, *The Rise of Anthropological Theory*, 290–372; King, *The Reinvention of Humanity*.

[162] Kroeber, *The Nature of Culture*; Lowie, *Culture and Ethnology*; Benedict, *Patterns of Culture*.

[163] Kroeber, *The Nature of Culture*, 22–23.

principle of *Omnis cellula e cellula* (every cell is derived from some other cell), the Boasian Robert Lowie formulated a similar cultural principle of *Omnis cultura ex cultura* (all culture is derived from other culture). Since racial and general psychological explanations had failed to solve the problems of culture, Lowie argued that it was now up to culture to solve the problems of culture.[164]

In the accounts of sociobiologists and evolutionary psychologists, Durkheim and the Boasians are often staged as the bogeymen of the Standard Social Science Model.[165] Purportedly, they were the ones who disconnected human behaviour from its biological basis, which led us towards dualism, irrationalism, and relapses into unscientific thinking.[166] Such criticisms are warranted to some extent. Durkheim and the Boasians did take their rejection of biological explanations too far. Durkheim claimed that even sexual jealousy, familial piety, and parental love are not innate but products of collective organization.[167] Boasian anthropology led to excesses, like Boas's famous pupil Margaret Mead, who painted a rosy picture of a stress-free sex life in Pacific Samoa that seemed to defy all biological constraints.[168] Later, it was found that her ethnographic data—as with more anthropological works at the time—were meagre and overly gullible.[169]

However, we should be careful not to dismiss the legacy of Durkheim and the Boasians altogether. These founding fathers of the social sciences did make essential observations that still stand. Culture may not be fully autonomous, but it *is* semi-autonomous. It has much of a dynamics of its own that cannot be reduced to the biological traits of individual people. True, after the advent of postmodern relativism in the 1970s, social sciences, and humanities segments fell victim to forms of irrationalism and obscurantism, but Durkheim and the Boasians were not yet part of that problem. They still endorsed scientific principles and disallowed

[164] Lowie, *Culture and Ethnology*, 46, 66, 90–95.

[165] Tooby and Cosmides, "The Psychological Foundations of Culture"; Pinker, *The Blank Slate*, 22–26; Slingerland, *What Science Offers the Humanities*, 82–83.

[166] Brown, "Human Nature and History," 138; Wilson, *Consilience*, 9.

[167] Durkheim, *Les Règles de La Méthode Sociologique*, 106.

[168] Mead, *Coming of Age in Samoa*.

[169] Freeman, *Margaret Mead and Samoa*; King, *The Reinvention of Humanity*, 161–67.

socio-cultural theories that contravene physical, biological, or psychological facts.[170] It was just that they thought they had good *scientific* reasons for their non-reductionist outlook.

What Durkheim and the Boasians referred to is an essential scientific principle known as "emergence."[171] It is often summarized as "the whole is more than the sum of its parts" or "more is different." Its principle is not unique to the relationship between biology and culture but can be found throughout scientific domains. Take the example of a frog jumping away from an approaching snake.[172] Modern biology accepts that the jump happens in accordance with physical and chemical laws. But it also assumes that we must look beyond those laws to understand how and why the frog jumps. Biologists thus delve into higher-level phenomena, like the complex interactions of the various organs at the moment of the jump, the development of the individual frog, or interactions between ancestral frogs and snakes in their evolutionary history. In other words, biology cannot be reduced to physics or chemistry.[173] Neither can culture be reduced to biology.

Now that evolutionary theorists are beginning to study culture more closely, its emergent properties are increasingly appreciated. In his defence of evolutionary psychology, Steven Pinker levelled some harsh criticisms against Durkheim and the Boasians.[174] But in a later book, in which he responds to the racial theories of current gene-culture coevolutionists, Pinker's argument suddenly turns Boasian, highlighting the impact of socio-cultural factors and culture's rapid changeability.[175] In his book on cultural evolution, Alex Mesoudi stresses how deeply children are socialized within their societies and how quickly migrants can adopt the cultural traits of their newly adopted homelands. Hence, "any explanation of human behaviour that ignores culture, or treats it in an unsatisfactory manner, will almost certainly be incomplete."[176]

The previous sections noted that Darwinism in the form of evolutionary psychology or gene-culture coevolution remains tied to the biological

[170] Durkheim, *Les formes élémentaires*, 25; Kroeber, *The Nature of Culture*; Lowie, *Culture and Ethnology*, 25–26.

[171] Holland, *Complexity*, 1–5; Rees, "Explaining the Universe," 42.

[172] Rose, *Lifelines*, 10–14.

[173] Dawkins, *The Blind Watchmaker*, 10–11.

[174] Pinker, *The Blank Slate*.

[175] Pinker, *The Better Angels of Our Nature*, 618–22.

[176] Mesoudi, *Cultural Evolution*, 2011, i–24, 1.

level. But when Darwinism is applied at the level of culture as an emergent semi-autonomous force, new explanatory vistas may be opened up. As many current evolutionary experts argue, culture gradually came into being through the processes of gene-culture coevolution. But when it arose, and when cultural information came to be transmitted from person to person, new variants continually appeared, and only the variants best adapted to their environments survived. The three critical ingredients of Darwinism, variation, selection, and reproduction, set a new evolutionary process in motion.[177] Cultural phenomena grew into other cultural phenomena through lineages of descent and began to take adaptive directions of their own. When Daniel Dennett called Darwinism "a universal acid," he especially thought of this idea of a semi-autonomous cultural Darwinian process, not least because it might shed new light on mysterious cultural practices like celibacy or martyrdom.[178]

However, most scholars of the socio-cultural disciplines and many biologists remain sceptical of the theory's added value. "What constitutes an evolutionary process as opposed to a 'merely' historical one? What explanatory work is done by claiming that culture has evolved?", historian Joseph Fracchia and geneticist Richard Lewontin wonder in a critical review.[179] As sceptics argue, the field of Darwinian cultural evolution grossly distorts and simplifies the workings of cultural processes. It allegedly fails to understand the importance of human agency and culture's contextual embeddedness.[180] Fracchia and Lewontin write, "This formulaic treatment is fully inappropriate to the labyrinthine pathways, the contingent complexity, the many nuances, and general messiness of history."[181] Moreover, as far as critics accept the theory of Darwinian cultural evolution, they think it merely restates the obvious; it rephrases common insights, parasitizes existing knowledge, or pointlessly re-describes "ordinary cause-and-effect sequences using the verbiage of natural selection."[182]

[177] Dawkins, *The Selfish Gene*, 189–201.

[178] Dennett, *Darwin's Dangerous Idea*; Dennett, *Breaking the Spell.*

[179] Fracchia and Lewontin, "Does Culture Evolve?," 57.

[180] Kuper, "If Memes Are the Answer, What Is the Question?"; Bloch, "A Well-Disposed Social Anthropologist's Problems with Memes"; Bryant, "An Evolutionary Social Science?"; Lewens, *Cultural Evolution*, 2015.

[181] Fracchia and Lewontin, "Does Culture Evolve?," 77–78.

[182] Lewens, "Cultural Evolution," 2012, 468; Pinker, "The False Allure of Group Selection."

This book does support the theory's innovative power: Darwinian cultural evolution can integrate, renew, and enhance our understanding of human culture and history. But considering all the scepticism, we need to move forward prudently. What exactly is it about socio-cultural phenomena that the theory wants to explain? Did established approaches explain it insufficiently, and in what way does Darwinism provide a better alternative? We begin with the first question about what the theory may explain, in other words, its explanandum. This requires a little detour as we start with neither socio-cultural nor biological phenomena. Instead, we first briefly immerse ourselves in the world of physical nature and ask ourselves an essential question: Do physical phenomena have functions?

CHAPTER 3

Functions

3.1 The Riddles of Functionality

I have had the pleasure of working for a few years at an interdisciplinary studies institute, allowing me to discuss various scientific topics with students. There were ample subjects that stimulated lively debates, like the biological underpinnings of gender, the pitfalls of cognitive relativism, and the alleged merits of free-market capitalism. But interestingly, none of the topics stimulated discussions as lively as whether physical phenomena can be functional. What, for instance, I asked them in an interdisciplinary course on evolutionary theory, about the Sun? Does the Sun have a function, and if so, what is it? Or what about mountains? Are they in any way purposeful, and what would their purpose be? Or the tectonic movement of the Indian continent?

Whenever I asked these questions, the students were divided. To some, it seemed evident that such phenomena were purposeful. The Sun serves the functions of making planets orbit, burning protons in its inner core, making plants grow, and tanning skins on warm sunny days. We would not discuss it here if the Sun had not been doing these things. Mountains can serve the purposes of stopping winds, letting rivers flow down, or giving us the pleasures of winter sport, while the movement of the Indian plate creates the Himalayan mountain range. Other students were willing to admit that the Sun has the *effect* of making planets orbit or letting plants

S. Hofhuis, *The Evolutionary History of Witch-hunting*,
https://doi.org/10.1007/978-3-031-66936-1_3

grow, while mountains, indeed, stop winds. Yet, this is not what such physical phenomena are *for*. It is not their *purpose*. After all, the Sun already existed before it made plants grow and will continue to exist long after all plants on Earth have disappeared. Neither do continental plates move because of their effects on mountain ranges. These things do not fulfil functions—they are just *there*.

The students who did take a functionalist stance on the existence of physical things were tapping into a pervasive tendency of human thought, often labelled as "teleology." It assumes that things have goals.[1] Amongst cognitive scientists today, it is argued that teleological thinking may even be an inborn trait of the human mind.[2] Religions are traditionally saturated with notions of supernatural agents infusing nature with purposes, and teleological interpretations of physical nature have also been embraced in science for many centuries, not least by the great Greek philosopher Aristotle.[3] "Nature does nothing in vain," Aristotle wrote, as "everything by nature is for the sake of something."[4] The ancient philosopher wanted to know what things are made of, how they are organized, and what sets them in motion. But on top of that, he was eager to understand what things were eventually *for*.[5] In Aristotle's world, all natural phenomena possess an internal drive towards completion concerning their capacities, their "final cause." The celestial bodies, for instance, have an inner potential of orbiting in perfect circles around the Earth, while the motions of the Moon and the Sun also drive the seasons on our planet.[6] This is not just what these bodies do; it is also their function to do it.

However, it is the anti-teleological students who find modern science on their side. The rejection of teleological explanations in physical nature is often considered a vital breakthrough of the seventeenth-century scientific revolution.[7] The English philosopher Francis Bacon, one of this revolution's heralds, wanted to have none of Aristotle's final causes: "The inquisition of Final Causes is barren, and like a virgin consecrated to God

[1] Perlman, "The Modern Philosophical Resurrection of Teleology"; Leroi, *The Lagoon*, 84.

[2] Banerjee and Bloom, "Why Did This Happen to Me?"

[3] Leroi, *The Lagoon*; Leunissen, *Explanation and Teleology in Aristotle's Science of Nature*; Broadie, "The Ancient Greeks."

[4] Johnson, *Aristotle on Teleology*, 80.

[5] Johnson, 42–55.

[6] Leroi, *The Lagoon*, 305–41; Johnson, *Aristotle on Teleology*, 131–58.

[7] Depew and Weber, *Darwinism Evolving*, 88–89; Haig, *From Darwin to Derrida*, xxiii–xxviii, 1–5.

produces nothing."[8] In the new worldview that gradually took shape in the works of Descartes, Galilei, and Newton, it would no longer be accepted that physical bodies move because of an internal drive towards completion or a final goal. Instead, objects like planets or stones are inert and undirected. They only keep doing what they are already doing unless an external force acts upon them. The rejection of teleology should not be placed too abruptly in time, as some critiques preceded the scientific revolution while many goal-oriented notions continued to linger long after it.[9] But the break with Aristotle in the physical sciences was still profound and has made teleology one of physical science's ultimate sins: things must be explained from their causes and not from allegedly functional effects.

Yet, during the discussions with the students, I then moved on to living nature. What about the long neck of the giraffe? Does it have a function, and if so, what is it? Or what about the pumping of a human heart? While the question regarding physical functions triggered debates, this time, students moved towards consensus quickly. They widely accepted that these things have functions. The giraffe's long neck is for reaching high-hanging leaves, and the heart is for circulating blood. Unlike physical phenomena, these things not only have effects but are also predisposed and adapted to have those effects. Not just some isolated parts of living organisms are functional, as their whole bodies, from the eyes, veins, leaves, bones, roots, and legs to toes, are organized in complex ways to achieve certain ends. The persistence of organisms depends on their effects. Of course, Aristotle also interpreted living nature teleologically.[10] "Every instrument is for the sake of something," he wrote, and among his concrete examples were wings for flying, lungs for breathing, and flowers for attracting insects.[11] He furthermore observed that specific instruments fulfilled several functions simultaneously, and an organ that particularly impressed him was the elephant's trunk.[12] The philosopher discerned the functions of grabbing food, snorkelling, defence, trumpeting, and uprooting trees. Aristotle experts sometimes argue that living nature was the philosopher's greatest passion and crucially shaped his teleological outlook.[13]

[8] Bacon, *The Works of Francis Bacon*, 4: Translations of the Philosophical Works 1:365.
[9] Johnson, *Aristotle on Teleology*, 25–30; Broadie, "The Ancient Greeks," 23.
[10] Leroi, *The Lagoon*; Johnson, *Aristotle on Teleology*, 159–210.
[11] Johnson, *Aristotle on Teleology*, 172.
[12] Leroi, *The Lagoon*, 137–40.
[13] Broadie, "The Ancient Greeks," 30–31.

While Aristotle's physics has been superseded in almost every respect, significant parts of his biology still stand. The life sciences today are still infused with functional analyses and purposeful language. When reading current scientific literature, one is likely to find sentences about somatic cell divisions being "used to make mortal tissues" or photosynthetic membranes having "an ingenious structure to fulfil all its functions."[14] Such purposeful language has often been contested in biology.[15] Francis Bacon's attack saw final causes in living nature as "remoras and hindrances to stay and slug the ship from further sailing."[16] Criticism continues today.[17] Nevertheless, goal-oriented reasoning has remained pervasive, and as an early twentieth-century biologist said: "Teleology is a lady without whom no biologist can live. Yet he is ashamed to show himself with her in public."[18] Today's biological teleology might differ from Aristotle's, but embracing the concept as such is still widespread and has even experienced a revival.[19]

It is widely recognized that living nature is different.[20] In contrast to physical phenomena, living organisms are full of functional complexity, making teleological reasoning indispensable. When reflecting on the difference, Daniel Dennett argues that the physical sciences can be satisfied with "how come" questions, asking for the process narrative of how things causally came about. In contrast, biology needs "what for" questions on top of that.[21] Many philosophers dealing with issues regarding functions have also developed a workable model for what a function is: when a phenomenon's existence has depended, or depends, on the effects that it has, it may count as functional.[22] Francis Bacon and kindred spirits thus failed to drive out the demon of Aristotle's final causes.

After finishing the discussion with students about biological functions, I still had one domain to go: the world of the socio-cultural. What about a hammer? Does it have a function, and if so, what function? This time,

[14] Hanke, "Teleology," 143–44.

[15] Leroi, *The Lagoon*, 345–78.

[16] Bacon, *The Advancement of Learning and the New Atlantis*, 105.

[17] Hanke, "Teleology," 143.

[18] Haig, *From Darwin to Derrida*, 359.

[19] Haig, 5–16; Perlman, "The Modern Philosophical Resurrection of Teleology."

[20] Dawkins, *The Blind Watchmaker*, 1–10; Godfrey-Smith, *Philosophy of Biology*, 50–65.

[21] Dennett, *From Bacteria to Bach and Back*, 33–54.

[22] Godfrey-Smith, *Philosophy of Biology*, 62; Perlman, "The Modern Philosophical Resurrection of Teleology."

consensus followed, and the question sometimes even stimulated some amusement. Of course, a hammer has a function: banging nails! Other artefacts like clocks, combs, or teapots triggered similarly prompt responses. More discussion ensued when complex socio-cultural institutions, like the police force or money, were discussed. Is the function of the police to uphold the law, as most students maintained, or is it a vehicle for oppression, as a few others suggested? The proposed functions of money diverged from exchanging goods and getting rich to preserving resources. But whether such phenomena fulfil any functions was not even up for debate. Socio-cultural phenomena are well adapted *in order to* have specific effects, and their persistence depends on their effects. In other words, they perfectly fit the function ascription mentioned above.

Unsurprisingly, Aristotle also interpreted socio-cultural phenomena like ships, statues, or houses from a teleological angle. He, for instance, wrote about coats: "There is an excellence that belongs to a coat, for a coat has a particular function and use."[23] Aristotle also reflected on the functions of more complex socio-cultural phenomena, like the institution of the state, which, in his view, has not only an origin and a development but also a purpose, an optimal size, and mechanisms for self-maintenance.[24] The idea that socio-cultural phenomena can be functional has remained paramount in the socio-cultural disciplines. One of sociology and anthropology's most influential theoretical movements even labelled itself "functionalism."[25] Emile Durkheim, who laid much of the groundwork for this approach, specified that the "social facts" that he analysed not only have effects but that these effects also ensure the persistence of those facts: "Without a doubt, the effect cannot exist without its cause, but this, in its turn, needs its effect."[26] Similarly, history books are full of military battles being fought to win a war, taxes being raised to increase the state's revenues, and palaces being built to glorify a ruler. Certain forms of purposeful reasoning have attracted criticisms, and functionalism has become contested as a school of thought. Teleology suffers from a bad reputation, as it is often associated with the idea that history is moving towards a predetermined goal, like Judgement Day, or the communist worker's paradise. However, no school of thought would reject the existence of functions in the socio-cultural realm entirely.

[23] Johnson, *Aristotle on Teleology*, 318.
[24] Leroi, *The Lagoon*, 314–15.
[25] Kincaid, *Philosophical Foundations of the Social Sciences*, 101–14.
[26] Durkheim, *Les Règles de La Méthode Sociologique*, 95.

The world of living nature and the world of the socio-cultural have something essential in common. In contrast to physical nature, things in both domains are purposefully organized. This observation was not lost on Auguste Comte, the founder of sociology as a discipline. He distinguished the "sciences of the raw bodies," which included the disciplines of physical nature, from the "sciences of the organized bodies," which concerned biology and sociology.[27] The insight now reappears in Daniel Dennett's work as he points out that the additional "what for" questions bring the biological and socio-cultural sciences together while setting them apart from the physical sciences.[28] This observation implies that both domains of science face similar questions and may benefit from each other's answers. The first question both domains need to address is how functions come about. After all, functions cannot be taken for granted, as the physical realm does not provide them. Aristotle already stressed that functional complexity cannot result from luck or chance.[29] Section 3.2 will discuss in greater detail how biologists have struggled with the origin of functions and in what way Darwinian theory offered a breakthrough. Section 3.3 subsequently explores how the socio-cultural disciplines struggled with the origin of functions and what some of their answers have looked like.

The existence of functions also raises a second question. What function do functions fulfil? Take the giraffe's long neck. As we saw, the purpose of the neck is to reach high-hanging leaves. But what, then, would the function of getting those leaves be? The answer is obvious: eating the leaves. Yet, this is only the beginning of a chain of questions that becomes progressively complex, as the next question would be what function eating has. Here, the answer remains feasible: ensuring the survival of the individual giraffe. But what is the function of the giraffe's survival? An answer might be that it serves reproduction, but even this does not resolve the explanatory chain because what purpose would reproduction serve? The problem had also dawned on Aristotle, and to him, the end of all functional explanations was that "being is better than not being, and living is better than not living. For this reason, there is the reproduction of animals."[30] But even this should not satisfy us because why would being

[27] Heerikhuizen, "Sociaal evolutionisme in de vormende jaren van de sociologie," 85–86.
[28] Wilson, *This View of Life*, 38–40; Dennett, *From Bacteria to Bach and Back*, 33–54.
[29] Johnson, *Aristotle on Teleology*, 100.
[30] Johnson, 175.

be better than not being, and what precisely is being reproduced? In Sect. 3.4, we will see how this has been, and is, a significant point of contention in evolutionary biology.

Questions regarding the purpose of purposes have also bedevilled scholars of the socio-cultural disciplines. Hammers fulfil the function of banging nails, but what would the function of banging nails be? The answer seems obvious: turning elements into a more extensive artefact. But here, things also get tricky when we ask ourselves what the function of artefacts is. A possible answer might be ensuring our biological survival. Yet, many artefacts seem superfluous from that point of view, so we are tempted to look for additional explanations. Is the function of artefacts to provide us with pleasure? To enhance individual people's status? Or to stimulate consumerism? If so, what would the functions of pleasure, status, or consumerism be? In Sect. 3.5, we will discuss some of the answers the socio-cultural disciplines have formulated regarding the purpose of socio-cultural purposes. Eventually, this will clear the ground for the comparison of the theory of Darwinian cultural evolution and its non-Darwinian counterparts.

3.2 The Origins of Biological Functions

The German thinker Immanuel Kant has a reputation as one of the greatest philosophers who ever lived—a well-deserved reputation, as he possessed the remarkable ability to synthesize and renew some of Western philosophy's most troublesome issues. Yet, a problem that he failed to resolve satisfactorily was the question of how biological functions come about.[31] The topic attracted Kant's attention because he understood the difference between organic and inorganic phenomena discussed above. Kant pointed out that rivers connect people in countries' interiors and that mountains preserve snow during summer. However, the existence of rivers and mountains is not causally or purposefully linked to such effects. Kant realized that living nature is different. Trees produce leaves, but this is not only an effect of the tree, as it is simultaneously the cause of its existence. Trees without leaves do not survive, which is also what trees produce leaves *for*. The purposeful organization of living organisms reminded Kant of things created by humans. The construction of a house, for instance, has the effect of the new residents paying rent, and this anticipated effect

[31] Kant, *Kritik Der Urteilskraft*, 305–458; Depew and Weber, *Darwinism Evolving*, 102–4.

is also the cause of the house being built.[32] It brought Kant to the conclusion that living organisms must be understood *as if* they had been purposefully created. It was "an indispensable need" to understand living things in teleological terms.[33]

Kant was not the first person to recognize that living organisms show a semblance of intentional design. The "argument from design" traditionally served as one of the essential proofs of the existence of a divine creator, and the ascent of modern science initially strengthened rather than weakened it. Well into the nineteenth century, many naturalists continued to work within the tradition of "natural theology," which both examined and celebrated nature's functions as proof of the benevolence and intelligence of our creator.[34] The English clergyman William Paley typically pointed out that the complicated design of a watch shows the existence of a designer—the watchmaker—as "there cannot be design without a designer."[35] Similarly, the marvellous designs of organisms proved the existence of a designer, God. One of Paley's followers, John Shute Duncan, referred to the giraffe's "lofty neck [which] enables him to feed on branches eighteen or twenty feet above the ground."[36]

Immanuel Kant, however, rejected this explanation.[37] He was not the first to do so, as Aristotle had thought that nature's functionality resulted from principles intrinsic to nature, and later thinkers like Hume and Voltaire remarked that an appeal to an intentional designer only shifted the explanatory problem. How did the designer come about, and how did he design?[38] Yet, if an intentional designer did not create living nature's purposes, how could science account for them? Kant knew that modern science disallowed teleological explanations in physical nature, but still, organisms were teleologically organized. Kant failed to provide his readers with a solution and asked them to accept that our minds inescapably interpret living organisms as purposeful but that we would always remain ignorant of how non-intentional purposes are scientifically possible.[39] He

[32] Kant, *Kritik Der Urteilskraft*, 318–27.

[33] Kant, 349–50.

[34] Bowler, *Evolution*, 38–41.

[35] Paley, *Natural Theology*, 10.

[36] Duncan, *Botanical Theology*, 67.

[37] Kant, *Kritik Der Urteilskraft*, 322–32, 51.

[38] Johnson, *Aristotle on Teleology*, 4–6; Hume, *Dialogues Concerning Natural Religion*, 63–66; Voltaire, *Oeuvres completes de Voltaire*, 19–33.

[39] Kant, *Kritik Der Urteilskraft*, 351–68.

projected: "It is pretty certain that in terms of merely mechanical principles of nature we cannot even adequately become familiar with, much less explain, organized beings and how they are internally possible." Kant thought it absurd to believe that someday, another Newton would appear, explaining how even a mere blade of grass is produced.[40]

Fortunately, not all scientists have been willing to settle with Kant's self-imposed defeatism. Naturalists increasingly began to flirt with the idea of species undergoing processes of fundamental change over many generations. Around 1800, the French naturalist Jean-Baptiste de Lamarck was the first to work the issue out in much scientific detail.[41] His account also included a materialistic explanation of how biological functions arise. Lamarck noticed that organisms are functionally organized to survive within their environments and thought such adaptations resulted from "the influence of circumstances."[42] Organisms change their habits in response to environments, and the acquired habits are transmitted to the descendants. The ancestors of giraffes had continually been striving for higher hanging leaves, and resulting from this habit, the trait had become hereditary.[43] Lamarck's proposal has made him famous, as this change mechanism is still known as "Lamarckism."

However, while Lamarck's proposal was significant, it should also be put into perspective. First, it was not the only mechanism of change he conjectured. More prominently, he postulated an internal force, "the power of life," which drives species change towards greater complexity.[44] Lamarck does not seem to have thought of this mechanism as an adaptive response to environments but as something that caused changes irrespective of conditions of existence. Lamarck was not the only naturalist who played down the importance of adaptedness to environments.[45] In contrast to functionalists like Aristotle or Paley, some researchers stressed that the internal structures of organisms often seemed unrelated to circumstances. Despite their radically different environments, all mammals have similar vertebrate body plans, ranging from whales to bats to horses. Such internal unities of type suggested that adaptations were epiphenomena.

[40] Kant, 352; The translation is derived from: Kant, *Critique of Judgment*, 282–83.
[41] Bowler, *Evolution*, 48–95; Pietro Corsi, "Jean-Baptiste Lamarck."
[42] Lamarck, *Philosophie zoologique*, 1:218–68.
[43] Lamarck, 1:256.
[44] Lamarck, 1:133; Pietro Corsi, "Jean-Baptiste Lamarck," 13–14.
[45] Depew and Weber, *Darwinism Evolving*, 48; Bowler, *Evolution*, 121–24.

Moreover, Lamarck's explanation for nature's functions was not built upon substantial evidence, which was at least one of the reasons why much of the scientific establishment remained committed to the idea of species being fixed in time. In retrospect, we can also question how plausible Lamarck's explanation for adaptive change was.[46] While the length of the giraffe's neck looks like a credible candidate, other adaptations, like camouflage, are less likely outcomes of any such principle. Lamarck's mechanism also faces the problem of many acquired characters not offering improvement. If broken legs and smallpox scars are transmitted, Lamarckian inheritance might quickly become a mechanism of deterioration rather than adaptation.

The world had to wait for a genuinely new proposal until 1859, when Charles Darwin published his *On the origin of species by means of natural selection*. Unlike Lamarck, Darwin provided substantial evidence for the transformation of species, ranging from biogeographical data, fossils, and animals' mutual affinities to rudimentary organs. This time, the scientific establishment gave in. What Darwin described as "descent by modification" became known as "evolution," and the idea became an established scientific fact.[47] In addition, Darwin delved more deeply into *how* change occurs. He allowed some room for Lamarck's adaptive mechanism of inherited acquired traits but thought it insufficient for all nature's adaptations. So, Darwin looked for something else.[48]

In his subsequent breakthrough, Darwin drew essential inspiration from the work of pigeon fanciers.[49] He noticed that pigeon breeders transformed the animals in various directions through artificial selection after multiple generations. When new and desired characteristics appeared, like specific wing lengths or colours of plumage, the fanciers preserved the traits by fostering the offspring of the animals carrying the traits. Darwin suspected a similar selective force operates in living nature, cumulatively maintaining chance variants that were adapted to environments. But what Darwin could initially not figure out was what kind of force this was.[50] Eventually, the work of the political economist Thomas Malthus gave him a start. Malthus stressed how a discrepancy between high population

[46] Dawkins, *The Blind Watchmaker*, 290–303.

[47] Bowler, *Evolution*, 141–223; Depew and Weber, *Darwinism Evolving*, 57–160.

[48] Darwin, *The Origin of Species*, 175–79.

[49] Bowler, *Evolution*, 160–61; Darwin, *The Origin of Species*, 71–100.

[50] Bowler, *Evolution*, 141–72; Zimmer, *Evolution*, 3–31.

growth and insufficient food production could result in mass starvation. As Darwin later recollected in his autobiography: "It at once struck me that under these circumstances favourable variations would tend to be preserved, and unfavourable ones to be destroyed. (…). Here, then, I had at last got a theory by which to work."[51]

Darwin developed a potent core idea: the selection and accumulation of chance hits. Giraffes did not develop the traits of somewhat longer necks because the trees were high in the first place.[52] The variation, Darwin argued, is random. Giraffes were born with slight variations of neck types in all directions. This was an insufficient explanation for adaptation because undirected variants are mostly non-adaptive. However, what was not random was which of these giraffes eventually survived. In the ongoing "struggle for existence," the giraffes that *accidentally* possessed longer necks could eat more leaves. Subsequently, they had better chances of survival, resulting in the survival of the hereditary traits of the longer neck itself. If these accidental hits were then cumulatively preserved in reiterated selection rounds, the process would lead to the adaptation that we know so well. It was the external selective force of scarcity in the environment that drove the transformation of species. Paraphrasing the concept of artificial selection, Darwin described it as "natural selection."[53] This process, he wrote, is "daily and hourly scrutinizing throughout the world, every variation, even the slightest; rejecting which is bad, preserving and adding up all that is good; silently and insensibly working, whenever and wherever opportunity offers, at the improvement of each organic being."[54]

The theory is so simple that it can be explained in a single paragraph.[55] But it is a deceiving simplicity, as it is not by accident that it took so long before anyone figured it out. The three critical ingredients of the process—variation, selection, and reproduction—had already been widely recognized long before Darwin. Naturalists knew that organisms continually vary, that vast numbers die in a relentless struggle for life, and that traits are passed on from parents to offspring. However, the variations amongst organisms were traditionally understood as oscillations from the normal state, while the massive natural culling was seen as a conservative

[51] Depew and Weber, *Darwinism Evolving*, 71.
[52] Darwin, *On the Origin of Species*; Cameron and du Toit, "Winning by a Neck."
[53] Darwin, *The Origin of Species*, 133.
[54] Darwin, 133.
[55] Wilson, *This View of Life*, 233.

force that ensured organisms did not diverge from the mean.[56] Darwin's essential insight was recognizing that reiterated selection is also a *creative* force. Everyday small-scale variation provided the raw material for a selection process that, given vast periods, could build adaptive complexity from it.

Darwin's theory is so logical that it is sometimes considered tautological. American author Charles Fort wrote in 1919: "The fittest survive. What is meant by the fittest? Not the strongest, not the cleverest – weakness and stupidity everywhere survive. There is no way of determining fitness except in that a thing does survive. (…) Darwinism: that survivors survive."[57] This objection is flawed because there is a way to determine fitness other than that a thing survives. The giraffe's long neck does not count as "the fittest" simply because it survived but because it is adapted to its specific environment: a savannah with high-hanging leaves.[58] As soon as we find giraffes with costly necks that are longer than trees are high, while no other function appears to be involved, the theory would face a problem. The relentless struggle for life should have removed such energy-costing, non-adaptive features. Darwin's theory predicts a superabundance of adaptive traits and innumerable less adaptive variants that disappear into evolutionary oblivion.[59] Other potentially falsifiable elements are the vast periods required and the non-directedness of new variation—the latter continues to be debated today.[60]

The non-tautological character of the theory also becomes apparent from the fact that it took quite a while before the scientific community accepted it.[61] Darwin's evidence for evolution was overwhelming, but his mechanism still contained loose ends. There were doubts whether the Earth was old enough to account for his slow and gradual scenario. Critics wondered why adaptive traits would not blend in a larger population. The exact mechanisms of heredity were still shrouded in mystery, and what critics found particularly hard to swallow was the idea that undirected variants provided sufficient raw material for something as complex and creative as living nature. Well into the twentieth century, biologists continued

[56] Depew and Weber, *Darwinism Evolving*, 37; Bowler, *Evolution*, 67.
[57] Fort, *The Book of the Damned*, 18.
[58] Cameron and du Toit, "Winning by a Neck."
[59] Darwin, *The Origin of Species*, 227–30; Cordingley, *Viruses*, 287.
[60] Wagner, *Arrival of the Fittest*.
[61] Bowler, *Evolution*, 177–324; Depew and Weber, *Darwinism Evolving*, 167–242.

to look for alternatives.[62] Older ideas like natural theology, internal life forces, or Lamarckian inheritance lived on. The early twentieth century also witnessed the rise of a new movement, "mutationism," inspired by the rediscovery of Mendel's genetic laws. Many of the first geneticists assumed that living nature's transformative power did not so much lie in selecting small-scale continuous variation but in the robust appearance of discontinuous, significant genetic "mutations," not necessarily adaptive.

Only in the 1930s and 1940s did Darwinism and genetics merge into a new paradigm named the "Modern Synthesis."[63] What this model adopted from genetics is the assumption that hereditary traits are located on genes and that characteristics are often inherited in a particulate and discontinuous manner. This would help explain why adaptive variants did not blend in with the broader population. Yet, what made the Synthesis thoroughly Darwinian was that new variation was thought to result from random small-scale genetic mutation and recombination, subsequently being steered towards adaptation by natural selection.[64] Some of the architects of the Synthesis had built their assumptions upon mathematical models, showing how selection could plausibly make chance variants spread throughout a population of interbreeding organisms. At one point, the Synthesis even out-Darwinized Darwin, as Lamarckian inheritance was excluded.

The Modern Synthesis was one of the triumphs of twentieth-century science. Around 1900, the many fields of biology, like palaeontology, field biology, and genetics, had been working past one another, often endorsing mutually inconsistent claims. By the mid-twentieth century, the various branches of biology increasingly began to work within an overarching and more consistent Darwinian framework.[65] One of the things that helped make the Synthesis a success was its clear research agenda.[66] When confronted with complex and functional traits, biologists look for the traits' effects within the environment that ensure their continuation. The theory thus nicely combines the nomothetic and the idiographic. Organisms adapt to their environments *in general*, but since adaptedness is always context-dependent and different in each instance, the theory also

[62] Bowler, *The Eclipse of Darwinism.*

[63] Bowler, *Evolution*, 325–46; Depew and Weber, *Darwinism Evolving*, 243–329.

[64] Godfrey-Smith, *Philosophy of Biology*, 40–41.

[65] Dobzhansky, "Nothing in Biology Makes Sense except in the Light of Evolution."

[66] Mayr, *What Evolution Is*; Ridley, *Evolution.*

looks for the particular.[67] Evolutionary biologists know that single traits can also fulfil multiple functions simultaneously. The functions of the elephant's trunk are translated into various "selection pressures," or effects that trunks have within their environments that have caused, and still cause, the organ's survival.[68]

However, the adaptive research agenda does not imply that everything in living nature is adaptive. Unlike natural theology, the theory can also explain why certain traits are non-adaptive. Darwin already referred to rudimentary organs, like the snake's limbs, that had lost their former function but had not entirely disappeared.[69] The idea of common descent also provides a plausible scenario for the mysterious similarities between organisms' internal structures.[70] The various types of mammals adapted to multiple environments after they diverged but maintained much of their internal composition. Natural selection is a tinkering small-scale process, so it cannot drastically rearrange organisms, implying that things are not optimally adapted to current environments; they also carry the heritage of past environments.[71] Additionally, there is the option of things being a by-product of something else. Bones are white, but it would be a mistake to call this an adaptation, as the colour is a by-product of the fact that bones are made of calcium salts, which happen to be white.[72] Lastly, Darwin speculated on "variations neither useful nor injurious [that] would not be affected by natural selection." Evolutionary biology accepts that something like that occurs at the genetic molecular level.[73] Many mutations have no phenotypic consequences visible to natural selection and continue to exist as "genetic drift."

The role of randomness in the model also requires careful consideration. A common misinterpretation of the theory states that organisms result from chance. Richard Dawkins responds: "Mutation is random; natural selection is the opposite of random."[74] But even the randomness of mutations deserves closer attention.[75] In quantum mechanics,

[67] Godfrey-Smith, *Philosophy of Biology*, 52.
[68] Shoshani, "Understanding Proboscidean Evolution."
[69] Darwin, *The Origin of Species*, 428–32.
[70] Darwin, 233.
[71] Dawkins, *The Greatest Show on Earth*, 337–72.
[72] Weiner, *Handbook of Psychology, History of Psychology*, 50.
[73] Darwin, *The Origin of Species*, 131; Ridley, *Evolution*, 155–93.
[74] Dawkins, *The Blind Watchmaker*, 41.
[75] Depew and Weber, *Darwinism Evolving*, 113–14; Baer, "Mutation."

randomness implies that things have no causes, but the term has a different meaning in evolutionary biology. The discovery of DNA in the 1950s showed that genetic information is encoded in DNA molecules, which are typically copied with high accuracy. But sometimes copying errors occur, and it is in these "mutations" that modern biology recognizes the origin of new variation. Such mutations to DNA are non-random because they have specific causes, like UV radiation.[76] The randomness that the model refers to only concerns randomness regarding fitness outcome—the mutations giving rise to longer giraffe necks were random because they were unrelated to high trees on the savannah. *That* was a chance hit.

Here, the mechanistic explanation of physical science meets the teleological explanation of biology.[77] The Newtonian model assumes that external forces steer inert and undirected objects. In an important respect, Darwin's model is much like that: the external force of a selective environment steers undirected variation.[78] Yet, while the Darwinian process is purposeless, its outcome looks radically different: purposeful complexity. Darwin demonstrated that design without designer is possible. Functional complexity can arise without the involvement of an intentional actor. In 1790, Kant claimed that a reconciliation between mechanical science and teleology was a chimera, but as Ernst Haeckel commented in 1899: "Seventy years later, this impossible 'Newton of organic nature' appeared with Darwin, who brilliantly resolved the daunting task Kant had declared to be entirely insoluble."[79]

Richard Dawkins thinks that Darwinism is "too big a theory" to keep it confined to the domain of living nature. Like Dennett, he wants to take it into the socio-cultural realm.[80] But can the theory again express its "universal acid"-like ability to fundamentally transform traditional concepts? Might it also help us explain socio-cultural "design without designer"? This first raises whether such an explanandum exists in the socio-cultural world. We saw that Paley and Kant referred to watches and houses as apparent examples of intentional human design. But can we safely assume that human actors deliberately created all purposes in the socio-cultural

[76] Dawkins, *The Blind Watchmaker*, 307–12; Baer, "Mutation."
[77] Haeckel, *Die Welträtsel*, 121; Godfrey-Smith, *Philosophy of Biology*, 60.
[78] Depew and Weber, *Darwinism Evolving*, 6–9, 110.
[79] Haeckel, *Die Welträtsel*, 119.
[80] Dawkins, *The Selfish Gene*, 191.

world? This is not as obvious as it might first appear, as the origin of functions has also puzzled scholars of the socio-cultural. To understand their debates, we now begin with a classic social case study: the Hopi rain dance.

3.3 The Origins of Cultural Functions

In 1957, the American sociologist Robert King Merton decided to plead for the significance of the Hopi rain dance.[81] According to many of his contemporaries, the ritual was hardly more than a relic of the past, outdated by modern knowledge. The Hopi Indians believed their dances could help create rain, but as modern meteorology demonstrated, rain has different causes. The ritual could be considered to be a relic without any function at all. However, Robert Merton believed things were not as simple as that. The dance's avowed purpose was to evoke rain, but hidden underneath, unintended and unrecognized by the Hopi themselves, these dances had other effects of greater importance. The ceremony, as Merton stated, "may fulfill the latent function of reinforcing the group identity by providing a periodic occasion on which the scattered members of the group assemble to engage in a common activity."[82] The dance provided the foundation for group unity.

The Hopi rain dance became a famous case study because it was emblematic of the school of thought to which Merton belonged: functionalism.[83] Merton's account of the rain dance is characteristic because he assumed that a socio-cultural phenomenon could fulfil a function beyond the actors' understanding.[84] Emile Durkheim typically claimed that we humans are "the victims of an illusion which leads us to believe that we have ourselves produced what has been imposed upon us externally."[85] The functionalists acknowledged that many functions *are* intended and recognized—Merton defined these as "manifest functions"—but the latent functions interested them most. In the functionalists' explanations of specific socio-cultural phenomena, relatively little attention is paid to the avowed purposes of the actors. As Bronislaw Malinowski argued, the function of magic was not to be found in the intended effects "on the soil

[81] Merton, *Social Theory and Social Structure*, 73–138.

[82] Merton, 118–19.

[83] Wallace and Wolf, *Contemporary Sociological Theory*, 15–66; Ritzer, *Contemporary Sociological Theory*, 13–18, 64–84.

[84] Radcliffe-Brown, *Structure and Function*, 142.

[85] Durkheim, *Les Règles de La Méthode Sociologique*, 7.

and spirits, on the spider and the full moon. This is a native belief which, important as it is, does not directly bind us." Instead, it was "the indirect effects of the words upon the psychology and physiology of the native organism, and hence upon social organization, which probably gives us the best clue to the nature of magical thinking."[86] Whereas religious believers may be convinced that their worship of gods serves a higher purpose, many functionalists have argued that a hidden function is more tangible: religion reinforces the integration and preservation of society.[87]

The notion of unintended and unrecognized functions was not confined to functionalism. During the 1970s, the cultural materialist anthropologist Marvin Harris argued that seemingly irrational beliefs form hidden adaptations to material-ecological environments. The Hindus may believe they refrain from eating cows to protect their karma, but an underlying explanation, as argued by Harris, was the preservation of an indispensable draught animal and source of manure.[88] The Aztec Indians may have been genuinely convinced that their sacrifice and consumption of humans ensured the rise of the Sun, but a hidden function of these rituals was different. Due to the lack of domesticated animals as a food source, the Aztecs suffered from protein deficiencies, compensated by human flesh.[89]

The former popularity of the functionalist explanatory model should not come as a surprise. Many socio-cultural practices appear adaptive in ways the actors did not intend or recognize. What is particularly striking is that the avowed purposes often do not coincide with apparently functional consequences. However, Merton articulated these views over half a century ago, and, ironically, later on, Merton's analysis *itself* became a relic of the past. The functionalist school lost its dominant position in social theory long ago. While the functionalists claimed to have uncovered all kinds of latent functions around the middle of the twentieth century, this form of reasoning became unpopular in the last sixty years, and the influence of functionalism waned. Other schools of thought won the day, from practice theory, social constructivism, and symbolic anthropology to rational choice theory. While these movements had many mutual disagreements,

[86] Malinowski, *Coral Hardens and Their Magic*, 241.

[87] Durkheim, *Les formes élémentaires*; Radcliffe-Brown, *Structure and Function*, 153–77.

[88] Harris, *Cows, Pigs, Wars, & Witches*, 11–34.

[89] Harris, *Cannibals and Kings*, 127–69.

an aversion to functionalism tied them together. They all found this idea of unintended and unrecognized functions problematic.[90]

To understand why they objected, let us first return to the Hopi rain dance, and we shall primarily focus on the critiques formulated by the school of practice theory.[91] As mentioned, the avowed purpose of the ritual was to make rain, and the latent function of social cohesion, Robert Merton argued, was unintended and remained unrecognized by the Hopi themselves. But as critics asked, would it be true that making rain, for the Hopi, was the only thing this ritual was about? A young Hopi man on his way to the rain dance was possibly also eager to marry and, therefore, curious about the possible candidates he might encounter. Similarly, he might urgently need to talk about a little bridge over the creek that needed repairing. Perhaps the relationship with a neighbouring tribe was under severe pressure; hence, he looked forward to the communal ceremony that provided a sense of shared strength. If the Hopi had been asked directly what this dance was for, he may have answered rain. However, underneath, he likely knew this dance was about more than just that.

If the arguments of practice theorists such as Pierre Bourdieu and Anthony Giddens are to be considered, these forms of tacit knowledge play a crucial role in human social life. Apart from the directly conscious and avowed purposes of our behaviour, we also possess, as Bourdieu defined it, a *sens pratique*—a sense of practice—or, as formulated by Giddens, a practical consciousness.[92] In our daily routines, from labour, ceremony, and leisure to how we dress, talk, or move our bodies, humans possess much implicit understanding. If asked about this practical knowledge directly, we cannot translate it into discursive expression, but the role of this *sens pratique* in the day-to-day construction and reproduction of the structures of society is seen as crucial. Hence, Merton's assumption that the rain dance was about more than just rain could have been correct. His mistake, however, was to conclude that these functions existed beyond Hopi knowledge.

Bourdieu and Giddens highlighted a crucial weakness in the functionalist argument. After reading these social theorists, the precise boundary

[90] Dahrendorf, "Out of Utopia"; Homans, *Bringing Men Back In*; Berger and Luckmann, *The Social Construction of Reality*; Gouldner, *The Coming Crisis of Western Sociology*; Giddens, *The Constitution of Society*; Stark, "Micro Foundations of Religion," 272.

[91] Bourdieu, *Le Sens Pratique*; Giddens, *The Constitution of Society*; Jenkins, *Pierre Bourdieu*.

[92] Giddens, *The Constitution of Society*, xxii.

between manifest and latent functions appears simplistic. The functional features of social phenomena require consideration of unexpressed levels of knowledge. Research on the Hopi has also shown that from these Native Americans' perspective, the practical and the spiritual are inextricably interrelated.[93] What provides Bourdieu's and Giddens's argument with additional strength is that their ideas correspond to, or are at least not contradicted by, the evidence provided by scientists in other fields. Psychologists and cognitive neuroscientists also claim that the human mind makes decisions that people are unaware of, let alone express discursively.[94] This claim means that an ultimate scientific explanation for the origin of unexpressed functions, as Bourdieu and Giddens envisioned them, appears possible.

What if we ask the same question about the origin of latent functions as Merton defined them if latent knowledge was not the cause? Merton remained unclear about this issue and did not offer an explanation beyond a rather vague appeal to unintended consequences.[95] However, this appeal is hardly convincing because it remains unclear why unintended consequences should lead to functions. Given the undirected nature of unintended consequences, chaos would be a more likely outcome than functions. The functionalists ultimately explained social phenomena by referring to their effects, which critics found unsatisfactory because science is supposed to explain things from their causes. In other words, the functionalists presented us with an explanandum but failed to provide a convincing explanans.[96] As Anthony Giddens said about Merton's account:

> This treats the Hopi as ignorant of what they are up to. Participants in most ceremonials, after all, have some sense of the various ends they might serve. Moreover, showing that the rain dance fosters social solidarity *cannot possibly* be an explanation of the actions of the Hopi unless at least some of them sense this is so and act accordingly.[97]

[93] Loftin, *Religion and Hopi Life in the Twentieth Century*.

[94] Dijksterhuis and Nordgren, "A Theory of Unconscious Thought"; Haggard P, "Human Volition."

[95] Merton, *Social Theory and Social Structure*, 73–138.

[96] Elster, *Explaining Technical Change*; Pettit, "Functional Explanation and Virtual Selection."

[97] Giddens and Pierson, *Conversation with Anthony Giddens*, 92. Emphasis added.

This weakness made the functionalists an easy target for their critics. By talking about social institutions that contain functions beyond people's knowledge or social systems that have needs, they could be accused of treating society as if it possessed a mind or consciousness of its own, while only people possess such capabilities. Marvin Harris spoke of a culture that adapted to the environment, whereas again, culture should be considered a means through which *individuals* adapt to their environment.[98] The functionalists were blamed for reification.[99] Abstract concepts such as "social systems" or "culture" appeared to live a life of their own. Sometimes, this reification seemed almost mystical. However, a satisfactory counter-argument remained elusive.

The diverse opponents of functionalism did not deny that society or culture could be a coercive force acting on individuals. The social constructivist Peter Berger stated that we produce society, but that society also produces us.[100] The symbolic anthropologist Clifford Geertz described culture as "a set of control mechanisms (…) for the governing of behavior."[101] Yet, the crucial distinction between functionalism and many of its critics remained the umbilical cord between socio-cultural functions and the knowledge that actors have of them. "Man is an animal suspended in webs of significance he *himself* has spun," Clifford Geertz wrote.[102] Or Peter Berger: "Society is a product of man. It has no other being except that which is bestowed upon it by human activity and consciousness."[103]

However, though such critiques of functionalism are legitimate in many ways, their opponents' arguments also contain unsatisfactory elements. The functionalists did not offer a convincing explanans, but does this necessarily imply that their explanandum is invalid? Despite these criticisms, socio-cultural phenomena appear to possess a functional integration and adaptive ability that transcends human actors' conscious and subconscious knowledge. From state structures and economic systems to religions or languages, socio-cultural structures often appear more functionally organized than possible had they been designed by human hand.[104] Individual actors' role in forming functional systems may often seem marginal at best

[98] Giddens, *The Constitution of Society*; Keesing and Strathern, *Cultural Anthropology*, 128.
[99] Stein and Stein, *The Anthropology of Religion, Magic, and Witchcraft*, 20.
[100] Hamilton, *The Sociology of Religion*, IV:50.
[101] Geertz, *The Interpretation of Cultures*, 43.
[102] Geertz, 5. Emphasis added.
[103] Hamilton, *The Sociology of Religion*, IV:50.
[104] Hodgson and Knudsen, *Darwin's Conjecture*, 6.

and hardly inspired by knowledge of the eventual consequences of their actions. What remains particularly striking is that actors' avowed purposes and seemingly functional consequences often do not coincide. The Hopi belief was most likely sincere, just as Hindus genuinely believe in the importance of cows for personal karma. The authenticity of the beliefs seems crucial because otherwise, these traditions would lose much of their prowess. However, how do people often develop factually false convictions, yet apparently functional in their results?

A possible explanation is that some Hopi were well aware that common dancing does not create rain but grasped the social function of others, sharing the belief that it does. What if they then purposely encouraged the ritual for that reason? Though socio-cultural intelligent design should not be rejected out of hand, it is questionable whether human actors possess that much sociological ingenuity in deceiving their companions. It is also questionable whether an appeal to the *sens pratique* will always suffice. As we discussed, it is possible that the Hopi, at the back of their minds, did have a specific knowledge that their dance was about more than just rain. Yet, whether the Hopi in their *sens pratique* also realized that it was functional for them to have a *genuine* belief in the magical powers of their dance and whether this belief was ingeniously introduced and transmitted from the *sens pratique* for that reason is something different. The capabilities of our practical consciousness could be easily overestimated this way. This explanatory route is also unsatisfactorily complex. Even if one would consider the scenarios of deliberate design, or *sens pratique* design plausible, we would still face problems regarding evidence. How can we empirically examine what the Hopi were consciously or subconsciously aware of when creating their practice? We have no sources that enable an investigation.

In other words, the counter-functionalist argument also appears to contain its mysteries, which is most likely why the functionalist reasoning turned out to be more challenging to eradicate than expected. As two anthropologists wrote about their colleagues: "Functionalism continues to be a powerful idea, and most anthropologists are probably more functionalist than they generally acknowledge."[105] The weaknesses in the counter-functionalist argument should make us wonder whether adaptive forces beyond our knowledge could nonetheless guide the evolution of

[105] McGee and Warms, *Anthropological Theory*, 154; Wallace and Wolf, *Contemporary Sociological Theory*, 57–66.

socio-cultural functions. But before we discuss what these forces could be, we need to address the arguably even more complicated issue of the ultimate purposes of biological and cultural adaptations.

3.4 The Purposes of Biological Functions

"Cui bono?" is a question that Daniel Dennett does not get tired of asking. Who benefits? It is a stock Latin phrase often used by lawyers, but Dennett also considers the question relevant when examining biological and socio-cultural worlds. When confronted with traits that seem functional, it is crucial to ask ourselves what they are ultimately functional for. But as Dennett warns us, the beneficiaries of functions can be elusive, so we need to cast our nets widely.[106] One of the reasons for Dennett embracing Darwinian theory so enthusiastically is that he believes it helps uncover some of the less apparent beneficiaries. Selfish gene theory made a lasting impression on him, and he is eager to extend its insights into the socio-cultural domain. In the next chapter, we will discuss Dennett's proposals. But first, let us explore how evolutionary biology has addressed the *cui bono* question.

What adaptations in living nature are essentially functional for was a question that already occupied Darwin. Overall, he thought the struggle for life was most intense between individuals of the same species, for instance, when they competed for food within the same area.[107] Traits that survive, like camouflage or claws, are generally for the benefit of the individual organism. Yet, Darwin realized that this model ran into some trouble.[108] What, for instance, about the sting of a bee? The trait seems to protect the bee's community, but it is an outright disaster for the individual bee, as using the sting often kills it. Many bees are even sterile and devote their entire lives to the community's well-being. Such traits look like evolutionary dead-ends. The solution, Darwin presumed, lay in applying natural selection beyond the individual.[109] Darwin hinted that the sacrifices of some members are compensated by the enhanced survival of other members of the community, who subsequently produce offspring with similar altruistic traits.

[106] Dennett, *Breaking the Spell*, 62–63.

[107] Darwin, *The Origin of Species*, 126; Depew and Weber, *Darwinism Evolving*, 75.

[108] Darwin, *The Origin of Species*, 258–63; Darwin, *The Descent of Man*, 149–57.

[109] Darwin, *The Origin of Species*, 135.

Current evolutionary biology shares Darwin's primary focus on individual selection since the bulk of adaptations indeed seem to benefit individual organisms.[110] However, numerous examples of organisms making selfless sacrifices for others indicate that individual selection cannot be the whole story. Some forms of altruism are still relatively easy to explain, like reciprocal altruism, where animals do each other small favours in return.[111] More challenging are animals taking significant risks for others. A classic example is prey animals, which send alarm calls when predators are in sight.[112] The trait looks adaptive because it helps others look for shelter, but not so much from the caller's point of view: the alarm might divert the predator's attention to the caller. Another example is parental care, with organisms making sacrifices for their offspring.[113] Social species of bees and termites, ants, or wasps are surely the most severe challenge.[114] Such species are characterized by a division of labour in which some individuals, most notably the queens, specialize in reproduction. Many others, the "workers," fulfil cooperative tasks such as food collection, rearing the young, or protecting the colony. Examples can be beguiling, like ants who developed thickened and enlarged heads to block the nest's entrance, or so-called kamikaze termites that explode to glue enemies of the colony in poisonous secretions.

It is not only *between* organisms that altruism can take self-destructive forms—it also occurs *within* organisms. Large organisms are made up of vast collections of cells that all descend from other cells. Most organisms are unicellular, but in the evolution of life, large assemblages of cooperating cells have evolved, the multicellular organisms, which are also characterized by an internal division of labour.[115] Some germ cells reproduce into the far future, while most somatic cells function as evolutionary dead-ends. The somatic cells specialize, think of skin, brain, or liver cells, and their existence is entirely devoted to the well-being of the whole. Like the kamikaze termites, somatic cells can also kill themselves in "programmed

[110] Okasha, "Units and Levels of Selection," 203.

[111] Dawkins, *The Selfish Gene*, 183–88, 202–33.

[112] Dawkins, 168–69; Kay, Lehmann, and Keller, "Kin Selection and Altruism."

[113] Dawkins, *The Selfish Gene*, 6.

[114] Keller, Chapuisat, and Losos, "Evolution of Eusociality"; Queller and Strassmann, "Kin Selection and Inclusive Fitness."

[115] Dawkins, *The Selfish Gene*, 258–59; Smith and Szathmary, *The Major Transitions in Evolution*, 201–16.

cell death," which occurs as soon as their existence burdens the organism.[116] These mechanisms are not perfect because now and then, some cells start reproducing for their own, which we call cancer.[117] But cancer is unusual since individual cells typically work harmoniously together for the higher-level benefit of the organism.

The explanatory problem of altruism is as follows: alarm-sounding animals, kamikaze termites, and somatic cells hinder their own survival, so their traits should be doomed. A new round of debate over this problem was triggered in the 1960s when the Scottish zoologist Vero Wynne-Edwards argued that animals altruistically regulate individual reproduction to not overexploit the environment's carrying capacity.[118] He explained that groups of altruists outcompeted groups of non-altruists, as the latter overexploited their resources and thus sealed their fate. However, many biologists were no longer willing to buy such explanations.[119] People like George C. Williams and William D. Hamilton pointed out that group selection, or species selection, suffers from the problem of internal subversion. Groups of altruists can be invaded by free riders who reap the benefits of the group without contributing to it themselves—think of an animal that profits from alarm calls but never raises alarms. Mathematical modelling illustrated that selection between individuals is fast and continuing, while selection between groups is weak and slow. Even when free riders make the group go downhill, there is still plenty of time for them to prosper at the expense of altruistic fellows.[120] Yet, how to explain higher-level altruism? Williams's and Hamilton's key innovation was not to look at the higher level of groups or species but to descend to the level of the gene.

This generation of evolutionary biologists realized, building on insights from the Modern Synthesis, that Darwinian selection is essentially about things being transmitted or not being transmitted. While individuals, groups, or species are not transmitted, ultimately, it is the survival of *genes* that counts—genes here often being defined as segments of DNA that survive the shuffle of sexual recombination.[121] To answer the *cui bono* question, these biologists looked at adaptations from the perspective of

[116] Wilson, *This View of Life*, 80.

[117] Yong, *I Contain Multitudes*, 87; Okasha, "Units and Levels of Selection," 200.

[118] Dawkins, *The Selfish Gene*, 7–10; Okasha, "Units and Levels of Selection," 200–201.

[119] Godfrey-Smith, *Philosophy of Biology*, 42–45, 120–28; Depew and Weber, *Darwinism Evolving*, 368–74.

[120] Dawkins, *The Selfish Gene*, 7–8.

[121] Depew and Weber, *Darwinism Evolving*, 369–70.

gene reproduction. Hamilton made the argument even "more vivid by attributing to the genes, temporarily, intelligence and a certain freedom of choice."[122] Suddenly, many things began to make more sense. Genes reside inside organisms, so genes typically "want" their organism to flourish—this would explain why most adaptations benefit individuals. However, since organisms are only temporary carriers, genes must ensure that copies of them will live on in other organisms. The most obvious way of doing so is to create new organisms that carry the same genes and to make the parents care about their offspring—this would explain parental altruism.[123] However, the same genes reside not only in offspring but also in other close relatives, like brothers and sisters or nephews and nieces. So, if altruism aids close relatives, there is a certain likelihood that the recipient of the altruistic action will be an altruist. When relatives are saved, copies of the genes are saved, and altruism might spread.[124]

The theory is known as "kin selection" or "inclusive fitness" and predicts that non-reciprocal altruism should focus on close relatives. Empirical research has provided abundant confirmation. Alarm calls predominantly focus on kin, and the underlying rationale would be that one carrier of the involved genes is put at risk to save the lives of a few other carriers.[125] The most startling evidence was found in social insects. These colonies are, in fact, huge families in which individual females are usually more closely related to full sisters than their offspring.[126] It is thus not a good strategy to let these females have offspring themselves. Instead, they should devote their lives to the reproduction of the mother, the queen, and protect her life at all costs. This would explain kamikaze-like actions: the deed kills the organism but enhances the reproductive chances of the genes involved. Further proof for kin selection was provided by multicellular organisms, as cells belonging to the same organism are genetic clones.[127] The somatic cells are like the "workers" who make the organism function as a whole to ensure the continuation of the genetically identical germline.

[122] Hamilton, "Altruism and Related Phenomena," 195.

[123] Dawkins, *The Selfish Gene*, 89–93.

[124] Dawkins, 88–107; Okasha, "Units and Levels of Selection," 202.

[125] Strassmann and Queller, "Cooperation and Conflict"; Kay, Lehmann, and Keller, "Kin Selection and Altruism."

[126] Hamilton, "Altruism and Related Phenomena"; Queller and Strassmann, "Kin Selection and Inclusive Fitness," 217.

[127] Okasha, "Units and Levels of Selection," 205.

In 1976, Richard Dawkins encapsulated the theory into his famous metaphor of "the selfish gene." In Dawkins's view, genes are the "replicators" of evolution, whose true purpose "is to survive, no more, no less."[128] In a mutual struggle for survival, genes have evolved ever more complex ways of making copies of themselves, and Dawkins considers organisms as just a way of achieving that goal. Organisms are the "vehicles" or "survival machines" of genes, and their whole performance is in aid "of one thing and one thing only, the spreading of DNA."[129] Usually, the interests of genes and organisms coincide: genes that programme gazelles to run away from lions enhance their own survival and the gazelle. But when the reproductive interests of genes and organisms do *not* coincide, selfishness comes to the fore. In that case, genes make organisms do things that are not good for organisms but for the good of the genes' own survival.

Selfish gene theory makes logical sense because genes that instruct organisms to make more copies of genes will be most successful at precisely that. Nevertheless, the theory keeps giving rise to two common misconceptions. The first states that Dawkins argues that *organisms* are selfish.[130] To some extent, Dawkins himself is to blame here, as one sentence from *The Selfish Gene* states that "we are born selfish."[131] But this claim contradicts much of his book because it is by placing selfishness at the level of genes that Dawkins wants to explain why organisms, under certain conditions, are altruistic. A second misconception is that Dawkins argues that genes are *literally* selfish. The biologist Steven Rose laments in a critique that "genes cannot be selfish; it is people, not neurons nor yet brains or minds, who think, remember, and who show emotion."[132] But Dawkins repeatedly emphasized that genes are molecules without any literal intentions, emotions, or intelligence.[133] The reason for writing about genes *as if* they possess the emotion of selfishness is that it makes it easier for us to comprehend what is happening. This is Darwinism: something can look and behave *as if* an intelligent agency is involved, while there is not.

The fruitfulness of Dawkins's metaphor is further illustrated when we consider interactions between organisms and other life forms that reside

[128] Dawkins, *The Selfish Gene*, 15, 45.
[129] Dawkins, *The Blind Watchmaker*, 111; Dawkins, *The Selfish Gene*, xxi.
[130] Midgley, "Gene-Juggling," 439.
[131] Dawkins, *The Selfish Gene*, 3.
[132] Rose, "The Biology of the Future and the Future of Biology," 141.
[133] Dawkins, *The Selfish Gene*, 19, 89, 139.

within them.[134] Multicellular organisms host a whole ecosystem of bacteria, fungi, protists, and viruses, known as the microbiome. Dawkins argues that the relationship between organisms and such creatures is somewhat similar to the relationship between organisms and selfish genes. Usually, the survival of microbes or viruses depends on the survival of their hosts, so they do not "want" their hosts to be harmed. We often think of viruses and bacteria as harmful, but life scientists have increasingly concluded that most are neutral or beneficial.[135] The survival of organisms can even depend on these tiny creatures, as they fulfil a wide range of positive coevolutionary functions. Human guts, for instance, are inhabited by numerous microbes that play essential roles in the digestion of food.[136]

However, we should not be naïve. Eventually, it is their own survival that the microbes or viruses' DNA (or RNA) care about. As soon as the reproductive interests of guests and hosts diverge, selfishness comes to the fore again.[137] Especially when microbes or viruses quickly move from organism to organism, there is little hindrance to causing severe damage to their temporary hosts as a by-product of their spread. We call this contagious disease. Such pathogens can, in some instances, even "manipulate" or "hijack" the bodies and behaviour of their organisms, turning them into vehicles for their own reproduction. The rhinovirus, for instance, the common cold, makes our noses run, which has no function for us.[138] But it *is* very functional for the rhinovirus because through the snot, it gets on our fingers, then on objects, and when other people touch those objects, the virus enters new bodies. A more frightening example is the rabies virus.[139] When it enters a dog, it manipulates its brain, turning normally peaceful animals into restless wanderers with foaming mouths who aggressively look for other animals to bite. We will not get anywhere if we ask what function this behaviour has for the dog. It is pretty dangerous for dogs because they might be killed for it. But when we ask what function it

[134] Dawkins, 234–66; Dawkins, *The Extended Phenotype*, 209–27; Cordingley, *Viruses*, 16–18.

[135] Yong, *I Contain Multitudes*; Wilson, *This View of Life*, 58–65.

[136] Yong, *I Contain Multitudes*; Strassmann and Queller, "Cooperation and Conflict," 672.

[137] Dawkins, *The Selfish Gene*, 234–66; Yong, *I Contain Multitudes*, 76–83; Bailey, "Evolution of Apparently Nonadaptive Behavior," 713.

[138] Zimmer, *A Planet of Viruses*, 10.

[139] Dawkins, *The Selfish Gene*, 246–47; Yong, *I Contain Multitudes*, 76.

has for rabies, it all suddenly makes sense: through the bites, the virus is transmitted. Organisms can be used as the marionettes of their invisible microbial or viral puppeteers.[140]

The relationship between organisms and their guests is ambiguous.[141] Complex organisms cannot survive without their microbiome, but exploitation and manipulation perpetually lurk in the dark. Some parasites can even be good and bad at the same time. Wasps, for instance, cannot lay eggs without the Wolbachia bacteria. Still, whenever it serves the bacterium's purposes, it turns male wasps into females, thereby harming the reproductive interests of wasps. The science writer Ed Yong describes microbes as having evolved "a continuum of lifestyles, between 'bad' parasites and 'good' mutualism," with flexible interactions, as "our allies can disappoint us and our enemies can rally to our side."[142] Again, we should be mindful of the power of metaphor here. Microbes or viruses do not "manipulate" or "hijack" their organisms in any literal sense; the anthropomorphic language only makes it easier to understand what is happening and predict novel phenomena.[143]

Yet, selfish gene theory is also criticized on more serious methodological and empirical grounds. To a large extent, the theory is built upon the Modern Synthesis, which, as we saw, ruled out Lamarckian inheritance and privileged the gene as the ultimate unit of heredity. However, numerous voices in biology maintain that this has taken things too far. It is claimed that we need an "expanded" or "extended synthesis" that offers a richer picture of evolution.[144] The first critique maintains that selection occurs at more levels than genes alone. According to a counter-model named "multilevel selection theory," genes have no special status, as other entities can also become targets of selection.[145] "One of the beauties of multilevel selection theory is that it employs the same principles at all levels of the biological hierarchy," David Sloan Wilson tells us.[146] Especially group selection is finding new support. Proponents admit that earlier

[140] Bailey, "Evolution of Apparently Nonadaptive Behavior," 713.

[141] Yong, *I Contain Multitudes*; Cordingley, *Viruses*, 39–45.

[142] Yong, *I Contain Multitudes*, 78–82.

[143] Cordingley, *Viruses*, 2.

[144] Pigliucci and Müller, *Evolution, the Extended Synthesis*; Laland et al., "The Extended Evolutionary Synthesis"; Gould, *The Structure of Evolutionary Theory*; Jablonka and Lamb, *Evolution in Four Dimensions.*

[145] Wilson, *The Social Conquest of Earth*; Wilson, *This View of Life*, 76–92.

[146] Wilson, *Darwin's Cathedral*, 2002, 141.

group selection models were naive, but it is said that the gene selectionists made a statistical error in their turn.[147] Selfish organisms may indeed outcompete altruistic organisms *within* groups. Still, when groups with a relatively high number of altruists vastly out-reproduce groups with egoists, the undermining effects can be compensated.

The response of mainstream biology has been mixed. According to many, the debate may revolve around a false opposition, as group selection and kin selection may come down to the same thing. In some cases, it is just more practical to look at the level of groups, the level of individuals, or the level of genes.[148] Others are more critical, maintaining that it is unclear what concrete empirical phenomena group selection would precisely help explain.[149] Supporters of group selection tend to re-describe well-known cases of inclusive fitness, like social insects or multicellularity, as resulting from group selection, sometimes without even mentioning the broad support for the alternative kin selection model.[150] But if group selection is an effective force, we should expect to find, as the evolutionary biologist Jerry Coyne puts it, "altruistic behavior in nature between individuals who were completely unrelated. That is, individuals would often sacrifice their lives (or reproduction) to help those who don't share their genes." As Coyne adds, "that is exactly what we *don't* see."[151] Moreover, multilevel selection blurs the essential distinction between things that are transmitted and those that are not.[152] Richard Dawkins: "There is one thing, and only one thing, that singles genes out as unique, and that is that they are the things that go on to the next generation."[153] Effective group selection would imply that genes are transmitted that make organisms behave in group-functional ways. But as we already saw regarding the individual organism, "selfish" interests of genes do not necessarily conflict with the interests of higher levels of organization. Enhancing group-ish behaviour in an environment of effective group selection may be the selfish genes' strategy of living on to the next generation. Even if the

[147] Wilson, 12–14; Turchin, *Ultrasociety*, 82–88.

[148] Okasha, "Units and Levels of Selection"; Sterelny and Griffiths, *Sex and Death*, 151–79.

[149] Bourke, "The Validity and Value of Inclusive Fitness Theory"; Coyne, "The Demise of Group Selection."

[150] Wilson, *This View of Life*, 77–81; Wilson, *The Social Conquest of Earth*.

[151] Coyne, "Genes First."

[152] Hodgson and Knudsen, *Darwin's Conjecture*, 158.

[153] Rose, Richard Dawkins interviews Steven Rose.

empirical evidence for group selection grows more robust, it is still questionable to what extent selfish gene theory is undermined.[154]

Another strand of critique appears more challenging, however, as it states that *transmission* takes place at more levels than genes alone. It is mainly the field of epigenetics that attracts attention here.[155] The term epigenetics refers to everything that happens to genetic information during the development of individual cells. All cells in multicellular organisms are genetic clones, but due to mechanisms of epigenetics in which some genes are switched on or off, cells are remarkably different. As such, epigenetics is uncontested. What stimulates debates is that recent studies suggest that epigenetic changes can be transmitted to offspring. Events during an organism's lifetime would affect which genes are switched on or off in subsequent generations without any changes to the genetic code.

Proponents of the extended synthesis also investigate further non-genetic transmission mechanisms, such as animal culture and a presumably neglected process described as "niche construction."[156] The latter refers to what happens when organisms actively seek their environments and subsequently change them, like rabbits that dig holes. Such behaviour is framed as the "construction" of a new environmental "niche," which generates new environmental selection pressures and adaptive change. We saw that Dawkins tends to depict genes as active "replicators," while organisms are presented as passive robot-like "vehicles." Epigenetics and niche construction suggest that such language might be up for renewal. Changes to the organismic vehicle are potentially transmitted without any modifications to the genetic replicator, and the active behaviour of the vehicle may change what future genes are selected. Instead of speaking of the selection of atomistic "genes," Jablonka and Lamb thus propose to use more flexible terms like "heritably varying traits," or developmental "networks."[157] These advances have a ring of Lamarckism, and this comparison is indeed sought explicitly.[158]

[154] Stewart-Williams, *The Ape That Understood the Universe*, 33.

[155] Maderspacher, "Epigenetics"; Jablonka and Lamb, *Evolution in Four Dimensions*, 111–52.

[156] Odling-Smee, "Niche Construction"; Laland et al., "The Extended Evolutionary Synthesis," 4.

[157] Jablonka and Lamb, *Evolution in Four Dimensions*, 77, 369–70.

[158] Jablonka and Lamb, 355–56.

Still, mainstream evolutionary biology remains sceptical about the extended synthesis, as many claims are rendered overblown.[159] The possible evolutionary significance of epigenetics is indeed attracting attention, but the evidence for robust adaptive effects over many generations is still flimsy.[160] The mechanisms of niche construction are far more widely accepted, but critics say this is precisely the problem: the concept hardly offers anything new. Niche construction would just be a new label for a wide variety of well-known phenomena.[161] Moreover, even if heredity would include whole developmental networks, we may again question whether it undermines the critical logic of selfish gene theory, as its logic does not necessarily rely on genes as hereditary units.[162] We could also expect developmental networks to become good at reproducing themselves, making them metaphorically "selfish."

Importantly, Dawkins highlighted that the logic of his theory does not depend on genes. In an argument named "universal Darwinism," he maintains that the mechanism of Darwinian selection is substrate-neutral and does not even automatically depend on organic molecules.[163] "Individuals and groups are like the clouds in the sky or dust storms in the desert," he wrote, while genes, "like diamonds are forever."[164] But significantly, Dawkins added that genes are not forever in the same way as diamonds. Like the clouds in the sky, DNA molecules come and go. What makes genes robust over generations is not their particular physical structures but the *information* encoded in those structures.[165] Any form of hereditary information that undergoes processes of non-random selection of undirected variation will result in adaptive complexity. This observation helped spark Dawkins's social-cultural application of Darwinism. It is to this application, as well as to those of others, that we turn in Chap. 4. But only after we have addressed how socio-cultural scholars answered their *cui bono* questions.

[159] Futuyma, "Evolutionary Biology Today"; Charlesworth, Barton, and Charlesworth, "The Sources of Adaptive Variation"; Coyne, "Once Again."

[160] Maderspacher, "Epigenetics," 426; Futuyma, "Evolutionary Biology Today," 6–7; Charlesworth, Barton, and Charlesworth, "The Sources of Adaptive Variation."

[161] Futuyma, "Evolutionary Biology Today," 1.

[162] Hodgson and Knudsen, *Darwin's Conjecture*, 114–15.

[163] Dawkins, "Universal Darwinism"; Godfrey-Smith, *Philosophy of Biology*, 45–49.

[164] Dawkins, *The Selfish Gene*, 34–35.

[165] Dawkins, *The Blind Watchmaker*, 170.

3.5 The Purposes of Cultural Functions

The question of who or what benefits from socio-cultural functions has been even more a point of contention in the social sciences and the humanities than how those functions come about. The debate strikes right at the heart of essential disagreements, not least because of its ideological ramifications. More so than the biological debate, the socio-cultural discussion is rather messy, with schools and disciplines talking past each other and answers being sought in many directions. To gain some overview, three significant perspectives will be discerned. The first explains socio-cultural phenomena as functional for maintaining social systems, the second for power structures, and the third focuses on fulfilling individual needs. This classification is somewhat arbitrary and should be understood as an ideal type. Various schools of thought also drew inspiration from each other, and the work of prominent social theorists is more multifaceted than can be discussed here. Still, the threefold division helps to bring some order to the chaos.

The first perspective is essentially about social order.[166] Why does society not fall apart? How do individuals maintain social systems that last over many generations? According to scholars working within this orientation, humanity's socio-cultural phenomena are doing the trick. Through symbols, rituals, traditions, institutions, and beliefs, people learn to stick together and act as a part of a greater whole. Society is seen as a vast organism, with humans being like the cells or the organs, all playing their part. This view is very old, but its modern scientific expression is found in functionalism. We already noted that Durkheim analysed social facts as existing externally to the individual, so he also sought to answer the *cui bono* question outside the individual. It would not be people's needs that socio-cultural practices fulfil but the "general needs of the social organism."[167] Durkheim was especially interested in the adoration of collective sacred symbols, like a totem, the cross, or a national flag. It creates a state of effervescence and excitement, symbolizing how the individual has no meaning or existence other than as part of something greater and

[166] Ritzer, *Contemporary Sociological Theory*, 13–19, 64–85; Wallace and Wolf, *Contemporary Sociological Theory*, 15–65.

[167] Durkheim, *Les Règles de La Méthode Sociologique*, 95.

more enduring, *other* than the self. This is also how he interpreted the rituals of self-mutilation at aboriginal funerals: the individuals are harmed, but society is made "more alive and more active than ever."[168]

The American sociologist Talcott Parsons developed the analogy of the social organism in more detail.[169] Social systems, he argued, consist of differentiated and interrelated parts that make the system function as a whole. When a biological organism meets challenges, like a temperature rise, the organism's internal mechanisms immediately retain the system's self-maintenance and equilibrium. Likewise, when individuals diverge from their roles, social systems immediately respond by putting these individuals back on track. In most cases, this is not even necessary, since people have internalized the norms and values of their society from a young age. Social cohesion thereby results from consensus. Many societies are characterized by gross dissimilarities in wealth and power, but functionalism developed an answer. Kingsley Davis and Wilbert Moore argued that social inequality is "an unconsciously evolved device by which societies ensure that the most important positions are conscientiously filled by the most qualified persons."[170] Durkheimian functionalism also exerted influence on historians, like the French Annales historians writing about structural collective sentiments.[171] The outlook has generated some internal criticisms, though.[172] In 1949, Robert Merton remarked that specific social usages or sentiments may be "functional for some groups and dysfunctional for other groups in the same society."[173] He furthermore highlighted how religion is not only a stabilizing force but also a thoroughly divisive one. We should not automatically assume that all existing institutions are necessary and, thus, by default, good.

Merton's self-critique foreshadowed the criticisms of functionalism that would unfold in the following decades. In the tumultuous 1960s, it was widely branded a conservative ideology under the veneer of objective science.[174] By analysing institutions as positively functional for the stability of

[168] Durkheim, *Les formes élémentaires*, 574.

[169] Parsons, *The Social System*; Ritzer, *Contemporary Sociological Theory*, 68–80.

[170] Davis and Moore, "Some Principles of Stratification," 184.

[171] Burke, *The French Historical Revolution*, 43–86; Bloch, *La société féodale.*

[172] Merton, *Social Theory and Social Structure*, 79–114, 484–88; Wallace and Wolf, *Contemporary Sociological Theory*, 48–55.

[173] Merton, *Social Theory and Social Structure*, 81.

[174] Wallace and Wolf, *Contemporary Sociological Theory*, 42–44, 64–66; Hamilton, *The Sociology of Religion*, 105.

society, the functionalists were accused of supporting and enhancing those institutions. The functionalist account of social stratification was vilified for legitimizing inequality and exploitation. The focus around this time thus shifted to the second critical perspective on the *cui bono* question: the maintenance of power structures. This perspective, too, has old precedents, and its modern academic form is sometimes called "conflict theory."[175] Here, society is not seen as a place where people work harmoniously together due to shared values but as a power arena in which groups continually compete for dominance. Norms and values are the weapons that favour those in power, and as far as social order exists, it primarily results from coercion.

This second perspective is less homogeneous an orientation than functionalism and is characterized—perhaps unsurprisingly—by more internal strife. One tradition, sometimes dubbed "analytic conflict theory," draws essential inspiration from Max Weber and aims to analyse society as objectively as possible. Power is seen as attractive but scarce, so conflicts of interest are permanent and inevitable. Power relations are multifaceted and entail a broad mixture of economic, political, and symbolic alliances. The other essential tradition takes an arguably even more negative stance on current affairs but expresses greater confidence in the future. After the mechanisms of oppression have been exposed, there is the promise of creating a far more harmonious society in which the human potential will be fulfilled. Intellectuals bear a moral responsibility to critique society and change it for the better. The dividing lines between the powerful and the oppressed are drawn in much sharper lines.

The arch-example of the latter approach is Karl Marx.[176] "The history of all hitherto existing society is the history of class struggle," he wrote.[177] The essence of power relations is thereby seen as material, essentially coming down to who owns the means of production. The "higher" domains of human social life, like morality or the arts, merely represent material power relations. Hence, religion is not a glue keeping society together but helps maintain inequality and exploitation. It is, as Marx famously put it: "the sigh of the oppressed being, the heart of a heartless world, and the

[175] Turner, "Conflict Theory"; Ritzer, *Contemporary Sociological Theory*, 85–91; Wallace and Wolf, *Contemporary Sociological Theory*, 68–157.

[176] Collins, "Reassessments of Sociological History," 157–68.

[177] Wiersing, *Geschichte des historischen Denkens*, 404.

soul of soulless conditions. It is the opium of the people."[178] Marx presaged that internal deterioration would prompt capitalism's self-destruction, leading to the establishment of communism. Marx's prophecies did not materialize, but Marxist theory remained resilient.[179] Thinkers like Antonio Gramsci and Theodor Adorno shifted Marx's focus to how structures of oppression are reproduced at symbolic levels of socio-cultural life. More loosely inspired by Marx, while also sharing tenets with the analytic tradition, is Pierre Bourdieu. He views all domains of socio-cultural life as consisting of particular "fields," such as the fields of the arts, religion, or sports, in which people continually defend their hierarchical position or try to improve it.[180] Trivial things, like preferences for art styles or beverages, are, in fact, the battlegrounds on which positions of power are distributed. It is also through subtle and implicit means that positions of dominance are maintained.

The interest in the symbolic aspects of oppression is also central to the other essential movement within this conflict theory tradition: poststructuralism, or as it is more broadly known, postmodernism.[181] Unlike Marxism, postmodernism rejects all pretensions of doing "hard" science. Nevertheless, its view of power as the essential force governing social life and the sharp dividing lines that it draws between collective groups of oppressors and oppressed can be similar. Instead of the bourgeoisie versus the proletariat, it is antagonisms between straight people and sexual minorities, white people and people of colour, or cisgenders and transgenders that take centre stage. Instead of the "worker's paradise," their ideals regarding diversity operate as history's grand destinies. Postmodernism is especially interested in how power is encoded in our use of language, or as it is often called, discourse. It is believed that power structures can be dismantled through a "deconstruction" of discourse.

Postmodernism has become particularly productive in feminist theory and gender theory. Authors like Luce Irigaray and Joan Scott have argued that patriarchal norms are deeply entrenched in the subconscious traditions of the West.[182] Males have always been a standard point of refer-

[178] Marx, *Der junge Marx*, 109.

[179] Ritzer, *Contemporary Sociological Theory*, 103–17; Farganis, *Readings in Social Theory*, 397–412.

[180] Bourdieu, *La distinction*; Ritzer, *Contemporary Sociological Theory*, 172–83.

[181] Thompson, *Postmodernism and History*; Ritzer, *Contemporary Sociological Theory*, 215–52.

[182] Irigaray, *Speculum*; Scott, "Gender."

ence—think of a word like "mankind"—while women were marked as "the Other." Judith Butler highlights how even a dichotomy between "men" and "women" already represents power by excluding gender-fluidity.[183] Although he disliked the designation, the French thinker Michel Foucault is often considered the most influential postmodernist. His main aim has been to deconstruct standard narratives about liberal modernity being a movement towards increased liberation from external constraints.[184] In Foucault's view, modernity has only made the mechanisms of control over human bodies more pervasive and exploitative. New strategies of subjugation and disciplining were developed in prisons and subsequently entered mental asylums, schools, and hospitals.

Power-oriented approaches have become tremendously influential, but they also face valid criticisms.[185] How does society stick together if shared values and consensus are mere illusions? Is coercion truly a sufficient explanation for the persistence of the social fabric? It is also questioned whether human interactions are as zero-sum as conflict-oriented theorists assume. While functionalism was accused of legitimizing the injustices of the status quo, this orientation is suspected of setting people against each other. Moreover, power and conflict-oriented models often suffer from a theoretical problem similar to functionalism: how did the alleged oppressive structures precisely take shape? Some conflict theorists assume that functional power structures come into being beyond people's intentions and recognition.[186] Karl Marx saw collective classes as the key actors of history—not individuals—and thought that oppressive mechanisms primarily take shape "behind people's backs."[187] A central concept to later Marxist theory was "false consciousness," referring to people's misguided understanding of their real class interests.[188] Likewise, Michel Foucault significantly downplayed the creative role of individuals, assuming that invisible sets of rules constitute discourses of power.[189] On the other hand, Pierre Bourdieu was eager to anchor the reproduction of power structures in the actions and practices of concrete individuals. He did not see

[183] Butler, *Undoing Gender*.

[184] Foucault, *Surveiller et Punir*; Downing, *The Cambridge Introduction to Michel Foucault*.

[185] Wallace and Wolf, *Contemporary Sociological Theory*, 155–57.

[186] Hamilton, *The Sociology of Religion*, 83, 103; Sanderson, *The Evolution of Human Sociality*, 67–68.

[187] Marx, *Das Kapital: Kritik der politischen Oekonomie*, I:6, 85–86, 378.

[188] Ritzer, *Contemporary Sociological Theory*, 25.

[189] Downing, *The Cambridge Introduction to Michel Foucault*.

strategies of oppression as consciously preplanned but did locate them in the more tacit domains of the *sens pratique*.[190] Analytic conflict theorists like Max Weber or Randall Collins were even more pronounced in finding the exertion of power in the micro-interactions of actual people. This brings their work close to our third central perspective on the *cui bono* question: individual needs.[191]

This last perspective is probably the most multifaceted and can be found in many theoretical schools. Still, a common thread is that socio-cultural phenomena are not thought to benefit abstract entities like "systems" or "classes" but flesh and blood people.[192] Its micro-oriented focus on individuals almost automatically leads to a closer affinity with psychology. Proponents do not deny that socio-cultural phenomena can fulfil social functions or that culture can be used in conflicts, but such functions are thought to emerge from the actions and needs of the individuals who created and maintained those phenomena. Neither is it denied that socio-cultural phenomena possess a complexity that exceeds the intentions of any actors involved. But again, such complexity is thought to accumulate from the micro-interactions of individuals.

Interestingly, this perspective also counts one prominent functionalist in its ranks.[193] Bronislaw Malinowski typically claimed that socio-cultural phenomena hang together in functional and unintended ways, but he thought that "the individual, both in social theory and in the reality of cultural life, is the starting point and the end."[194] When explaining rituals of magic, Malinowski not only focused on the social function, but additionally addressed psychological and emotional dimensions: people live in uncertainty about things beyond their control, such as rain, and a magical ritual, like a rain dance, provides a consoling sense of empowerment.[195] After functionalism's decline, individual-oriented approaches gained new momentum, leading to the popularity of schools like symbolic anthropology, social constructivism, and symbolic interactionism. These schools

[190] Bourdieu, *La distinction*; Bourdieu, *Le Sens Pratique*.

[191] Weber, *Wirtschaft Und Gesellschaft*, 6; Wallace and Wolf, *Contemporary Sociological Theory*, 150.

[192] Farganis, *Readings in Social Theory*, 263–368; Wallace and Wolf, *Contemporary Sociological Theory*, 197–380; Hamilton, *The Sociology of Religion*, 157–215.

[193] Malinowski, "The Group and the Individual in Functional Analysis."

[194] Malinowski, 964.

[195] Malinowski, *A Scientific Theory of Culture*, 198–99; Malinowski, *Magic, Science and Religion*, 90.

wanted to stay as close as possible to the meaning that actors themselves give to things. The anthropologist Marshall Sahlins thought that the unity of cultural order is constituted by meaning. It was "this meaningful system that defines all functionality."[196] When considering religion, Clifford Geertz even wanted to dispense with questions regarding functions altogether, rendering it a chimera. Yet, surprisingly and quite inconsistently, Geertz launched his functional interpretation of religion only a few sentences later: religion serves as a "framework in which a wide range of experience – intellectual, emotional, moral – can be given meaningful form."[197] Geertz did not share Malinowski's optimistic outlook, remarking that "over its career, religion has probably disturbed men as much as it has cheered them." Religion does not help us avoid suffering but teaches us *how* to suffer and resist the meaninglessness of profound and unremovable pain.[198]

These approaches focusing on symbols and meaning have little pretensions of doing "hard" science. But over the past decades, the individual-oriented perspective has also given rise to an approach that is staunchly scientific in its ambitions: rational choice theory. This school of thought is closely linked to economics but has also gained ground in other social sciences.[199] A fundamental assumption is that human actors are rational profit seekers who make cost-benefit analyses based on available knowledge. Humans deliberately choose those things that seem to serve their purposes most effectively in each instance. The approach is commonly applied to the exchange of economic goods but has also been used in explaining the exchange of more intangible goods, like approval, respect, or happiness. Anything people can desire is phrased as a reward that rational agents will pursue. The world of religion is understood as a marketplace where religious goods are on offer, with consumers making calculations of utility.[200] There is continuing competition, such as cults competing with cults in Ancient Rome or churches competing with churches in Modern America. Some suppliers sometimes gain a monopoly, like the Roman Catholic Church in Medieval Europe, but the position of religions always remains unstable. When other suppliers offer better bargains, people will look for solace elsewhere. Proponents of this approach

[196] Sahlins, *Culture and Practical Reason*, 206.

[197] Geertz, *The Interpretation of Cultures*, 123.

[198] Geertz, 103–5.

[199] Wallace and Wolf, *Contemporary Sociological Theory*, 302–80; Farganis, *Readings in Social Theory*, 263–302.

[200] Iannaccone, "Voodoo Economics?"; Stark, "Micro Foundations of Religion."

pretend their perspective highlights dimensions of religious life that others failed to see, like financial motives behind religious choices and social, emotional, intellectual, or spiritual benefits.

This orientation also faced criticism. From the Durkheimian perspective, socio-cultural phenomena are functionally integrated in ways that extend the meaning or intentions of actors.[201] A conflict theorist might remark that the impact of power is underestimated. People do not always rationally or meaningfully *choose* what cultural forms to adopt, as they are often *imposed* on them.[202] Furthermore, the symbolic approaches, as well as rational choice theory, claim to stay close to the meaning that actors themselves give to things, but this is not entirely convincing.[203] If sixteenth-century Catholics had been confronted with the idea of their religion being a way of "giving meaning to life" or "an attractive good on the religious marketplace," it probably would have sounded strange and heretical to them. In their view, Catholicism was more likely thought of as an eternal truth on which salvation depended.

To conclude, each of the three main perspectives has highlighted crucial dimensions of socio-cultural life. But they can lead us astray when applied too generally or too exclusively. Socio-cultural phenomena often do not seem to fulfil the functions they are supposed to fulfil, according to one of those perspectives. This indicates that such phenomena are highly multifaceted and escape any single-minded framework. A typical response, not least amongst historians, is to opt for an eclectic approach. One picks and chooses those elements from general frameworks that seem most fitted to the phenomenon one studies.[204] Take the example of beliefs in the afterlife. According to Durkheimians, such beliefs bolster social behaviour in individuals. When people think bad conduct is punished after death, they will think twice before behaving asocially.[205] Conversely, Marx saw the afterlife as something that maintains exploitation: it makes people less likely to revolt against earthly injustices. From the individualistic perspective, it has been highlighted that the belief offers emotional or intellectual support. Malinowski argued that humans cannot face the idea of complete

201 Kincaid, *Philosophical Foundations of the Social Sciences*, 142–90.

202 Hamilton, "Rational Choice Theory: A Critique."

203 Stark, "Micro Foundations of Religion," 266–74; Hamilton, *The Sociology of Religion*, 157–64, 216–18.

204 Ritzer, "Sociology: A Multi Paradigm Science"; Sil and Katzenstein, "Analytic Eclecticism in the Study of World Politics."

205 Davis, *Human Society*, 526–31.

cessation, so grasping at death, they reach for "the comforting belief in spiritual continuity and in the life after death."[206] Overall, there seems to be something to be said for all of these perspectives. So why not mix them?

Eclecticism is not without problems of its own, though. In a critique of theoretical pluralism, Dutch sociologist Nico Wilterdink notes, "On closer scrutiny, the toolkit of theories is quite a mess. The instruments are not ordered neatly next to each other, with each having a clear function of its own. In this higgledy-piggledy box, it is often unclear beforehand what tool can be used for what."[207] Marvin Harris once rudely commented that eclecticism abounds with hidden dangers: "In practice, it is often little more than a euphemism for confusion, the muddled acceptance of contradictory theories, the bankruptcy of creative thought, and the cloak of mediocrity."[208] Eclecticism is probably to be preferred over Harris's procrustean bed of "cultural materialism," but his critique is not entirely beside the point. Take, for instance, the question of whether latent functions exist. When examining the issue, the biologist David Sloan Wilson discovered that some social scientists take the existence of latent functions for granted while others vehemently oppose it. Wilson rightly concludes, "You know there is a problem when one man's heresy is another man's commonplace."[209] Eclecticism will not resolve this problem. Moreover, eclecticism borrows insights from more ambitious approaches, like functionalism or Marxism, but those approaches were only developed because its creators refused to settle with eclecticism's messiness. Eclecticism exploits the wells of grander theories without offering much creativity in return.

Hence, ideally, we should want to work with an approach that unites valuable insights from other perspectives into a more consistent framework while also being a source of creativity itself. Some sociologists, like Norbert Elias, Anthony Giddens, and Randall Collins later in his career, have tried to combine various perspectives into something more consistent.[210] But none of those attempts persevered. We shall now examine whether Darwinian theory is the panacea we might hope for.

[206] Malinowski, *Magic, Science and Religion*, 51.

[207] Wilterdink, *Omstreden Wetenschap*, 11.

[208] Harris, *The Rise of Anthropological Theory*, 285; Harris, *Cultural Materialism*, 287–314.

[209] Wilson, *Darwin's Cathedral*, 2002, 84.

[210] Elias, *Was ist Soziologie?*; Giddens, *The Constitution of Society*; Collins, *Interaction Ritual Chains*.

CHAPTER 4

The Theory

4.1 Darwinian Cultural Evolution

In the spring of 1955, a remarkable meeting took place between scholars from two different generations in the American city of New Haven.[1] With a colleague, the social scientist Donald Campbell visited Albert Galloway Keller, an emeritus professor from Yale. Like many of his academic generation, Campbell had a progressive orientation—he adored Roosevelt's New Deal and opposed anything conservative. On the other hand, Keller was one of the last living of an older generation of evolutionary theorists who had used Darwinian theory to understand culture and society and leaned towards the political right. Keller's evolutionary ideas had once been avant-garde, but in the 1920s and 1930s, they were already disregarded, and by 1955 seemed to disappear into oblivion. Yet, despite the ideological differences, Donald Campbell and his colleague were curious to meet the old Keller. They found his Darwinian perspective stimulating and were eager to learn more about it.

The two visitors were pleasantly surprised by Keller's hospitality. At the same time, it soon became apparent how widely their political views diverged. Campbell supposed that Keller's theory about Darwinian cultural selection justified Roosevelt's New Deal: old institutions had

[1] Campbell, "Variation and Selective Retention in Socio-Cultural Evolution," 24–25.

S. Hofhuis, *The Evolutionary History of Witch-hunting*,
https://doi.org/10.1007/978-3-031-66936-1_4

malfunctioned, and the New Deal created new and diverse institutions that stimulated innovations. But Keller started an exposé on how society was changing for the worse, how his favourite magazine had turned into a left-wing propaganda sheet, and, to Campbell's dismay, he even told jokes about Eleanor Roosevelt's teeth. Darwinian social scientists had a reputation for conservatism, and Campbell argued that the picture was more diverse. But here, he had to admit that Keller just proved the point. Nonetheless, these scholars came to speak extensively about their shared interest in Darwinian cultural evolution. From the perspective of the history of science, it was the bridge between an endpoint and a new beginning. Keller had been one of the last representatives of a school of thought that once looked promising but faded away, while Campbell represented a revival that would gradually unfold over the following decades.[2] This meeting in 1955 thus demonstrates something important: the idea of Darwinian cultural evolution was never entirely gone.

Darwin's theory concerned biological evolution. But soon after his book was published in 1859, scholars began to wonder whether its explanatory power might extend to the socio-cultural realm.[3] Darwin himself was particularly intrigued by apparent similarities between the evolution of living nature and human languages. He compared the competition between words in languages to the process of natural selection and toyed with the idea of non-intentional adaptations to languages. While Darwin remained first and foremost a biological researcher, others explored such similarities in more detail.[4] Walter Bagehot compared natural selection and the competitive struggle between political institutions and nations, Benjamin Kidd wrote about the selection of ethical principles, and William James contemplated the selection of ideas in the history of science.

These avenues of research looked promising, but what was being selected often needed to be clarified. Applications of Darwinism to the socio-cultural realm tended to result in confusing mixtures of the alleged selection of individuals, groups, nations, races, institutions, and ideas. What often remained particularly unclear was the distinction between the selection of biological variants and cultural variants.[5] Many Darwinian

[2] Carneiro, *Evolutionism in Cultural Anthropology*, 68–72, 91–95, 173–78.

[3] Carneiro, 68–72.

[4] Hodgson and Knudsen, *Darwin's Conjecture*, 8–11; Darwin, *The Descent of Man*, 108–13.

[5] Hodgson and Knudsen, *Darwin's Conjecture*, 5–8.

theories were ultimately concerned with the former.[6] The emergent properties of socio-cultural processes, as would later be explored by Durkheim and the Boasians, were still poorly understood. Typically, in his book *The Descent of Man,* Darwin expressed his worries about the foreseeable degeneration of the human race, quoting a social commentator named William Rathbone Greg: "the careless, squalid, uninspiring Irish multiplies like rabbits," whereas the "frugal foreseeing, self-respecting, ambitious Scot (...) leaves few behind him."[7] Darwin agreed and interpreted these alleged differences as a consequence of biological predispositions, and not of cultural variations or stereotypes. Darwin expressed his concerns about modern society, which supposedly selected the wrong biological types; people with undesirable traits outbreed people with desirable traits.[8] In the late nineteenth and early twentieth centuries, such Darwinian concerns became a powerful influence on social thought. Today, this is what often goes under the name of social Darwinism.[9]

At the same time, some contemporaries did recognize that the selection of biological and socio-cultural units were two potentially different things. The philosopher David George Ritchie observed that human experiences could be transmitted independently from the continuity of biological traits and that the mechanism of natural selection could thus be "applied in a new sphere."[10] He wrote in 1896: "There is going on a natural selection of ideas, customs, institutions, irrespective of the natural selection of individuals and of races."[11] Later on, in 1915, Albert Keller made it the key focus of attention in his book *Societal Evolution.*[12] The wisdom of hindsight now allows us to conclude that it formed a promising line of investigation that deserved closer scrutiny.

Yet, the socio-cultural disciplines increasingly distanced themselves from Darwinian approaches, as they came to be associated with apologies for reactionary politics, biological reductionism, rough capitalism, and

[6] See, for example: Ammon, *Die Gesellschaftsordnung und ihre natürlichen Grundlagen*; Lapouge, *Les sélections sociales.*

[7] Darwin, *The Descent of Man*, 164.

[8] Darwin, 158–72.

[9] Hawkins, *Social Darwinism in European and American Thought 1860–1945.*

[10] Ritchie, *Darwin and Hegel and Other Philosophical Studies*, 2:47–49, 99–100.

[11] Ritchie, "Social Evolution," 171.

[12] Keller, *Societal Evolution.*

racially supremacist colonialism.[13] Unfortunately, such associations were not entirely unwarranted. Especially concerns about the biological degradation of the race that had taken an ugly turn. Darwin backtracked when it came to political measures, as he believed that any forced attempts to stop the "unfit" from propagation would conflict with our instinct of sympathy, "the noblest part of our nature."[14] However, others had fewer inhibitions, and in the twentieth century, Darwinism became one of the key inspirations for the notorious socio-political programme of eugenics. In various countries, such as the USA and Sweden, this programme resulted in mass sterilizations of the alleged unfit, and in Nazi Germany, people were killed for this reason.[15] Such crimes contaminated all socio-cultural applications of Darwinism. The nuance that a scientific theory about the selection of cultural units is different from eugenics easily got lost.

Cultural Darwinism also became too strongly associated with a school of thought often labelled as "evolutionism." This perspective treats socio-cultural development as a form of progression in several stages.[16] The idea can be traced back to the Enlightenment at least and in the second half of the nineteenth century, found new scientific support in authors like Herbert Spencer, Edward Tylor, and Lewis Hunt Morgan.[17] While this approach never convinced many historians, in the first half of the twentieth century, it was radically rejected by many social scientists. Critics like the Boasians vilified evolutionism for its conjectures, imprecisions, and racist and colonialist biases. But again, we need to highlight distinctions. The idea of Darwinian cultural evolution is potentially different because it can imply more capricious and historically contingent processes.[18] Franz Boas himself recognized this, but his comments were to no avail; the bad reputations of social Darwinism and evolutionism weighed heavily on cultural Darwinism, which waned accordingly.[19]

[13] Hofstadter, *Social Darwinism in American Thought, 1860–1915*; Carneiro, *Evolutionism in Cultural Anthropology*, 75–98.

[14] Darwin, *The Descent of Man*, 159.

[15] Levine, *Eugenics. A Very Short Introduction*.

[16] Carneiro, *Evolutionism in Cultural Anthropology*; Wilterdink, "The Concept of Social Evolution."

[17] Harris, *The Rise of Anthropological Theory*, 8–53, 250–604.

[18] Carneiro, *Evolutionism in Cultural Anthropology*, 171–79; Mesoudi, *Cultural Evolution*, 2011, 37–40.

[19] Lewis, "Boas, Darwin, Science, and Anthropology."

Campbell decided to pick up the idea in the 1950s, describing it as a model of "variation and selective retention in socio-cultural evolution." He recognized its scientific potential, as he wrote in 1965:

> Elaborate adaptive social systems, such as the priesthood dominating irrigation civilizations, could have emerged, just as termite societies, without any self-conscious planning or foresightful action. It provides a plausible model for social systems that are 'wiser' than the individuals who make societies, or the rational social science of the ruling elite.[20]

In other words, Campbell proposed a possible answer to the question posed above: How do non-intentional cultural adaptations come about? Nonetheless, few colleagues were willing to adopt his ideas.[21] Despite this opposition, several isolated scholars further elaborated upon the theory, taking it in new directions. One of them was Richard Dawkins.[22] In his book *The Selfish Gene,* he added a bold little chapter on the cultural application of Darwinian evolution. As Dawkins claimed, there are cultural equivalents of the gene. By shortening the Greek word for imitation, *mimeme*, and making it sound like gene, he created a new word for this cultural item: the meme. In Dawkins's view, all elements of culture that are transferred as separate units could count as memes, and amongst the examples he mentioned were "tunes, ideas, catch-phrases, clothes, fashions, ways of making pots or building arches."[23] Like the genes, Dawkins saw these memes as unconscious, blind replicators that undergo a Darwinian process of variation, selection, and replication, thus evolving into things that looked like they were designed.

What made Dawkins's little chapter special was that he also imported the concept of "selfish replicators" into the cultural domain. Genetic evolution had initially created bigger brains to store memes that helped human propagation. But once these new memetic replicators arose, a new kind of evolution was set in motion. Many of Dawkins's biological colleagues still tended to look for sociobiological answers when explaining human behaviour, but Dawkins presumed that novel replicators could

[20] Campbell, "Variation and Selective Retention in Socio-Cultural Evolution," 28.
[21] Hodgson and Knudsen, *Darwin's Conjecture*, 18–23.
[22] Dawkins, *The Selfish Gene*, 189–201.
[23] Dawkins, 192.

develop novel purposes. The reproductive interests of genes and memes do not necessarily coincide. "What we have not previously considered is that a cultural trait may have evolved in the way it has, simply because it is advantageous to itself."[24] Dawkins's key proposal was that memes could spread like selfish genes or infectious parasites while subverting the interests of their human carriers. In other words, our behaviour is not only controlled by selfish genes but also by "selfish memes."

Dawkins's little chapter went largely unnoticed for quite a while, but in the 1990s, it gained more traction.[25] Several people became intrigued by the idea of "selfish" cultural phenomena designed by natural selection and spreading in a pathogen-like way at the expense of their human hosts. It was mainly in this idea that Daniel Dennett saw Darwinism's most significant potential of renewing the socio-cultural disciplines.[26] In 1997, a so-called *Journal of Memetics* was launched to turn meme theory into an accepted scientific discipline. In 1999, the English psychologist Susan Blackmore wrote a best-seller named *The Meme Machine*.[27] Moreover, Dawkins's meme became an enormous and lasting success in popular culture. Today, the internet is full of so-called internet memes, small texts, images, or videos, often humorous, that spread across the web.

However, the academic reception of "memetics" was a disaster. Many biologists were sceptical, and responses in the socio-cultural disciplines were almost entirely hostile.[28] The French anthropologist Maurice Bloch said that memes as body snatchers are "as intriguing, as frightening and as likely as invasion by little green men from Mars."[29] Critics rejected meme theory as either a vacuous re-description of things we knew all along or as a fad that completely misrepresents the workings of socio-cultural processes. The suggestion that cultural units have a selfish "will of their own" seemed to reduce human agents to passive vehicles while reifying culture to a bizarre extent. The term "meme" also encountered scepticism, as hardly anyone saw the added value of chopping culture into little distinct

[24] Dawkins, 200.

[25] Robert Aunger, *Darwinizing Culture*.

[26] Dennett, *Darwin's Dangerous Idea*, 335–69.

[27] Blackmore, *The Meme Machine*.

[28] Millikan, *Varieties of Meaning*; Lewens, *Cultural Evolution*, 2015; Fracchia and Lewontin, "Does Culture Evolve?"; Kuper, "If Memes Are the Answer, What Is the Question?"; Wilterdink, "Darwinisme in de Sociale Wetenschappen," 451–52; Kronfeldner, *Darwinian Creativity and Memetics*.

[29] Bloch, "A Well-Disposed Social Anthropologist's Problems with Memes," 201.

units. These critiques were not without effects because meme theory failed to establish itself as a scientific field. The *Journal of Memetics* was discontinued in 2005.

At the same time, however, and more quietly at first, the idea of Darwinian cultural evolution stimulated approaches that would become a more durable success.[30] In the 1980s and 1990s, the academic duos Luigi Cavalli-Sforza and Marcus Feldman and Peter Richerson and Robert Boyd developed a form of cultural Darwinism that did manage to establish itself as an academic field. These authors had mathematical natural science backgrounds, and their approach used mathematical evolutionary modelling from biology. Their quantifications of how cultural traits spread within a population reaffirmed the insight that cultural transmission can occur semi-independently from biological transmission. Simultaneously, they avoided the pitfalls of memetic terminology. Instead of "memes," the common term "culture" was adopted, and as a unit of selection, Richerson and Boyd preferred the more flexible notion of "cultural variant." From around the 2000s onwards, this application of Darwinian theory started to gain ground. A new field of cultural evolution is proliferating and has earned its own research budgets, conferences, institutes, and major publications.[31] Its claims may not sound as grandiose as those from memetics, but the ambitions are still significant. The psychologist Alex Mesoudi expresses the prospect of "synthesizing" the balkanized socio-cultural sciences, just like Darwinian theory once synthesized the disciplines of biology.[32]

A few aspects of this growing interdisciplinary field stand out. First, its precursors have succeeded in setting a new standard of quantitative modelling.[33] The methods used are highly diverse, ranging from field studies and laboratory experiments to the collection of comparative datasets. Still, nearly all of them are anchored in mathematical models, often adopted from population biology. Evolutionary biologists, for instance, use phylogenetic models to reconstruct evolutionary relationships between species, and such models are now being applied to reconstruct evolutionary connections between culturally transmitted traits such as fairy tales or

[30] Cavalli-Sforza and Feldman, *Cultural Transmission and Evolution*; Boyd and Richerson, *Culture and the Evolutionary Process*; Richerson and Boyd, *Not By Genes Alone*.

[31] Lewens, *Cultural Evolution*, 2015; Mesoudi, "Pursuing Darwin's Curious Parallel."

[32] Mesoudi, *Cultural Evolution*, 2011, 205–20.

[33] Mesoudi, "Cultural Evolution," 2016.

languages.[34] Consequently, the relationship with qualitative and interpretative approaches in socio-cultural disciplines remains largely unexplored territory.[35]

Second, while the field distances itself in some respects from the biologically reductionist programmes of sociobiology and evolutionary psychology, it still seeks to build models that are based on realistic assumptions about individual cognition. The field's approaches often presuppose the existence of innate and adaptive biological "learning biases" that result from gene-culture coevolution. Those biases make some cultural variants more likely to be adopted and transmitted than others.[36] Some exponents are moving even further from evolutionary psychology, claiming that these learning biases can be a product of cultural learning.[37]

Third, many key authors endorse Campbell's and Dawkins's claims about the evolution of non-intentional cultural adaptations.[38] Joseph Henrich states about the cultural evolutionary process that: "(it) is – in some crucial sense – smarter than we are. Over generations, often outside of conscious awareness, individuals' choices, learned preferences, lucky mistakes, and occasional insights aggregate to produce cultural adaptations."[39] The biologist David Sloan Wilson writes that "cultural evolution often results in practices that work without anyone knowing why they work." He explicitly draws comparisons to sociological and anthropological functionalism.[40] However, Dawkins's claim about "selfish replicators" has not found much support.[41]

The field of cultural evolution is thriving, so it has become challenging to summarize all the current trends. So, here, we focus on our two key questions: how do cultural adaptations come into being, and what are they purposeful for? Darwinian cultural evolution offers two potentially new answers: first, variation and selective retention in cultural evolution result in unintended and unrecognized cultural adaptations, second, cultural

[34] Tehrani J.J, "The Phylogeny of Little Red Riding Hood."

[35] Mesoudi, *Cultural Evolution*, 2011, 18–20.

[36] Richerson and Boyd, *Not By Genes Alone*, 119–26; Mesoudi, *Cultural Evolution*, 2011, 55–84.

[37] Mesoudi, "Cultural Evolution," 2016, 492.

[38] Wilson, *Darwin's Cathedral*, 2010; Henrich, *The Secret of Our Success*; Kincaid, "Functional Explanation and Evolutionary Social Science," 2007.

[39] Henrich, *The Secret of Our Success*, 34.

[40] Wilson, *This View of Life*, 207.

[41] Lewens, *Cultural Evolution*, 2015.

adaptations are well-built to ensure their self-reproduction, sometimes even at the expense of their human hosts. We shall now take a precise look at these answers. Do they have genuine explanatory power, and what might concrete examples of such cultural adaptations be?

4.2 Darwinian Origins of Cultural Functions

Robert Merton provided no convincing explanation for how latent functions arise. Unintended consequences produce chaos, rather than functions. But what if we assume that people introduce all sorts of cultural variants with unintended consequences, most of which die out, but now and then, variants appear that are functional by accident? What if such cultural chance hits are then cumulatively preserved in repeated selection rounds? In the case of the Hopi rain dance, this would come down to a scenario in which the newly introduced dance enhanced social cohesion only accidentally at first. But when Hopi groups dancing faced ecological or military challenges, the cohesion resulting from their ritual improved the group's persistence, also leading to the persistence of the dance itself. Earlier, we saw that a phenomenon can be seen as functional when its existence has depended, or depends, on the effects that it has. Within this framework, the rain dance can thus be said to have served a function of social cohesion, even if the actors themselves were incognizant of the effect.

What again did Anthony Giddens write? "Showing that the rain dance fosters social solidarity *cannot possibly* be an explanation of the actions of the Hopi unless at least some of them sense this is so and act accordingly."[42] Like a William Paley of sociology, Anthony Giddens argued that there cannot be design without a designer. Yet, through a Darwinian process, the evolution of socio-cultural design without a designer becomes a real option. This model implies that Merton's unintended consequences provide the raw material for a process of selection that builds its adaptations from it. Anthony Giddens overlooked something crucial.

The Darwinian scenario is quite simple but like biological Darwinism, deceptively so. All ingredients of the process have been recognized innumerable times before. Merton was not the first to highlight unintended consequences—John Locke, Georg Hegel, Max Weber, Norbert Elias, and many others preceded him.[43] Competition between cultural forms has

[42] Giddens and Pierson, *Conversation with Anthony Giddens*, 92. Emphasis added.

[43] Vernon, "Unintended Consequences."

also been widely recognized, as analytic conflict theory and rational choice theory help to illustrate. That culture is socially reproduced from person to person has often been marked as culture's essence. What mainstream socio-cultural scholarship has failed to recognize correctly, however, is the counter-intuitive outcome of these elements *in combination*: unintended socio-cultural functions. Admittedly, some earlier authors came close to something like it occasionally.[44] Notions of selection and survival were often in the background in functionalist accounts.[45] Talcott Parsons maintained that when people do not fulfil their social roles, "society would very soon cease to exist."[46] Marvin Harris said things about "selection processes" and "differential survival" with a Darwinian flavour.[47] Yet, such isolated remarks were not developed into a consistent Darwinian framework, and so it was thus that the functionalists and Harris failed to defuse the counter-functionalist critique.

Now that the field of cultural evolution is reviving the old functionalist notions, this counter-functionalist position is reappearing. Historical sociologist Joseph Bryant notes that societies "have no power to 'act' or 'adapt' in any fashion" since human actors reflexively and non-randomly construct and define their social spheres.[48] The anthropologist Tim Ingold laments that the new Darwinians create a wrongheaded "topsy-turvy world" in which humans do not adapt through culture but where culture adapts through humans.[49] Yet, these authors seem to have missed Darwinism's true potential. A Darwinian model can help us understand how socio-cultural phenomena have the power to "act" or "adapt" through humans. Such critiques illustrate how the theory not only parasitizes existing knowledge but offers something new and essential.

The model does raise theoretical questions. What, for instance, about history's idiographic complexity? Some current versions of cultural evolution are so very abstract, mathematical, and coarse-grained that it will indeed be challenging to reconcile these versions with things that many historians hold dear. However, idiographic complexity does not have to be a problem at all: the Canadian sociologist Marion Blute rightly observes

[44] Whitehouse, "Why Is a Science of the Sociocultural so Difficult?," 422.

[45] Kincaid, "Functional Explanation and Evolutionary Social Science," 2007, 214–19; Malinowski, *Magic, Science and Religion*, 40.

[46] Quoted in: Wallace and Wolf, *Contemporary Sociological Theory*, 44.

[47] Harris, *Cultural Materialism*, 60.

[48] Bryant, "An Evolutionary Social Science?," 476.

[49] Ingold, "The Trouble with 'Evolutionary Biology,'" 16.

that the theory has the potential to resolve the old divide between the idiographic and the nomothetic. Cultural phenomena adapt to their environments in general, but since adaptedness is context-dependent and different in each instance, the theory simultaneously looks for the particular.[50] Like the multifunctional elephant's trunk, cultural phenomena can be considered subject to changing and multifaceted selection pressures. This assumption that cultural traits are "adaptive" should not be taken too far, though, assuming that everything cultural has a function.[51] It was already said that Darwinian theory offers room for non-functional traits. Sociocultural phenomena may contain adaptations to past environments, they can be a by-product of something else, and some new cultural variants may be neutral in their effects on their survival, thus leading to cultural equivalents of genetic drift.[52]

The most pressing issue, however, as already resurfaced in Bryant's and Ingold's comments, is the role of human agency. While biological evolution may indeed be a process of "design without designer," the evolution of cultural adaptations, at first sight, looks more like "design *with* designers" in the form of creative and reflexive human agents. In so far as Joseph Bryant accepts the notion of culture adapting to environments at all, he thinks it is "accomplished through the symbolic interaction of acculturated, minded selves."[53] Even Steven Pinker, who thoroughly supports Darwinian theory, maintains that the model requires "iterations of copying of random errors," which purportedly does not hold in the cultural realm. Instead, our highly non-random brains deliberately produce cultural "mutations" in response to felt needs.[54]

What to make of this critique? First of all, we should remind ourselves that the randomness of the model does not refer to the causes of "mutations" but only to their fitness outcome. Pinker may be right that cultural variants usually do not occur randomly, but the effects that make them survive can still be accidental and unrecognized.[55] But of course, human intentional design still makes a significant difference here, as any model of

[50] Blute, "History versus Science," 345.

[51] Wilson, *Darwin's Cathedral*, 2002, 203; Kincaid, "Functional Explanation and Evolutionary Social Science," 2007, 225.

[52] Campbell, "On the Conflicts between Biological and Social Evolution," 196; Mesoudi, *Cultural Evolution*, 2011, 76–79.

[53] Bryant, "An Evolutionary Social Science?" 475.

[54] Pinker, "The False Allure of Group Selection."

[55] Wilson, *This View of Life*, 196; Smail, *On Deep History and the Brain*, 91.

Darwinian cultural evolution should recognize. However, the difference does not yet refute the Darwinian mechanism because intentional design and blind evolution may complement each other.[56] We could think of it as a spectrum, with conscious, intentional design at one end and blind Darwinian design at the other. The evolution of adaptations might often be a complex mixture of intended, subconscious, and unintended elements, with the levels of recognition also differing from person to person. The extent to which the Darwinian mechanism operates must be addressed empirically.

This empirical dimension immediately leads to the counter-argument that Darwinism is a tautology. In Anthony Giddens's words, "It is of little value indeed to claim that those societies or types of society which have survived for a period of time, because they survived, must have survived."[57] Apart from contradicting the objection that the theory is incorrect, this fails to recognize how the model contains multiple non-tautological elements.[58] The Hopi rain dance scenario has several premises: (1) The Hopi introduced innumerable variants, most of which did not make it. (2) The variant of the rain dance strengthened social cohesion. (3) The increased survival of these groups due to this social cohesion helped preserve the ritual itself. (4) The effect on social cohesion was unintended and unrecognized by the Hopi involved. These premises all have the potential of being false to various degrees.

Fortunately, the current field of cultural evolution has already presented many case studies. A fruitful domain is the resurgence of old functionalist theories of religion.[59] Trendsetting was David Sloan Wilson's 2002 book *Darwin's Cathedral*, which analyses how seemingly irrational and fictional religious convictions can have this worldly effect on enhancing social cohesion. Wilson, for instance, looked at the historical example of Calvinism, which emerged in sixteenth-century Geneva.[60] He analyses the European Reformation as a process of many experiments, few of which survived, with Calvinism possessing certain traits that made it a durable

[56] Wilson, *Darwin's Cathedral*, 2002, 76–79; Blute, *Darwinian Sociocultural Evolution*, 138–61.

[57] Giddens, *The Constitution of Society*, 235; Moore, *Visions of Culture*, 171.

[58] Kincaid, "Functional Explanation and Evolutionary Social Science," 2007, 223–26, 237.

[59] Sosis and Petersen et al., "Why Cultural Evolutionary Models"; Joseph Bulbulia et al., "The Cultural Evolution of Religion."

[60] Wilson, *Darwin's Cathedral*, 2002, 86–124.

success. Calvin's Catechism, for example, told its adherents to abandon self-will and to make sacrifices for the common good. Other current case studies include old functionalist classics like everyday celebrations, dances, and self-mutilation rituals.[61] It is thereby added that it can even be essential that the actors themselves do not understand the functional rationale of their supernatural belief systems because the sincerity of beliefs endows those systems with at least some of their effectiveness.[62]

In addition, there is a resurgence of the kind of eco-materialist explanations that Harris and others once explored. Joseph Henrich recalls how, during fieldwork in Fiji, he came across a taboo on the consumption of eels during pregnancies. The taboo makes sense from a scientific perspective, as eels contain elements potentially toxic to foetuses. However, when Henrich asked the Fijians about the function, they referred to it as "custom" and did not seem to be doing it for health reasons. Henrich claims that the underlying adaptiveness of the taboo is, as he calls it, one of culture's "secret tricks."[63] This line of investigation is promising and might also revitalize older case studies, like the Hindu holy cow or Aztec human sacrifice.

Economics is the discipline that has probably toyed with Darwinian cultural evolution most endurably.[64] Since the days of Adam Smith, economists have been familiar with the idea of "an invisible hand" that leads liberal economic processes towards non-designed functional outcomes. The famous Austrian economist Friedrich Hayek elaborated upon that, using selectionist notions that were vaguely Darwinian in their outlook.[65] Hayek saw "the emergence of order as the result of adaptive evolution," which resulted from a "survival of the successful."[66] In addition, he cautioned politicians not to interfere in this process since the evolutionary mechanisms were far better able to lead to adaptive outcomes than deliberate design; this would help explain why capitalist economies outperformed communist ones. More recently, others have looked at the selection of firms and organizations, claiming that businesspeople often do not fully understand which institutions or routines made their companies successful

[61] Henrich, *The Secret of Our Success*; Whitehouse, "Dying for the Group."
[62] Henrich, *The Secret of Our Success*, 106, 165; Norenzayan, *Big Gods*, 54.
[63] Henrich, *The Secret of Our Success*, 100–103, 145.
[64] Ridley, *The Evolution of Everything*, 96–117; Frank, *The Darwin Economy*.
[65] Beck and Witt, "Austrian Economics and the Evolutionary Paradigm*."
[66] Hayek, *The Constitution of Liberty*, 57–59.

or unsuccessful.[67] Another domain is the evolution of language.[68] Some linguists maintain that our language systems contain internal structures and usefulness that result from unintended Darwinian selection.

These many case studies look promising, but we must note empirical issues. Take the example of the Fijian food taboo on eels. Henrich's blind scenario sounds plausible, but it is difficult to rule out that past Fijians consciously or subconsciously intended and recognized the taboo's positive effect on pregnancies. Unfortunately, Fijian history has not left behind many sources that would enable a detailed investigation into how the taboo evolved and what people themselves understood about its effects. This problem includes other examples, like the Hopi rain dance or Aztec cannibalism. Importantly, this lack of evidence is not just a problem for the blind Darwinian scenario but also competing scenarios. In many cases, we cannot know how functions came about.

Historians are amongst those best able to offer solace because they traditionally study processes that leave many sources behind. Wilson's case study of Calvinism looks more promising because there are ample sources that would allow a more detailed reconstruction of how Calvinist doctrines enhanced the cooperative survival of Calvinist groups and show what involved actors understood about this rationale. Notably, Wilson has not yet done this in detail. "We can safely conclude that conscious, intentional thought explains some but not all of the functionality of Calvinism," he insists, but his empirical evidence is meagre.[69] Wilson consulted Calvin's catechism and ordinances, but their implementation was still relatively intentional and top-down. He highlights how Calvinists themselves continually encouraged social behaviour and compared their communities to organisms.[70] This, however, indicates manifest functions rather than latent ones.

Another example is a thought-provoking book by Joseph Henrich, which delves into how people in Western Europe have developed a distinctive culturally induced psychological pattern of individualism, nonconformism, and trust in strangers over the past two millennia.[71] Henrich attributes a decisive early role to the medieval Roman Catholic Church. By stimulating monogamy and by banning marriages between relatives, the

[67] Nelson and Winter, *An Evolutionary Theory of Economic Change.*
[68] Kirby, "Language as an Adaptive System."
[69] Wilson, *Darwin's Cathedral*, 2002, 121.
[70] Wilson, 86–124; Wilson, *Evolution for Everyone*, 233–54.
[71] Henrich, *The WEIRDest People in the World.*

Church broke the old shackles of kin-based institutions. The Church reaped the benefits, as it became a more effective rival for people's loyalty at the expense of tribal loyalties. Possessions of extinct families also fell into ecclesiastical hands. Importantly, Henrich claims that all this happened accidentally and unintentionally.[72] But did medieval clergymen really have no clue whatsoever about their marriage policies breaking tribal power and enhancing the Church's supremacy? Answering that question would require a thorough qualitative engagement with medieval sources. Henrich's book does not address this empirical dimension.

The absence of more detailed research in this direction is probably due to the field's quantitative orientation. This focus has many advantages—the value of mathematical modelling for Darwinian theory is undisputed. Still, if we want to put the idea of "design without designer" to the test, the field simultaneously needs to move beyond that. The design hypothesis demands investigations into how involved actors comprehended the functionality of their socio-cultural phenomena and compares those actors' understanding with the actual adaptive effects that helped those phenomena survive throughout particular periods. This requires *Verstehen*, hermeneutics, thick description, and the meaningful interpretation of historical documents. What the field needs right now is a qualitative turn. In the following chapters, we shall see how European witch-hunting provides a promising case study.

Yet, two theoretical steps first need to be taken. We still have to look at the purposes cultural adaptations fulfil. But there is also another topic that demands attention. The proposed cultural Darwinian model assumes that some effects of cultural phenomena are real, like social cohesion or protein intake, while other effects are non-real, like supernaturally induced rainfall or sunrise. This assertion, in turn, presupposes that modern science potentially better understands certain aspects of reality than alternative viewpoints. However, this will make many socio-cultural scholars dig their heels in today. Over the past half-century, various forms of epistemic relativism have gained influence, rendering distinctions between fact and fiction, science and non-science, or objective and subjective obsolete. If warranted, this would make the whole Darwinian model break down immediately. This relativist movement is often called "postmodernism." We will now take a closer look at it, especially by addressing the work of the author Bruno Latour.

[72] Henrich, 161–67, 281.

4.3 Intermezzo: Epistemic Relativism

"We won't try to discipline you, to make you fit into our categories; we will let you deploy your own worlds, and only later will we ask you to explain how you came about settling them." This is how the French philosopher and social theorist Bruno Latour suggests social scientists approach the actors they study. The task of defining and ordering the social, he argues, "should be left to the actors themselves, not taken up by the analyst."[73] Social scientists should not treat actors as mere "informants" while attributing a broader and more complete understanding of things to themselves.[74] Latour even renders the idea of scientists having a better understanding of a reality "out there" unjustified altogether.[75]

Taken to its logical extreme, this viewpoint puts a bomb under the cultural Darwinian model. Accordingly, when I present the approach of Darwinian cultural evolution to colleagues from the social sciences or the humanities, Latour is the name they mention the most when making objections. Reading Latour would clarify why academic scholars should not aim to distinguish real from illusory effects as I do with the Hopi rain dance. That Latour's name is often mentioned is hardly surprising since he belongs to the most cited scholars in the humanities.[76] Latour is also described as the "most illustrious champion" of the vast science and technology studies field, which aims to examine science in its broader context.[77] Hence, considering his reputation, we must contemplate whether Latour's epistemic position indeed convincingly undermines the distinction between real and non-real effects.

Latour is often labelled as one of the representatives of postmodern relativism. The question of precisely what this movement is can engender much confusion and controversy, as the term's meaning is as slippery as the movement's academic status is contested.[78] In academic parlance, postmodern relativism refers typically to the thought of a loose amalgam

[73] Latour, *Reassembling the Social*, 23.

[74] Latour, 32–33.

[75] Latour and Woolgar, *Laboratory Life*, 180–83.

[76] "Most Cited Authors of Books in the Humanities, 2007 | Times Higher Education (THE)."

[77] Fuller, *The Philosophy of Science and Technology Studies*, 4.

[78] For overviews of the movement, see: Searle, "Rationality and Realism, What Is at Stake?"; Gross and Levitt, *Higher Superstition*; Boghossian, *Fear of Knowledge*; Butler, *Postmodernism*.

of French thinkers, including Jacques Derrida, Michel Foucault, Jean-Francois Lyotard, Jacques Lacan, Julia Kristeva, Gilles Deleuze, and Bruno Latour, although some of these authors, not least Latour, reject the label themselves. From the 1960s onwards, their influence on the social sciences and the humanities has been substantial. Their ideas gained traction, especially in the Anglo-Saxon world, with Judith Butler, Donna Haraway, and sometimes Richard Rorty being mentioned as influential advocates.[79] The controversies regarding the movement came to a head in the so-called science wars of the 1990s, but disagreements continue to this day.

In the eyes of its supporters, postmodernism is a creative movement that has helped us understand that reality cannot be represented through human conceptual categories. This implies that no account of reality can be epistemologically more accurate than any other account.[80] Of course, relativist notions about knowledge being historically conditioned, with no one having privileged access to the truth, are much older.[81] Yet, in the late 1970s, when grand narratives of modernity appeared to have broken down, people's minds were ripe for a novel surge of relativist ideas. Postmodernism then showed scepticism towards unifying metatheories and instead aimed to be acceptant of difference and variety. Its thinkers were especially interested in language, which was seen as a system of signs that all refer to each other. Language constructs reality, implying that science is just one language system amongst others and has no privileged access to an objective world. In addition, postmodern theorists portrayed modern science as being inextricably linked with power. By deconstructing science's language and its claims of objectivity, established positions of power can be challenged. While the movement is not necessarily in itself an ideology, it enables ideologies like feminism, LGBTQ+ activism, antiracism, multiculturalism, or postcolonialism. Postmodern theory would thereby carry the promise of creating a more accepting social environment in which oppressed groups or cultures are shielded from the charge of holding false beliefs.[82]

[79] Wilterdink, "The Sociogenesis of Postmodernism"; Leezenberg and Vries, *History and Philosophy of the Humanities*, 293–354.

[80] Brown, *Postmodernism for Historians*, 7–11; Thompson, *Postmodernism and History*, 91–94.

[81] Baghramian and Coliva, *Relativism*.

[82] Boghossian, *Fear of Knowledge*, 130; Law, "STS as Method," 46–47.

Opponents, on the other hand, describe the postmodern epistemic position as folly.[83] The movement is said to have broken down standards of academic quality, so "higher superstition," "intellectual impostures," or "the vacuous rhetoric of mountebanks and charlatans" are some of the designations.[84] The language postmodernists use is said to be opaque and obscurantist; it has the pretension of being profound but comes down to a vacuous mixture of trivialities and absurdities—Noam Chomsky spoke of "truism, error, or gibberish."[85] Another point concerns the denial of scientific facts. Postmodern relativists are accused of dismissing scientific discoveries about the world, seeing those discoveries as mere socio-cultural constructs. But if the postmodernists mean what they say, how can they substantiate their claims about the world?[86] "Fact-constructivism is such a bizarre view that it is hard to believe that anyone actually endorses it. And yet, it seems that many do," Paul Boghossian states.[87]

In the eyes of these opponents, postmodernists respond to such criticisms unsatisfyingly. Questions about the precise meaning of their words are usually met with evasion, irritation, and denunciations. Noam Chomsky observes that instead of clarifying what they exactly mean to say, "the response is cries of anger: to raise these questions shows 'elitism,' 'anti-intellectualism,' and other crimes." The counter-accusation of elitism is seen as particularly strange since it is the postmodernists who use unnecessarily complicated language. Not explaining their positions in plain and concrete language fulfils a function, however: people would immediately deflate the hot air they produce. Another method postmodernist thinkers are accused of using is suddenly pretending that their relativist claims are ironic rather than absurd. Their opponents retort: if postmodernists are joking around, why do they respond so angrily to criticisms, accept significant academic salaries, and want to be taken seriously on so many occasions?[88]

[83] Sokal and Bricmont, *Intellectual Impostures*; Sokal, *Beyond the Hoax*; Chomsky, "Noam Chomsky on Post-Modernism"; Dawkins, "Postmodernism Disrobed"; Pluckrose and Lindsay, *Cynical Theories*; Aya, "Reign of Error."

[84] Gross and Levitt, *Higher Superstition*; Sokal and Bricmont, *Intellectual Impostures*; Dawkins, "Postmodernism Disrobed," 141.

[85] Quote in: Aya, "Reign of Error," 33.

[86] Sokal and Bricmont, *Intellectual Impostures*, 49–96; Wallace and Wolf, *Contemporary Sociological Theory*, 421–26.

[87] Boghossian, *Fear of Knowledge*, 25.

[88] Dawkins, "Postmodernism Disrobed," 142.

The example we will mostly stick to here is Bruno Latour, whose thought directly affects the distinction between natural and non-real effects. Some critics describe the French scholar as one of the most dishonest exponents of postmodernism.[89] His work is said to be "vague, contradictory or simply unintelligible," consisting of "philosophical mystifications" that boil down to "a metaphysical conceit (in both senses of the word) of astounding proportions."[90] What would make Latour particularly tricky is that he masks his mysticism with a charming, seemingly science-friendly posture. His work has a Janus-faced character, springing from relativism to endorsements of scientific facts and back. Latour can deny on one occasion what he demonstrably said on an earlier one; previous relativist statements are then suddenly transformed into banalities nobody would disagree with.[91] Yet, Latour's "animism" and "new age mysticism" are considered far from innocent as they foster the post-truth predicament of alternative facts. Latour has leftist political views, but his critics warn that his methods open the door to fact-free Machiavellianism in the context of any ideology.[92]

Unsurprisingly, Latour and the people inspired by him dismiss such charges as vindictive.[93] "People never really understood what I was saying," Latour retorts.[94] "I was associated with that postmodern relativist stuff, I was put into that crowd by others."[95] His critics allegedly fail to see that Latour criticized the kind of postmodern relativism that perceives scientific findings as mere human constructions of the world. Latour is celebrated for offering a nuanced approach to transcending older human and non-human divisions.[96] Instead of solely focusing on nature determining science, as positivists would have it, or science constructing nature, as outright constructivists would have it, Latour has looked at the human

[89] Malm, *The Progress of This Storm*; Sokal and Bricmont, *Intellectual Impostures*.

[90] Stamenkovic, "The Contradictions and Dangers of Bruno Latour's Conception of Climate Science," 4; Gross and Levitt, *Higher Superstition*, 58.

[91] Sokal, "Pourquoi j'ai écrit ma parodie"; Gross and Levitt, *Higher Superstition*, 58–59.

[92] Stamenkovic, "The Contradictions and Dangers of Bruno Latour's Conception of Climate Science," 2, 9, 29.

[93] Luckhurst, "Bruno Latour's Scientifiction," 5–6; Kofman, "Bruno Latour, the Post-Truth Philosopher, Mounts a Defense of Science."

[94] Paulson, "The Critical Zone of Science and Politics:"

[95] Vrieze, "Bruno Latour, a Veteran of the 'Science Wars,' Has a New Mission."

[96] Dijstelbloem, "De democratie anders."

and non-human actors who become connected through networks. For example, we should not solely perceive a researcher like Louis Pasteur as an observer who made discoveries about microbes that already existed passively.[97] Pasteur, his instruments, the microbes, colleagues, social institutions, politicians, and other actants formed complex networks in which all actants changed due to their interactions. Latour maintains that he "was just arguing that humans and nonhumans are now mixing together more and more intimately."[98] It was merely a more scientific and honest view of how science is practised. Latour's assertion that specific networks become more robust than others would also give his work a non-relativist twist. In the eyes of his supporters, this renowned French thinker, in fact, "mounts a defense of science," which we badly need in an age of post-truth politics.[99]

In Latour's defence, it can indeed be said that he did innovative work on the complex interactions characteristic of scientific research. Latour became particularly famous for bringing the methods of ethnography, traditionally used to study non-Western peoples, into the scientific laboratory.[100] He aimed to look at scientific researchers as if they were a foreign tribe. During his sojourns in the laboratory, Latour noticed that the endurance of scientific "facts" does not simply hinge on their correspondence to the truth. To a large extent, a wide range of social, institutional, and technical factors decide the matter, not least the competitive struggles for dominance between scientists supporting or contesting those facts. Networks of alliances can make a scientific claim stand or fall. While the sociology of science preceded Latour, his empirical work did offer a new counterweight against overly simplistic notions of scientists as objective and rational agents of truth.

However, while we may question the rudeness of his critics, a calmer look at Latour's work does show them to be correct about his epistemology.[101] "Although Latour repeatedly denies being relativist, his writings speak for themselves," Stamenkovic rightly contends.[102] Epistemic relativism is defined as the view that "everything or every truth is relative to

[97] Latour, *The Pasteurization of France.*

[98] Paulson, "The Critical Zone of Science and Politics."

[99] Kofman, "Bruno Latour, the Post-Truth Philosopher, Mounts a Defense of Science."

[100] Vries, *Bruno Latour*, 21–52; Latour and Woolgar, *Laboratory Life.*

[101] Baghramian and Coliva, *Relativism*, 135–39.

[102] Stamenkovic, "The Contradictions and Dangers of Bruno Latour's Conception of Climate Science," 7.

some standards so that, when two or more people disagree about these issues, they may all be correct."[103] Unmistakably, Latour has made statements that fit this definition. Take this quote from 1988:

> There is no difference between the 'real' and the 'unreal,' the 'real' and the 'possible,' the 'real' and the 'imaginary.' Rather, there are all the differences experienced between those that resist for long and those that do not, those that resist courageously and those that do not, those that know how to ally or isolate themselves and those that do not.[104]

Simply put, whether something is true or false is beside the point; it only matters whether it endures. Hence, what is relativism if Latour's rejection of the quest for truth is not it? In a book on his actor-network approach from 2005, Latour also made striking claims about the nature of reality, typical for Science and Technology Studies.[105] He argued that we live in a "pluriverse" in which things are fundamentally multiple. He specifies: "This has nothing to do with the 'interpretive flexibility' allowed by 'multiple points of views' taken on the 'same' thing. It is the thing itself that has been allowed to be deployed as multiple."[106] The notion of "alternative facts" presupposes the idea that there are not just multiple perspectives on reality, with no one knowing the whole and complete truth—in itself a reasonable and defensible position—but that there are multiple versions of reality itself. That is precisely what Latour is saying here.

As mentioned, Latour purportedly takes a nuanced position regarding the debates about constructivism. But if this is so, how should we interpret this statement from 1987? "Since the settlement of a controversy is the cause of Nature's representation, not its consequence, we can never use this consequence, Nature, to explain how and why a controversy has been settled."[107] If taken seriously, it implies that when scientists agreed that the Moon has craters, this was not in any way a consequence of the fact that the Moon indeed has craters. A closer scrutiny of his actor-network perspective leads us to further strange assertions. As Latour argues, scientific examination of entities changes those entities to the extent that we cannot

[103] Tanesini, "Epistemic Relativism."
[104] Latour, *The Pasteurization of France*, 159.
[105] Law, "STS as Method," 43–44.
[106] Latour, *Reassembling the Social*, 116–17; Latour, *Face à Gaïa*, 50.
[107] Latour, *Science in Action*, 258.

gain knowledge about them as they existed *before* they were examined. Hence, we cannot make claims about things beyond our networks. In 1976, French scientists used modern tools to study the mummy of Ramses II and concluded that the pharaoh died of tuberculosis. Latour objected because this bacillus was only diagnosed and examined as such by the late nineteenth-century researcher Robert Koch. Its existence could, therefore, not be extrapolated into ancient Egypt.[108] "Before Koch, the bacillus had no real existence."[109] Hence, saying that Ramses II died of tuberculosis would be as anachronistic as saying that he died due to machine gun fire or the stress of a stock market crash. But is Latour's comparison not patently absurd? Machine guns did not exist in ancient Egypt, while everything that we know about bacilli confirms that they did already exist. If Latour is serious here, he might just as well blame Koch for unleashing tuberculosis on humanity!

Other authors associated with postmodern theory make similar epistemological claims. Some write more clearly than Latour, like Richard Rorty and Michel Foucault. Yet, epistemologically speaking, these authors also slide towards relativism. In Foucault's view, claims about reality or truth are disguised representations of power; "in fact power produces; it produces reality; it produces domains of objects and rituals of truth."[110] Richard Rorty proposed abandoning the aim of truth-finding as correspondence to reality altogether and replacing it with "pragmatism," which redefines truth as "what is good for us to believe."[111] As Rorty explained:

> We pragmatists cannot make sense of the idea that we should pursue truth for its own sake. We cannot regard truth as a goal of inquiry. The purpose of inquiry is to achieve agreement among humans about what to do, to bring about consensus on the ends to be achieved and the means to be used to achieve those ends.[112]

Relativism's critics correctly point to the problem of inconsistency and self-refutation. If Latour is serious about his claim that scientific research fundamentally changes the entities it examines, what about his own research? Did laboratories already exist before he began studying them?

[108] Latour, "Ramsès II est-il mort de la tuberculose ?"

[109] Latour, 84.

[110] Foucault, *Surveiller et Punir*, 196; Rabinow, *The Foucault Reader*, 51–75.

[111] Rorty, *Objectivity, Relativism, and Truth*, 22.

[112] Rorty, *Philosophy and Social Hope*, xxv.

On what grounds can he make claims about laboratories if the scientists who work there do not necessarily share those claims? Anyone who uses relativism consistently will wind up in quicksand. But Latour did not use relativism consistently—he only used it when it suited his purposes. When it concerns his own claims, relativism suddenly undergoes a vanishing act. Latour, for instance, refers to scientific insights on climate change.[113] But why not use his weapons against him? Climate change was only created when scientists examined it. Temperatures did not exist before human and technological actors measured them, disallowing long-term comparisons. In the Latourian pluriverse, climate change may also occur in some versions but not in alternative versions. Similarly, Richard Rorty repeatedly refers to aspects of Darwinian evolution to support his "pragmatist" viewpoints; however, Darwinian evolution is built upon a Mount Everest of controllable evidence that resulted from the consistent use of scientific principles like truth-finding, rationality, and empirical evidence.[114] By using Darwinian insights, Rorty selectively relies on scientific evidence while undermining the principles that made the evidence possible in the first place.

If epistemic relativism is self-refuting this way, why did it spread in the academic world? The answer to this question is multifaceted, but postmodern relativists focus on politics and ideology as crucial determinants of academic viewpoints. Let us, therefore, have a look at postmodernism's own ideological context.[115] The relationship between science and ideology has always been a problematic one. Science examines reality, and reality, to the dismay of any ideologist, tends not to comply with ideological desires. Ideologies need to find a way around that.[116] At the time of Galileo Galilei, the Roman Catholic church used a rather crude strategy: oppressing science. But throughout the modern era, as the prestige of science grew, this approach became increasingly untenable. New ways of dealing with this issue came into being. Communists embraced science's prestige unreservedly but often maintained that Marx and Lenin had reached ultimate scientific truths that could no longer be contested. That approach broke down in the 1970s and 1980s, and it is probably not by accident that

[113] Latour, *Face à Gaïa*.

[114] Rorty, *Philosophy and Social Hope*, 38, 64–69.

[115] Gross and Levitt, *Higher Superstition*; Aya, "Reign of Error."

[116] This analysis drew inspiration from: Harari, *Sapiens: A Brief History of Humankind*, 279–83.

postmodern relativism won the day amongst certain branches of the left with something more manoeuvrable. On the one hand, people like Latour endorse science; they accept its nomenclature, tap into its budgets, and rely on its authority whenever convenient. But when scientific research does not lead to the outcomes that were hoped for, a plethora of relativist tools can be used to get rid of those outcomes.[117] Latour-like relativism can thus operate as a flexible protective shield for ideological desires. "The promise of power to mold the world in accordance with one's sense of justice is far more qualified and ambiguous than was the case with hard-line Marxism," Gross and Levitt maintain.[118]

Many key relativists are straightforward about their ideological aims. In the passage quoted above, Richard Rorty illustrates how he makes the scientific aim of truth-finding subservient to his ideologically loaded objective of bringing about consensus. In addition, he intends to "demote the quest for knowledge from the status of end-in-itself to that of one more means towards greater human happiness."[119] In a similar vein, Foucault frankly admits that he has no intention of emancipating "truth from every system of power (which would be a chimera, for truth is already power)" but instead aims to detach "the power of truth from the forms of hegemony, social, economic, and cultural, within which it operates at the present time."[120] In other words, because the concept of truth is inherently political, we can just swap the political and ideological purposes for which we use it. Latour similarly conflates the political with the scientific, and stated in a recent interview that "I do not think it is important to be rational, as long as you are civilized."[121] The moralistic fallacy is not considered a fallacy.

One of the hallmarks of good science is that it invites everyone, at least potentially, to participate in the debate on an equal basis. If someone's claims are consistently rational and supported by sufficient empirical evidence, they deserve to be taken seriously, no matter how despicable or uncivilized they may seem initially. In the relativist pluriverse no such obligations exist. When someone's scientific arguments do not serve the proper purposes of "bringing about consensus" or "greater human

[117] Wilterdink, "The Sociogenesis of Postmodernism," 197, 206.
[118] Gross and Levitt, *Higher Superstition*, 42–43.
[119] Rorty, *Philosophy and Social Hope*, xii.
[120] Rabinow, *The Foucault Reader*, 74–75.
[121] Latour, *Politics of Nature*; Verschuer, "'Hoop is nu een laffe uitvlucht.'"

happiness," they can be rejected outright and without further ado.[122] In a detailed study, Helen Pluckrose and James Lindsay elucidate how the early generation of postmodern relativists paved the way for increasingly zealous ideologists who no longer tolerate counter-voices.[123]

But now, let us move back to the model of Darwinian cultural evolution. We saw that epistemic relativism, if warranted, would undermine the model and that a currently popular form of relativism can be found in the work of Bruno Latour. But we also saw Latour's epistemic position is too unclear and inconsistent to be considered a valid academic argument. Yet, having said that, let us also note that there is nothing wrong with Latour's call for taking actors seriously. The importance of carefully interpreting people's views of things is one of the major topics of this book. A certain degree of fallibilism is also justified. It can be meaningful to remind ourselves that the current scientific understanding of the world is incomplete. Other systems of thought may well see aspects of reality that modern science fails to see. Many of today's insights will probably, and hopefully, be overtaken by future and better insights. Philosophically speaking, we cannot rule out the possibility that someday, new evidence will validate unexpected causal links between common dancing and rainfall. Yet, because such evidence is currently unavailable, it would be unwise to let relativist contemplations distract us too much. We will now pick up where we left off for now. What does Darwinian cultural evolution have to say about the purposes of cultural adaptations?

4.4 Darwinian Purposes of Cultural Functions

"Man employs the extrasomatic tradition that we call culture in order to sustain and perpetuate his existence and give it full expression." This is how the cultural evolutionist Leslie White celebrated human culture in 1959. Its function, he thought, was to make life "secure and enduring for the human species."[124] Scholars sympathetic to culture "evolving" or being "adaptive" have more often sought culture's functionality in this direction.[125] After all, humans are the carriers of culture, so if culture enhances our biological survival, it should also thrive. When looking for

[122] Aya, "Reign of Error," 35; Chomsky, "'Rationality/Science,'" 52.
[123] Pluckrose and Lindsay, *Cynical Theories.*
[124] White, *The Evolution of Culture*, 8.
[125] Moore, *Visions of Culture*, 157–59; Malinowski, *A Scientific Theory of Culture*, 72.

culture's beneficiaries from a Darwinian point of view, this indeed looks like an obvious place to start.[126] Current theorists of gene-culture coevolution indeed argue that humans have evolved into culture-absorbing sponges, which strongly indicates genetic benefits. Why otherwise would culture have evolved in the first place? When looking at humans spread over the globe, it is not difficult to fathom *how* culture has enhanced human survival. When individuals adopt cultural information about their environment, it becomes much easier for them to adapt. Just think of the Inuit and how they have survived the hostile Arctic setting due to culturally learned assets of building igloos, hunting seals, and making clothes. Joseph Henrich designates culture as "the secret of our success."[127]

Yet, it was already mentioned above that cultural phenomena like martyrdom, celibacy, asceticism, or self-mutilation suggest that direct biological survival cannot be the whole story. Admittedly, the celibate Shakers have dissolved as a cultural unit, while Christian denominations that require their members to be "fruitful and multiply" are still among us.[128] But Shakerism was still reproductively successful for quite a while because many converts compensated for the lack of biological descendants.[129] Their celibacy may even have been advantageous, as it enhanced their reputation for moral purity. Moreover, innumerable details of human culture, like art styles, fashions, or television quizzes, seem arbitrary and superfluous from a biological perspective alone.[130] Aztec human sacrifice possibly offered the advantage of protein intake, but these rituals were characterized by ceremonial splendour and a meaningful complexity that indicated additional purposes. "Why build a temple, when all you need is a butcher's block?" Marshall Sahlins once retorted to Harris's eco-adaptationist model.[131] This message has not been lost on the current field of cultural evolution. It recognizes that the *cui bono* question needs to be addressed more broadly.

One name is thereby on people's lips again: Emile Durkheim. "The godfather of the field," he has been called. When encountering the work

[126] Stewart-Williams, *The Ape That Understood the Universe*, 231–43; Richerson and Boyd, *Not By Genes Alone*, 99–147.

[127] Henrich, *The Secret of Our Success*.

[128] Stewart-Williams, *The Ape That Understood the Universe*, 247.

[129] Sobo and Bell, *Celibacy, Culture, and Society*, 115.

[130] Moore, *Visions of Culture*, 205; Stewart-Williams, *The Ape That Understood the Universe*, 242.

[131] Sahlins, "Culture as Protein and Profit."

of this French sociologist David Sloan, Wilson thought Durkheim to be speaking to him "across the chasm of time."[132] Some of the mentioned case studies illustrated how the new proponents of cultural evolution identify the cultural group, or society, as the beneficiary of culture's hidden adaptions. The current field abounds in Durkheimian talk about religion as an "organic" and "interlocking" system, a "complex adaptation that serves to support human social life" or the "glue" that holds society together.[133] Power and oppression are considered an insufficient basis for society, so there is a perpetual need for ritual performance and social values, and the old comparison of societies to organisms is reconsidered.

The theoretical pedigree of this revitalization of the organismic framework partly lies in multilevel selection theory from evolutionary biology. The principles of Darwinian selection are applied at all levels of the human socio-cultural hierarchy, and a mechanism named "cultural group selection" is drawing attention. It assumes that culturally marked groups are entangled in continuing competition processes and that the ones that operate most smoothly as organismic units survive. "The costly institutions of complex societies manage to spread and propagate because the societies that possess them destroy those that don't," is how Peter Turchin frames it.[134] But the cultural group selectionists also look at less warlike mechanisms, like the differential survival of environmental challenges, the ability to attract individuals from other groups, and the adoption of practices and beliefs from more successful groups.[135]

This revitalization of the Durkheimian orientation is not welcomed by everyone, though. Critics find the differences between human societies and organisms too significant for this comparison to take hold.[136] Human individuals are not as loyal to social groups as cells are to bodies, and

[132] Slingerland and Sullivan, "Durkheim with Data," 314; Wilson, *Evolution for Everyone*, 239.

[133] Campbell, "On the Conflicts between Biological and Social Evolution," 202; Sosis and Petersen et al., "Why Cultural Evolutionary Models," 46, 48; Turchin, *Ultrasociety*, 207; Wilson, *Darwin's Cathedral*, 2002, 118.

[134] Turchin, *Ultrasociety*, 20.

[135] Henrich, *The Secret of Our Success*, 166–69; Richerson and Boyd, *Not By Genes Alone*, 195–236.

[136] Bryant, "An Evolutionary Social Science?," 462–63; Kincaid, "Functional Explanation and Evolutionary Social Science," 2007, 243–44; Wilterdink, "Te veel darwinisme, te weinig sociologie," 305.

human groups also lack the precise boundaries and reproductive moments that organisms have. Steven Pinker illustrates his misgivings by highlighting how the Roman Empire differs from an organism:

> It was the Roman Empire that took over most of the ancient world, not a group that splintered off from a group that splintered off from a group that splintered off from the Roman Empire, each baby Roman Empire very much like the parent Roman Empire except for a few random alterations, and the branch of progeny empires eventually outnumbering the others.[137]

The cultural group selectionists probably do not mean to compare cultural groups to organisms as literally as Pinker suggests, but their framework easily triggers such misinterpretations. Like its biological counterpart, this multilevel model does not distinguish between things that are transmitted and things that are not. In biology, genetic information is potentially transmitted, while groups are not. The same goes for culture: unlike groups, cultural information is possibly reproduced. When applied to cultural evolution, the multilevel framework tends to create a conceptual fog, eliciting confusing discussions about confusing mixtures of the selection of groups, genes, peoples, societies, or cultures, all operating as adaptive units.[138] What exactly is being selected here? What is being reproduced?

In response to this confusion, some authors render the whole idea of cultural group selection a non-starter.[139] Group selection does not even work in biology, so why bother about its socio-cultural application? But that is too hasty because as soon as we make a sharp distinction between things that are transmitted and things that are not, the idea of cultural group selection offers two fruitful lines of investigation. While cultural groups are not transmitted, the biological or cultural information that makes individuals behave as part of a culturally marked group may be subject to transmission and repeated selection rounds. The problem in biology was that group selection is undermined by free riders in the form of egoistic mutants or invaders who bring in new genes that the group cannot get rid of. However, what the cultural group selectionists bring forward—and this is their first fruitful line of investigation—is that culture

[137] Pinker, "The False Allure of Group Selection."

[138] For a confusing discussion, see the debate over: Pinker.

[139] Pinker; Coyne, "Genes First"; Stewart-Williams, *The Ape That Understood the Universe*, 201–12.

has created an evolutionary setting in which the selection of group-functional genetic predispositions has become more effective.[140]

What happens when a human free rider enters a cultural group characterized by an unusual internal cooperativeness and solidarity level? Humans are great cultural imitators, so the free rider might adopt the group's culture, thus becoming a group-ish altruist himself. In case the free rider was to persist in his exploitative behaviour, the cultural group might then decide to morally punish the detractor by excluding or even killing him—something that cultural groups quite often do.[141] Hence, culture might have opened up new possibilities for maintaining group cohesion, turning human groups into entities stable enough to be targeted by natural selection for group-functional genes. Gene-culture coevolution may have endowed humans with bio-psychological propensities for more group-ish behaviour. In evolutionary biology, it is unclear what animal traits group selection would precisely help explain, but our species offers more promising examples, as studies into religion, nationalism, or football clubs show. Humans possess notable tendencies of rallying around collective symbols, making sacrifices for the common good, and defending the group's interests against enemies, sometimes even at the price of death.[142] The cultural group selectionists remark that such group-ish biological traits simultaneously come up against certain limits. "These new tribal social instincts were superimposed onto human psychology without eliminating those that favor friends and kin," Richerson and Boyd maintain.[143] Nepotism has remained a continuing threat to the cohesion of any group.

At the level of such cultural pressures, the second key line of investigation of cultural group selection lies in the selection of group-functional *cultural* variants.[144] Pinker was correct that the Roman Empire did not reproduce like an organism, but what may still have been subject to repeated rounds of selection and transmission is the cultural variants that made Rome flourish, like its military strategies, religious patriotism, or political institutions. Here, we can think of selection processes between empires as wholes and the selection of subgroups within the Roman Empire, such as cities, military legions, religious cults, or patron-client

[140] Koopmans, "Het mysterie van de naastenliefde," 126; Turchin, *Ultrasociety*, 67–94; Richerson and Boyd, *Not By Genes Alone*, 195–236.

[141] Sauer, *Moral*, 63–104.

[142] Whitehouse, "Dying for the Group."

[143] Richerson and Boyd, *Not By Genes Alone*, 215.

[144] Norenzayan, *Big Gods*, 10; Wilson, *Darwin's Cathedral*, 2002, 28–37.

networks. In this sense, cultural group selection possibly explains how socio-cultural systems evolved their Durkheimian qualities of installing and maintaining values that make individuals behave as loyal members of cultural groups.

However, in our earlier discussion of the socio-cultural *cui bono* question, we noted that many cultural phenomena do not appear to enhance the functioning of societies as wholes. Let us remind ourselves of Dennett's dictum to cast our nets out widely, and in this light, the conflict tradition first warrants renewed attention. The role of power structures has already received some attention in the cultural Darwinian field.[145] William Durham created the term "selection by imposition," meaning that cultural variants can survive because some people force them onto others. The focus of the cultural group selectionists on dynamic warlike competition also gives the field a conflict-oriented twist. Yet, the presumed outcome of the competitive processes still fits the Durkheimian mould of more cohesive societies. So, it is not entirely unwarranted that the field is sometimes criticized for insufficient attention to power and oppression.[146] But let us notice here that there is nothing un-Darwinian about power and exploitation per se. Living nature teems with adaptations of organisms that exploit others. The survival of cultural traits may well, in many cases, depend on who is in charge, creating selection pressures on variants that serve the interests of the privileged at the expense of the oppressed.

The most natural ally of Darwinian theory is the analytic variant. Its dynamic model of everlasting power struggles over scarce resources looks quite Darwinian already. This may not be accidental, as some early twentieth-century exponents, like Max Weber, seem to have drawn inspiration from the social Darwinism of their day.[147] The utopian desires underlying Marxism and postmodernism seem more challenging to reconcile with a Darwinian framework, but still, these schools of thought may also contain elements worthy of integration. We saw that Marxists and postmodernists often aimed to uncover hidden power structures, and Darwinism may finally provide a flexible explanation for how such hidden structures emerge. One could, for instance, think of a scenario in which

[145] Wilson, *This View of Life*, 182–83; Durham, *Coevolution*, 198; Turchin, *Ultrasociety*; Blute, *Darwinian Sociocultural Evolution*, 80–130.

[146] Lewontin, "The Wars Over Evolution"; Lewens, *Darwin*, 208; Bryant, "An Evolutionary Social Science?," 467.

[147] Weber, *Wirtschaft Und Gesellschaft*, 21; Runciman, "Was Max Weber a Selectionist in Spite of Himself?"

elites who genuinely believed themselves to have been chosen by gods and who succeeded in convincing others of this belief had better average chances of preserving their supremacy during particular periods of history. Such processes might again be partially the accumulation of accidentally adaptive variants rather than intelligent design alone.[148]

Now that we are casting our nets out widely, the strengths of the individual-oriented approaches might also be reconsidered. Some of their elements look vaguely Darwinian, like the idea of micro-interactions adding up to complex phenomena or rational choice theory's focus on competition. Daniel Dennett and David Sloan Wilson also sympathize with the latter's assumption that cultural phenomena often fulfil individuals' needs. Yet, while rational choice theory mainly focuses on the demand side—what people want—the Darwinian model can help elucidate the supply side and how cultural phenomena adapt to fulfil those wants.[149] We should not only think of deliberate and conscious aims here. Cultural variants may, for instance, survive because they only accidentally appealed to people's desires or made sense within meaningful frameworks. In sum, Darwinian cultural evolution can offer an open-minded view. The fact that the determinants of reproductive success are diverse and context-dependent implies that the *cui bono* question might be addressed differently in each instance. The model provides us with the advantages of eclecticism while integrating insights from older approaches into a more consistent framework.

Yet we saw that the theory also has something genuinely new in store for us: the idea of selfish memes. Considering the fierce resistance to this proposal, how seriously should we take it as a scientific hypothesis? It has remained unnoticed that older non-Darwinian approaches sometimes came close to something quite like it. The Boasians stressed that culture has a semi-autonomous power, leading to talk about culture as having its own interests. "We may know all about the distribution of a tribe's form of marriage, ritual dances, and puberty initiations and yet understand nothing of the culture as a whole which has used these elements *to its own purpose*," the prominent Boasian Ruth Benedict wrote in 1934. Benedict treated culture as an agency which selects traits from the environment

[148] Turchin, *Ultrasociety*, 152, 160.
[149] Wilson, *Darwin's Cathedral*, 2002, 79–83; Dennett, *Breaking the Spell*, 182–84.

"which it can use, and discards those which it cannot." This process "need never be conscious during its whole course."[150]

Within the Darwinian framework, Dawkins was not the first to propose the idea of culture developing reproductive interests of its own either.[151] One year earlier, the anthropologist Ted Cloak wrote:

> A cultural instruction (...), although carried by an organism, can replicate itself interorganismically without waiting through the organism's life cycle. Indeed, it may replicate itself within minutes of being stored. So it should not be surprising to find (...) culture features that perform particular functions which are irrelevant, or even destructive, to the organisms whose organs help to make or do them.[152]

Cloak also preceded Dawkins in exploring the parallel of cultural things being "like an active parasite that controls some behaviour of its host." At best, he argued, culture behaves like a symbiont, but at worst, it manipulates and exploits us like its slaves.[153] Dawkins referred to Cloak and developed the suggestion into his overall context of "selfish replicators."[154] This is a potent idea because cultural forms that instruct people to make more copies of those cultural forms will be most successful at precisely that. If we accept that cultural phenomena adapt beyond our intentions and recognition, it is not a far stretch to presume that this might also lead to adaptations that do not align with our own interests, especially considering the pervasiveness of pathogens in biological evolution. So why has meme theory found so little support?

Some of this resistance stems from the terminology of memetics, and adopting a more conventional vocabulary might be a good idea. We already have the word culture, and it is unclear why memes would provide a better alternative. The term meme also has a connotation of discreteness, which, considering the often more fluid character of cultural items, might easily lead us astray.[155] Thus, I will stick to Richerson's and Boyd's "cultural variant" as the more flexible selection unit. Another issue that raises

[150] Benedict, *Patterns of Culture*, 33. Emphasis added.
[151] Cloak, "Is a Cultural Ethology Possible?"
[152] Cloak, 172.
[153] Cloak, 172.
[154] Dawkins, *The Selfish Gene*, 190.
[155] Richerson and Boyd, *Not By Genes Alone*, 88–91.

hackles is the ontological status of memes.[156] Genes are located in the molecular structures of DNA, but what exactly would memes be made of? This touches upon an old notion of culture being something intangible. Clifford Geertz once stated about culture that "though ideational, it does not exist in someone's head; though unphysical, it is not an occult entity."[157] This objection is easy to refute because all culture consists of materialized information in human brain cells or artefacts like computers or books. The maintenance of culture also continually requires the extraction of energy from the environment, which is precisely what keeps engendering the Darwinian process.[158] There is no such thing as a cultural free lunch.

Nevertheless, many queries remain. What would such "selfishly" reproducing cultural forms look like exactly? This requires careful consideration, as not all examples brought forward are convincing. Instead of selfish memes, Richerson and Boyd speak of maladaptive "rogue cultural variants" and use higher education as an example because, on average, highly educated people have fewer children.[159] However, the possibly negative effect of culture on genetic fertility is already quite a trivial observation, and higher education, in general, helps people to live healthier and happier lives. Or take the smoking of tobacco.[160] This harms human health, but that does not yet mean no human beneficiaries are involved. In today's world, we might think of people working for the tobacco industry who get richer by deliberately developing strategies to make consumers addicted. To avoid vacuity, the selfish culture hypothesis needs to become specific about what a cultural "pathogen" would entail. The Belgian philosopher Maarten Boudry argues that, for the "selfish meme" concept to gain traction, we need concrete examples of cultural phenomena that exhibit purposes of their own but which do not coincide or overlap with human purposes already familiar to us.[161] Ideally, we should find phenomena that contain "blindly" evolved adaptive elements that are well-built for ensuring their own survival while harming the purposes of the people transmitting them and the functioning of their societies.

[156] Dennett, *From Bacteria to Bach and Back*, 221–24.
[157] Geertz, *The Interpretation of Cultures*, 10.
[158] Dennett, *Breaking the Spell*, 80–81.
[159] Richerson and Boyd, *Not By Genes Alone*, 148–90, 156.
[160] Stewart-Williams, *The Ape That Understood the Universe*, 252.
[161] Boudry and Hofhuis, "Parasites of the Mind."

Luckily, the meme theorists also offer more promising case studies, like earworms, conspiracy theories, exasperating cultural arms races, woke ideology, or extreme overwork.[162] The case study that has been developed in the most detail is that of religion.[163] Because what should a "selfish" religion be expected to look like? This religion would be unwelcoming if adherents moved to another faith. So, we might expect the religious system to have evolved mechanisms of keeping people in, like threats of divine punishment or incentives for its members to penalize or kill apostates. Something else that might endanger the religion's survival is criticism or the ridicule of its sacred symbols. The faithful should thus be encouraged to stop that. All systems of ideas contain internal inconsistencies, but this might be overcome by telling adherents that questioning the doctrine is a sin and that observance requires uncommitted faith. It would also be helpful if members are stimulated to defend and spread the faith through peaceful means and resort to violence if that does not work.

None of this will sound unfamiliar to anyone acquainted with the history of Christianity and Islam. It is probably not by accident that the two most successful religions have so often contained doctrinal elements that forcefully enhanced their propagation. "*Only* Jesus can save you," Christians often do not get tired of telling us, and the first pillar of Islam is not about serving the community or following the leaders, as it says that "there is no God but God, and Mohammed is his messenger." It is the undivided loyalty to the religion *itself* that comes first.

Or take the celibacy of Roman Catholic priests.[164] Their sexual continence is a disaster for the priests' genes, as well as for their romantic and sexual pleasures. However, for Roman Catholicism, the prospects look less bleak, as devoted priests often have much cultural "offspring." And what about martyrdom?[165] "The more we are cut down by you, the more we grow in number; the blood of Christians is seed," the early Christian apologist Tertullian once told pagan authorities of Ancient Rome.[166] Tertullian assumed that the loss of some members was compensated by the

[162] Saad, *The Parasitic Mind*; Haring, *For A Successful Life*.

[163] Dennett, *Breaking the Spell*; Stewart-Williams, *The Ape That Understood the Universe*, 219–23, 250–68; Dawkins, *The God Delusion*, 222–40; Blackmore, *The Meme Machine*, 188–90.

[164] Blackmore, *The Meme Machine*, 138–39; Dennett, *Darwin's Dangerous Idea*, 367.

[165] Dennett, *Breaking the Spell*, 284–85; Stewart-Williams, *The Ape That Understood the Universe*, 302–3.

[166] Gwynn, *Christianity in the Later Roman Empire*, 7.

conversion of others, and modern research into early Christianity, to some extent, confirms his suspicions.[167] In 2015, the Algerian-French brothers Cherif and Said Kouachi massacred the editorial staff of the Charlie Hebdo magazine for publishing Mohammed cartoons. The brothers then willingly paid with their own lives as well. This was a catastrophe for the Kouachi brothers' life-fulfilling purposes but not necessarily for the spread of Islamism. Their actions may have been adaptive for protecting their interpretation of the faith from the undermining effects of public ridicule. While religion can stimulate extraordinary selflessness in individuals, the traits of religion *itself* can reveal egoism.

Let us remind ourselves of the kamikaze termites and the dogs suffering from rabies. We will not get anywhere if we ask what function their exploding or biting has for those animals. Yet, it suddenly makes a lot of sense when we ask what function it has for the biological information inside the dogs. Might extreme interpretations of religion function as rabies-like "viruses of the mind" that "hijack" and "manipulate" the behaviour of their human hosts? Viruses can camouflage their real nature, evade the defences of their carriers, inject themselves into cells, and then turn the hosts into vehicles for their own propagation. Might religion, to a certain degree, do something quite like that? This is what meme theorists are proposing. While genes gang up in genomes, memes are supposed to work together in what Susan Blackmore calls "memeplexes," which, under certain conditions, exploit their human carriers.[168] Dennett speaks of "a finely tuned amalgam of brilliant plays and stratagems, capable of holding people enthralled and loyal for their entire lives."[169] Of course, none of this is to suggest that religions or other cultural phenomena are literally selfish or that they can think in any way. Again, Darwinism potentially explains how certain phenomena can look as if an intelligent selfish agent is operating behind the scenes.

In a dismissal of meme theory, the English philosopher Tim Lewens wrote:

> There are plenty of mainstream views about cognition that remind us that people make decisions all the time that are not in their best interests. We do

[167] Bowersock, *Martyrdom and Rome*, 59–74.
[168] Blackmore, *The Meme Machine*, 19–20.
[169] Dennett, *Breaking the Spell*, 154.

> not need memetics to expose the widespread existence of various forms of irrationality, weakness of will, self-deception, false consciousness, subconsciously motivated action, and so forth.[170]

What Lewens seems to have missed here is that the theory is not so much about explaining why people can be cognitively vulnerable to self-destructive behaviour but, rather, how culture can become well adapted to *exploit* those vulnerabilities. Meme theory has offered something new, which can be illustrated through our example of the afterlife. Richard Dawkins admired how religion ingeniously uses the threat of hellfire to enforce people's commitment. He thinks that it "might almost have been planned deliberately by a Machiavellian priesthood trained in deep psychological indoctrination techniques." But Dawkins doubts whether priests were that smart. Instead, he proposes that "the idea of hell fire is, quite simply, *self perpetuating*."[171] Dawkins touches upon a dimension of afterlife beliefs that no one seems to have addressed before him.

Richard Dawkins uses meme theory mainly as a device for his strident efforts to rid people of what he calls "the god delusion."[172] But it is important to reemphasize that the idea of "selfish replicators" does not exclude other beneficiaries. On the contrary, we might generally expect the reproductive interests of cultural information to coincide with purposes at higher levels of organization. Dennett writes, "We should not forget that the vast majority of memes, like the vast majority of bacterial and viral symbionts, are neutral or even helpful."[173] For good reasons, Susan Blackmore and Daniel Dennett thus take a more multifaceted stance on religion, more prominently highlighting beneficial aspects, like group functions, altruism, consolation, or the magnificence of religious art. We should remind ourselves that it is primarily at the points where reproductive interests of culture and purposes at other levels do *not* coincide that the essential explanatory power of the theory comes to the surface.[174]

The debate over selfish meme theory has been ongoing for a long time and paddles in theoretical circles. But as the cultural evolution expert Alex

[170] Lewens, *Cultural Evolution*, 2015, 31.
[171] Dawkins, *The Selfish Gene*, 197–98.
[172] Dawkins, *The God Delusion*, 222–40.
[173] Dennett, *Breaking the Spell*, 184.
[174] Stewart-Williams, *The Ape That Understood the Universe*, 293–303; Dennett, *From Bacteria to Bach and Back*, 205–47.

Mesoudi rightly states, empirical research must address it.[175] Do historical and ethnological records provide examples of cultural phenomena that reproduced at the expense of their human hosts and societies? David Sloan Wilson acknowledges that evolution does not make everything nice and considers cultural "pathogens" possible, but thinks them very exceptional.[176] Paraphrasing Ed Yong, we might also expect to find a continuum of lifestyles between "bad" parasites and "good" mutualism, with cultural allies who disappoint us and enemies who rally to our side. The only way to find out is indeed by having an empirical look. Daniel Dennett wrote that some aspects of his theory sketch are pretty well established but that "getting down to specifics and generating further testable hypotheses is work for the future."[177] In this light, we will now enter the sinister world of European witch-hunting.

[175] This comment was made during the panel discussion of the conference *The generalized theory of evolution* (University of Duesseldorf 31-01-2018).

[176] Wilson, *This View of Life*, 157–59, 182.

[177] Dennett, *Breaking the Spell*, 310.

CHAPTER 5

Witch-Hunts

5.1 Witch Persecutions as a Case Study

In October 1782, a German publicist named Wilhelm Ludwig Wekhrlin uncovered a case of judicial miscarriage that was meant to have faded into the mists of history.[1] At the time, Wekhrlin had already made a name for himself as an audacious political satirist eager to challenge the authorities—something these authorities did not always take lightly, as various rulers had already expelled him from their territories because of it. By 1782, Wekhrlin had found new refuge in a part of what is now Bavaria, ruled by Prince Kraft Ernst von Oettingen-Wallerstein. This princedom offered the publicist a newly won freedom to write a journal. As a child of the Enlightenment, Wekhrlin wanted it to address "cases of intolerance, superstition, brutalities of criminal justice, and the violation of the weak."[2] In 1782, he received reports about a legal case in Switzerland that was right up his alley. Wekhrlin learned that authorities in the remote canton of Glarus had tried and decapitated a forty-eight-year-old woman named Anna Göldi. The official verdict was that she had poisoned the daughter of a prosperous family when she worked for the family as a servant. What attracted Wekhrlin's attention was that this allegation seemed to be a cover

[1] Wekhrlin, *Chronologen. Ein periodisches Werk*, 213–24; Hauser, *Der Justizmord an Anna Göldi*.

[2] Hauser, *Der Justizmord an Anna Göldi*, 14.

S. Hofhuis, *The Evolutionary History of Witch-hunting*, https://doi.org/10.1007/978-3-031-66936-1_5

for something else. Anna Göldi had been accused of aligning herself with the devil. After her interrogators had tortured her, she confessed that Satan had appeared to her as a black dog and gave her magical powers that she then used against the daughter of the family. The girl was supposed to have spat nails and to have suffered from convulsions. This was not just a case of poisoning; it was a witch trial! As Wekhrlin exclaimed in his journal: "Behold the dreaded fate of those people who find themselves at the mercy of such magistrates!"[3]

The government of Glarus responded furiously to Wekhrlin's allegations, as witch-hunting had already become highly controversial. They asked the Prince of Oettingen-Wallerstein to hand him over. Yet, while the Prince was somewhat dismayed about Wekhrlin's piece, he had no interest in taking things further and declined the Swiss request. This infuriated Glarus's government even further, as they thought that Wekhrlin's "vicious slander, falsehoods, insolence, and disrespect" should not go unpunished.[4] In a vindictive mood, they put a price of a hundred Cronenthaler on Wekhrlin's head, burned his journal in public, and sent around a search warrant that described the German satirist as a person with a thin face, thin legs, pale skin, and an unsightly appearance; just ugly in general. But Wekhrlin did not flinch. He continued publishing about the affair and expressed his hopes that the freedom of the press would be like "an iron wedge around the neck of tyrants."[5] News about the witch trial in Glarus indeed spread like wildfire throughout Germany and beyond. Additional reports from Switzerland mostly confirmed Wekhrlin's version of the story, and enlightened public opinion was appalled. The trial was rendered a judicial murder against an innocent suspect, and people found it difficult to believe that something this superstitious and backward could still happen in their "philosophical" century.[6] The public outcry was not without effect: Anna Göldi has gone into the history books as the last person to be tried and killed for witchcraft in Europe.[7] This Glarus trial has come to symbolize the end of one of Europe's most infamous and horrendous historical episodes.

[3] Wekhrlin, *Chronologen. Ein periodisches Werk*, 224.
[4] Hauser, *Der Justizmord an Anna Göldi*, 25.
[5] Hauser, 26.
[6] Hauser, 140–47; Korrodi-Aebli, "historicum.net."
[7] Goodare, *The European Witch-Hunt*, 320–22.

An infamous and horrendous episode it certainly had been. What looked like an absurdity to late Enlightenment polite society—allegations about harmful magic and meetings with the devil—had been taken extremely seriously by many for over three centuries.[8] Significant parts of European society had been willing to try, torture, and burn people for such allegations. Ironically, it had all started not very far away from where it would eventually end, the Alps of Switzerland.[9] In the early fifteenth century, at the foot of the Western Alps, Europe had first been confronted with stories about a new and dangerous sect of Satan worshippers. Its adherents were supposed to ally themselves with the devil and receive harmful magical powers in return. They came to be known as witches, and it was also in this region that the first hundreds of people were tried and burned for the purported crime. Initially, the rest of Europe did not take the danger very seriously. Still, the notion of the diabolical witch gradually spread over the next century, overcoming various strands of scepticism. Especially in the late sixteenth and early seventeenth centuries, in what we now call the early modern period, the hunt for witches reached an ominous peak. The exact numbers are impossible to reconstruct because much historical source material is lacking, but approximations today assume that around 50,000–60,000 people were killed on suspicion of diabolical witchcraft, with women being markedly overrepresented.[10]

Admittedly, this number is lower than has been assumed in the past. Earlier estimations amounted to nine million, and the actual number is less impressive in comparison with other causes of death at the time, like religious warfare or contagious disease. However, 50,000–60,000 is still significant and does not fully convey the phenomenon's scale. It does not include the tens of thousands of people who were tried and tortured as suspected witches but not killed. Nor does it have the people who were ostracized as witches by their local communities but who were not officially tried, the people who lived under fear of being accused as a witch, or those who were terrified about the harm that witchcraft might inflict on them.[11] What makes witch trials particularly horrific from a modern perspective is that we now know that witches did not exist. Or at least, they

[8] Rummel and Voltmer, *Hexen und Hexenverfolgung*; Levack, *The Witch-Hunt in Early Modern Europe*; Goodare, *The European Witch-Hunt*; Dillinger, *Hexen und Magie.*

[9] Behringer, *Witches and Witch-Hunts*, 60–70.

[10] Rummel and Voltmer, *Hexen und Hexenverfolgung*, 74–79; Levack, *The Witch-Hunt in Early Modern Europe*, 19–22.

[11] Levack, *The Witch-Hunt in Early Modern Europe*, 20–22.

did not exist in the way that people at the time often thought they existed.[12] So, apart from making the trials increasingly grim, it also makes them increasingly intriguing from a scientific point of view. Why did Europeans persecute innocent people for so long and on such a scale? Why was there a hunt for imaginary enemies?

What is relevant for this research is that the question of why witch-hunts occurred has attracted much theoretical attention.[13] Scholars have tried to explain them from various angles: romanticist, liberal, Marxist, functionalist, rational choice, feminist, and poststructuralist.[14] But despite all these efforts, understanding witch-hunting has remained a tough nut to crack. None of the approaches has become particularly successful. In the eyes of historians with an idiographic orientation, this offers further proof that large theoretical frameworks are bound to fail. Scholars should interpret the witch-hunting phenomenon within its particular and complex historical context.[15] Hence, if we want to find out whether the theory of Darwinian cultural evolution can live up to its promises, this case study poses the perfect challenge. Can the theory shed new light on a phenomenon covered from so many angles?

The topic of witch-hunting has already attracted some modest attention in debates over cultural evolution. In their argument about "rogue cultural variants" that resemble Dawkins's "selfish memes," Peter Richerson and Robert Boyd shortly refer to outbreaks of witch panics as a possible example of a supernatural belief system that spread in a harmful way amongst a population.[16] In contrast, the philosopher Mary Midgley uses witch trials as a telling example of why such approaches are bound to fail if taken to the socio-cultural realm.[17] Midgley concedes that witch-hunts may, at first sight, look like "selfish memes" that invaded communities lacking immunity to them. Still, she argues that such talk misses the point: "The meme story simply fails to give us any kind of explanation at all." Instead, she maintained that "what we need to understand in such a case is how people could begin to think and act in this way in spite of the beliefs, customs and ideals which had prevented them from doing so

[12] Levack, 12–19; Hutton, *The Witch*, 2016–17.
[13] Barry and Davies, *Palgrave Advances in Witchcraft Studies*.
[14] Davies and Barry, "Introduction."
[15] Gaskill, "The Pursuit of Reality"; Dillinger, *Hexen und Magie*.
[16] Richerson and Boyd, *Not By Genes Alone*, 167–69.
[17] Midgley, "Why Memes."

earlier." As Midgley calls it, it entails a "more awkward form of investigation" into "the habits of the heart."[18]

This disagreement bears upon the topics that concern us here. Are qualitative historical scholarship and cultural evolution mutually exclusive, or can they be integrated in novel and unexpected ways? Does the notion of adaptive, "selfish" cultural phenomena provide us with a fruitful line of investigation? These questions will be addressed here as follows. In Chap. 5, I introduce the topic of witch-hunting. Chapter 6 then looks at the various explanatory models developed and why those models failed to tackle the witch-hunting phenomenon satisfactorily. This failure helped shape an idiographically oriented "state of the art" in witchcraft historiography. Yet, this position is unsatisfying in its turn as it leaves some striking aspects of witch beliefs and trials unexplained. After I propose a new Darwinian scenario in Sect. 6.5, Chaps. 7 and 8 will examine this proposal that may help bridge those explanatory gaps.

But before all that, it must be acknowledged that Mary Midgley was right in at least one crucial respect. To understand the witch-hunts, we must undertake "this awkward form of investigation" into "the habits of the heart." European witch trials were driven by imaginations that can look irrational and bizarre today. Hence, interpreting and explaining these events demands a deeper understanding of the intricate symbolic world the involved actors lived in; it requires methods like *Verstehen*, hermeneutics, and thick description.[19] But while motivations for witch-hunting were dauntingly complex, one element unmistakably played an important part: fear. The expert Lyndal Roper comments that "we cannot understand why witches were hunted unless we take the fears of those who hunted witches seriously."[20] So, arguably, the best way to introduce this thought world is by looking at some of the critical anxieties of the age, as those anxieties helped shape the environment in which witch-hunting could flourish.

5.2 The Early Modern Environment

In the public mind, the period from the fifteenth to the seventeenth centuries is often associated with the Renaissance, voyages of discovery, the scientific revolution, the Reformation, a new individualism, and a novel

[18] Midgley, 83–84.
[19] Roper, *Witch Craze*, 6–12.
[20] Roper, 9.

zest for life. As it left the alleged darkness of the Middle Ages behind, this era marked the beginning of the modern world.[21] Those who adhere to such lofty views may be surprised that *this* was the period in which most witch-hunts occurred, not the Middle Ages. This looks like a striking paradox, and historians have indeed struggled with it for quite some time.[22] For instance, they treated witch-hunts as a marginal phenomenon not characteristic of its time and place or as a creation of the late Middle Ages that just happened to linger on.[23]

However, if we genuinely want to resolve the paradox, we must recognize that there might not be one. While much of the classical narrative about the period still stands—this indeed was the age of Erasmus, Montaigne, and Galilei—historians have noted for quite some time that this era also had more sinister sides.[24] It was not only the age of Renaissance sculpture, the telescope, and the printing press but also fierce intolerance, religious wars, increased interest in magic, obsessions with the diabolical, and profound anxieties. We may now see these centuries as the beginning of the modern world, but to many contemporaries, it looked like the last age before an imminent apocalypse.[25] The Second Coming of Christ seemed near, so enormous upheaval was expected. Numerous people presumed that the Anti-Christ was about to appear, followed by a resurrection of the dead, the last judgment, and the end of the world. Indications were supposed to be everywhere: strange weather, raging Turks, false prophets, monstrous births, and ominous comets. When a bishop of Vienna, Frederick Nausea (*nomen est omen*), reflected on his epoch in 1532, he concluded that "the last age is near at hand, that the end of the world is indeed at the door, that the present age is utterly ruined, and that the death of it (as we might say) is indeed approaching to confirm the prodigies in every way."[26]

[21] Woolfson, *Palgrave Advances in Renaissance Historiography.*

[22] Briggs, *Witches & Neighbours*, 13.

[23] Hansen, *Zauberwahn, Inquisition und Hexenproceß im Mittelalter*, 4, 535–36.

[24] Kinsman, *The Darker Vision of the Renaissance*; Delumeau, *La Peur en Occident*; Clark, *Thinking with Demons.*

[25] Spinks and Zika, *Disaster, Death and the Emotions in the Shadow of the Apocalypse, 1400–1700.*

[26] Clark, *Thinking with Demons*, 366.

Many people thought to experience epic times in which their faith would be sorely tested.[27] Righteous Christian communities saw themselves as embattled fortresses needing defence against dangers and enemies looming. The English historian Stuart Clark calls it an age of "cognitive extremism"; more so than in many other epochs of European history, people tended to divide the world into binary pairs or symbolic contraries, with religious values being strictly dichotomized.[28] During the decades around 1500, certain parts of the elites had embraced a somewhat more epicurean and optimistic Renaissance worldview, but over the sixteenth and early seventeenth centuries, the general mood seems to have darkened. Conflicts of the Reformation proved insoluble and grew increasingly bitter, stimulating a more gloomy and pessimistic view of the world.[29] On both sides of the religious divide, there were increasing demands for a morally rigorous, self-disciplined lifestyle and a rejection of worldly pleasures. Many rulers and subjects embraced the ideal of the godly state, characterized by an imminent state of emergency, a dogmatic religious sensibility, and a zealous concern for the purification of society of all evils.[30]

The ultimate source of evil in this paranoid mindset was the devil, also known as the Prince of Darkness, Satan, Lucifer, or the Evil One.[31] His origins were believed to lie in heaven, where he had served God as one of his archangels. But this initial peace and tranquillity were disturbed when Satan started a rebellion against God and was cast out of heaven. He now continued his existence in hell and was filled with an insatiable hatred of God and humanity and a thirst for revenge. Satan also sensed that the apocalypse was imminent and that time was running out, so his wish to harm was all the more pressing. The devil should not be understood as a real threat to divine powers, as God was omnipotent.[32] Dualistic notions about God and the devil being competing forces had been stamped out as one of the alleged heresies of the Middle Ages. On the contrary; God permitted diabolical evil because it served him as a device for testing and punishing sinful humankind. The extent of the devil's powers depended

[27] Delumeau, *La Peur en Occident*; Levack, *The Witch-Hunt in Early Modern Europe*, 100–121.

[28] Clark, *Thinking with Demons*, 39.

[29] Levack, *The Witch-Hunt in Early Modern Europe*, 100–121; Behringer, *Witches and Witch-Hunts*, 119–20; Goodare, *The European Witch-Hunt*, 155–88.

[30] Roper, *Witch Craze*, 18.

[31] Delumeau, *La Peur en Occident*, 232–53; Goodare, *The European Witch-Hunt*, 55–70.

[32] Dillinger, *Hexen und Magie*, 42–53.

on human cooperation, but because human sinfulness was ubiquitous, Satan had ample opportunities to exert his evil ways. The suffragan bishop of Trier, Peter Binsfeld, wrote in 1589 that "when the devil's temptations are withstood, he becomes as weak as an ant, but when his temptations are met, he will grow strong as a lion."[33]

What was additionally frightening was that legions of demons assisted the devil.[34] Like Satan, these beings had once lived peacefully as angels in heaven, but since they had joined the devil's rebellion, they were now a part of his malign campaign of tricking, seducing, and punishing humankind. The angelic origin of the devil and his demons implied that they had maintained many of their former traits. What they were precisely capable of was a topic of intellectual debate in a domain of scholarship that historians now call demonology. It was generally assumed that the devil and demons were spiritual beings without physical bodies and were thus unable to eat, drink, or procreate. They were also incapable of overriding people's free will, performing miracles, or changing substances. Yet, they did have special powers, like travelling at incredible speeds, knowing about the future, and creating illusions resembling miracles. Amongst those illusions was metamorphosis; they could appear in people's lives in disguise as apparently normal humans or animals. When people at the time were confronted with beings they did not know, there were thus reasons for distrust. For instance, when Martin Luther was faced with an unknown dog in his bed in the Wartburg Castle, he assumed it was a demon and flung it out of the window.[35]

This mindset of "cognitive extremism," with demonic forces looming all around, also tainted social relations. Opposing groups were often not seen as just opponents but as agents of the devil.[36] The Middle Ages had already witnessed persecutions and burnings of so-called heretics, like the Cathars, Hussites, and Waldensians, who had diverged from the dominant Christian doctrines.[37] It was said that such groups held secretive gatherings where they did outrageous things, such as frenzied naked dancing, desecrating holy symbols, or plotting against Christianity. Similar fears existed regarding the non-Christian minority of the Jews. They allegedly

[33] Binsfeld, *Tractat Von Bekantnuß der Zauberer unnd Hexen*, 43.

[34] Eire, *Reformations*, 618–59.

[35] Eire, 650.

[36] Delumeau, *La Peur en Occident*, 254–304.

[37] Moore, *The Formation of a Persecuting Society*; Oberste, *Ketzerei Und Inquisition Im Mittelalter*.

performed cannibalistic rituals with Christian children at their assemblages named Sabbaths. Medieval scholarly works about these groups had titles like *Malleus Haereticorum* (The Hammer of Heretics) or *Malleus Judaeorum* (The Hammer of Jews), and the institution of the Holy Inquisition was installed to combat them.

In the fifteenth and sixteenth centuries, this zeal seems to have intensified. Jews were increasingly expelled, and by 1570, not many Jews were left in Western Europe.[38] The religious splintering of the continent as a consequence of the Reformation also created new antagonisms between Catholicism and various branches of Protestantism. Western Christendom was confronted with religiously divided societies, which many considered profoundly undesirable. Endangering the unity of the "Corpus Christianum," it was bound to provoke God's wrath.[39] Catholics described Lutheranism as "the German plague," and a sixteenth-century French preacher warned that if Protestant deviants were not removed, God would "exterminate us all."[40] The hatred of Protestantism stimulated a vigorous institutional response known today as the Counter-Reformation. Protestants, in their turn, stereotyped the Catholic church as "Satan's synagogue" and the Pope as "the whore of Babylon."[41] At times, people found ways of living peacefully together across religious divides. Still, such arrangements were fragile and could easily give way to brutal conflicts, like the wars of religion in France, England, The Netherlands, Switzerland, and Germany.

Another phenomenon that was distrusted within this male-dominated world was women.[42] There were positive notions about women as chaste and obedient servants and companions of their husbands, and Catholics venerated the Holy Mary. But these ideals sharply contrasted with many actual women, the earthlier Eves, who were often perceived as weak of mind and character with an overall tendency towards moral transgressions.[43] Religious and secular authorities alike shared notions about women as imperfect men who were less rational or intellectual and prone

[38] Kaplan, *Divided by Faith*, 294–332.

[39] Kaplan, 35.

[40] Schilling, *Aufbruch und Krise: Deutschland 1517–1648*, 207; Kaplan, *Divided by Faith*, 75.

[41] Rublack, *Reformation Europe*; Kaplan, *Divided by Faith*, 35.

[42] Delumeau, *La Peur en Occident*, 305–46; Wunder, *Er ist die Sonn', sie ist der Mond*; Wiesner-Hanks, *Women and Gender in Early Modern Europe*.

[43] Roper, *Witch Craze*, 136.

to traits like petulance, credulity, and madness. Such stereotypes had deep roots, often dating back to classical antiquity or biblical times. Amongst the legitimizations for misogynous stereotypes was the widely held belief in bodily humours—a theory that assumed that four fluids in our bodies determine our temperaments. Supposedly, women had an overabundance of the fluid black bile associated with the Earth, which created forces in the body that moved downwards, explaining women's wider hips and more emotional tendencies. A crucial consequence of women being more earthly and emotional was that they were believed to have a stronger sex drive.[44] The female sexual appetite was often thought insatiable and challenging to control. This was a problem because extramarital sex was forbidden, and women's abilities of seduction gave them a means to exert power over men. Given this context, women were likely seen as the devil's likely tool.

Another relevant set of anxieties is popular beliefs.[45] What ordinary people living in the countryside precisely believed is difficult for historians to reconstruct because these people were largely illiterate. But the accounts of the educated suggest that ordinary people thought to inhabit a world teeming with non-human spiritual beings, like nature spirits, ghosts, revenants, giants, fairies, and goblins.[46] Some of those creatures could dwell in and around the house, but they were mostly found beyond the villages and the cultivated lands in the untamed and dangerous world of woods, lakes, moors, and mountains. Stories were told about their gatherings in the middle of the night at remote places like hilltops or moors.[47] Interactions between humans and such beings were possible. For instance, they flew into people's cellars to drink barrels of wine or appeared as "persons" to people in need and promised help, but when help was accepted, they suddenly turned into spirits.[48] A popular folkloric trope concerned the so-called Wild Hunt—the belief that fairies, the dead, or supernatural hunters collectively flew around during the night under the leadership of pagan gods like Wotan, Holda, or Diana.[49] Such supernatural beings could be beneficial, neutral, or harmful.

[44] Levack, *The Witch-Hunt in Early Modern Europe*, 132.
[45] Goodare, *The European Witch-Hunt*, 121–54; Hutton, *The Witch*, 74–95, 120–46.
[46] Delumeau, *La Peur en Occident*, 75–97; Dillinger, *Hexen und Magie*, 37–42.
[47] Dillinger, *Böse Leute*, 139; Wilson, *The Magical Universe*, 10.
[48] Dillinger, "Hexenverfolgungen in der Grafschaft Hohenberg," 25–27.
[49] Hutton, *The Witch*, 125–43.

It was a "magical universe."[50] The world was alleged to be full of hidden spiritual forces that impacted daily life. While these forces were often unpredictable, they were not inanimate or indifferent; they could be influenced. People used magical practices, like spells, charms, amulets, or rituals, to stimulate fruitful harvests, good health, or safe journeys. Some people were considered experts, the so-called cunning people, who were thought to have unusual abilities to influence and contact the supernatural world.[51] By the end of the Middle Ages, the Christian religion may have penetrated the minds of ordinary people, but it had to coexist with a rich heritage of local customs and beliefs that often preserved older elements. Parts of the educated classes shared this interest in magic. From the twelfth century onwards, many ancient and Arabic texts were translated into Latin, and they were full of theories about magic, which triggered a fascination for the occult sciences amongst Renaissance intellectuals.[52] Religious authorities often distrusted intellectual and ordinary views and practices regarding magic, as they appeared to diverge from official doctrines and require demonic powers. But magical acts were so widespread that attempts to stamp them out tended to misfire. Moreover, official Christianity also used practices that looked like magic, such as ringing church bells against storms or using holy water for protection.[53]

What various levels of society at specific periods were far less lenient on, however, was the use of magic in supposedly harmful ways, which usually goes under the name of witchcraft. The fear of witchcraft did not make Europeans unique. Historians and anthropologists have found anxieties about it throughout premodern societies in world history.[54] Yet, while belief in witchcraft was widespread, the concept becomes slippery as soon as scholars today try to define it with precision. Does it mean that certain people are supposed to harm others through magical means only unconsciously? Is the power of witchcraft acquired or innate? And does it involve the invocation of supernatural forces or the unlocking of natural secrets? Answers to such questions widely diverge if they make sense to people at all. Despite intercultural differences, the universal use of the term witchcraft is helpful. In premodern societies, suspicions within communities can

[50] Wilson, *The Magical Universe*.

[51] Briggs, *Witches & Neighbours*, 171.

[52] Clark, *Thinking with Demons*, 214–32; Dillinger, *Hexen und Magie*, 25–27.

[53] Wilson, *The Magical Universe*, xxii; Thomas, *Religion and the Decline of Magic*, 27–57.

[54] Behringer, *Witches and Witch-Hunts*, 1–46, 196–248; Hutton, *The Witch*, 3–43; Stein and Stein, *The Anthropology of Religion, Magic, and Witchcraft*, 213–29.

be found about individuals who harm others through hidden and uncanny means.[55] What is also widespread is the belief that action can be taken to resist these witches, for instance, through countermagic or by persecuting and murdering them.

Premodern Europe was no exception. Already long before the age of the witch trials, it was familiar with what historical scholars nowadays call "village witchcraft."[56] Within local communities, people often had built up reputations for maleficent magic and were dreaded for it. Magical countermeasures were continually taken against them.[57] Such fears should not immediately be dismissed as illusory because reputations for maleficent magic might have been grounded in reality to some degree.[58] There are clues that specific individuals did try to harm others through magical means and that they may even have cultivated their reputations for it. After all, it could make others careful not to mistreat them. Psychosomatic mechanisms possibly also made such attempts have real consequences, for instance, when the stress of being cursed had real damaging effects on people's health. The levels of witch persecution in Europe fluctuated. In Ancient Rome, there were panics about women who allegedly spread disease through magical means, with hundreds being executed.[59] Worries about harmful magic also continued in the Middle Ages, and trials still sporadically occurred.[60] However, Medieval political and ecclesiastical authorities were, in general, not very interested in the persecution of witchcraft and tended to dismiss beliefs in it as superstitious. Only in the fourteenth century fears started to grow again, particularly regarding alleged sorcery against political opponents. This helped set the stage for a new era of witch-hunting.[61]

Late medieval and early modern European communities also faced various actual and purported dangers. Europe was experiencing the harsh effects of the so-called Little Ice Age.[62] After a relatively warm period in the Middle Ages, the average temperatures gradually began to drop from

[55] Hutton, *The Witch*, 10–35.

[56] Goodare, *The European Witch-Hunt*, 88–121.

[57] Wilson, *The Magical Universe*, xxvi; Davies, "The World of Popular Magic," 170.

[58] Wiesner-Hanks, *Women and Gender in Early Modern Europe*, 260; Briggs, *Witches & Neighbours*, 55.

[59] Behringer, *Witches and Witch-Hunts*, 48–50.

[60] Hutton, *The Witch*, 59–73, 147–79.

[61] Goodare, *The European Witch-Hunt*, 35–38.

[62] Behringer, *Kulturgeschichte des Klimas*; Parker, *Global Crisis*, 3–25.

around 1300, particularly from the 1550s onwards, and the effects were severe. There were strange weather conditions, like extreme winters, sudden frost in spring, or cold and wet summers. Being part of a Malthusian economy of vulnerable ecologies, communities often lived close to the subsistence level, so a failed harvest or livestock loss could mean the end. This also added to an already existing problem of overpopulation. After the devastations of the plague in the fourteenth century, the populations of Europe started to grow again and by the late sixteenth century, the growth had reached the limits of what natural resources could harness. There were rising food prices, epidemics, inflation, and famines, and by the 1590s, agricultural and industrial production had reached its lowest level in three centuries.[63] This stimulated an atmosphere in which the health and fertility of people and livestock were a matter of intense concern.

Individuals depended on robust social bonds within their local communities to face these challenges. People often lived in the same place for generations, interacting daily to exchange services and goods. They knew each other well.[64] During times of scarcity, such social cohesion could provide much-needed solidarity, but there was also a darker side; enmities and suspicions were often building up within communities.[65] Unpopular individuals could become the objects of hostile gossiping. When confronted with perceived transgressors, communities could lash out harshly against them. The punishment of wrongdoers was increasingly taken out of the hands of those local communities, being monopolized by the state. But that did not necessarily soften the disciplinary measures.[66] During the sixteenth century, the number of public executions by authorities grew significantly and usually went together with much physical torment. Public punishments were almost folk festivals, attracting many visitors. The German historian Richard von Dülmen has called them the "theatre of horror."[67]

Considering all these aspects of the late medieval and early modern period, it is not so strange, after all, that this was the age of European witch-hunting. When historians nowadays place the trials in their

[63] Kamen, *The Iron Century*; Schilling, *Die neue Zeit*, 264–82.
[64] Kamen, *Early Modern European Society*, 9–11.
[65] Briggs, *Witches & Neighbours*, 138–40; Goodare, *The European Witch-Hunt*, 88–120.
[66] Kamen, *Early Modern European Society*, 188–96.
[67] Dülmen, *Theater des Schreckens.*

historical context, they treat them as an integral part of European society.[68] However, as petrifying as all these actual or alleged dangers were, they did not yet create witch persecutions. What created the possibility of witch-hunting was that all these anxieties, which had often existed independently from each other, gradually came together into a cluster of beliefs that made things even more frightening. Historians have labelled it "the cumulative concept of witchcraft." We now turn to this concept.

5.3 The Cumulative Concept of Witchcraft

The notion of a "cumulative concept of witchcraft" was first proposed by the German historian Joseph Hansen.[69] According to his early twentieth-century argument, a cumulative diabolical interpretation of witchcraft gradually took shape in the late Middle Ages in the Alps. It then spread throughout Europe, constituting the basis of the witch-hunts. A moment of particular importance in his account was a papal bull against diabolical witchcraft in 1484, followed in 1486 by the publication of the *Malleus Maleficarum* (The Hammer of Witches). This extensive book was authored by the clergymen Heinrich Kramer and Jacob Sprenger—although the latter's involvement is contested—and it soon became a standard scholarly reference.[70] Many aspects of Hansen's theories about witch-hunting have become outdated, but his notion of a "cumulative concept of witchcraft" is still widely in use among experts today.[71] American historian Brian Levack describes the concept as "a composite or cumulative notion: composite because it consisted of ideas that formerly were not connected, and cumulative because the synthesis of these ideas occurred gradually, each new addition building upon or supplementing others."[72] While the set of elements might vary, five key ideas are often on the list.

The first essential ingredient concerned the idea of a "diabolical pact."[73] Witches were supposed to be ordinary Christians who had allied themselves with the devil and were now serving his malicious campaign against

[68] Briggs, *Witches & Neighbours*, 13.

[69] Hansen, *Zauberwahn, Inquisition und Hexenproceß im Mittelalter.*

[70] Behringer and Jerouscheck, "'Das unheilvollste Buch der Weltliteratur'?," 33; Broedel, *The "Malleus Maleficarum" and the Construction of Witchcraft*, 19.

[71] Rummel and Voltmer, *Hexen und Hexenverfolgung*, 5; Behringer, *Witches and Witch-Hunts*, 57–59; Goodare, *The European Witch-Hunt*, 50–51.

[72] Levack, *The Witch-Hunt in Early Modern Europe*, 29.

[73] Levack, 29–37; Dillinger, *Hexen und Magie*, 42–47.

God and humanity. Since demonic forces could not override people's free will, the pact always resulted from a person's own decision, making it an inexcusable crime that deserved severe punishment. It was a deliberate form of apostasy and heresy, and if witches did not confess and repent, their souls would be lost to an afterlife of eternal agony. However, Satan needed to make ingenious efforts to tempt Christians into this alliance. A Westphalian witch-hunter, Heinrich von Schultheis, compared the devil to a bird catcher who used the sweet tones of his pipe to lure birds in.[74] As a master of lies, Satan continuously looked for moral weak spots, and many experts stressed that one category of people was especially likely to fall for that: women.[75] According to King James VI of Scotland, female witches outnumbered male witches by twenty to one. He explained: "The reason is easy, for that sex is frailer than man is, so it is easier to be entrapped in these gross snares of the Devil."[76]

Satan was also aware that a vulnerable spot was to be found in the female sex drive, which brings us to the second key element of the cumulative concept: "diabolical intercourse." The alleged diabolical lovemaking often went directly along with the pact, and a typical scenario of female initiation into witchcraft could happen as follows.[77] It all began when women were somewhere alone, at home or outside, in a situation of emotional or economic distress. A stranger then appeared who was kind, comforting, and interested in the causes of their troubles. He was a young and handsome man with stylish clothes, such as trousers of satin or velvet, and he appeared virile, sexually experienced, and willing to offer help. The stranger began to promise these women lovely things like wealth and prosperity; he brought them money and jewellery and tried to seduce them to have intercourse with him.[78] However, there was a catch because this stranger did not offer his favours for free; he demanded that they abjure God.

This was no ordinary stranger. Hidden behind the appealing disguise was a demon or even the devil himself. But while Satan was a marvellous

[74] Schultheis, *Eine Außführliche Instruction*, 442.

[75] Levack, *The Witch-Hunt in Early Modern Europe*, 135–41; Binsfeld, *Tractat Von Bekantnuß der Zauberer unnd Hexen*, 14–15, 42.

[76] King James I and Tyson, *The Demonology of King James I*, 128.

[77] Dillinger, *Böse Leute*, 112–17; Roper, *Witch Craze*, 82–103; Goodare, *The European Witch-Hunt*, 298–305; Pohl, *Zauberglaube und Hexenangst im Kurfürstentum Mainz*, 245–58.

[78] Roper, *Witch Craze*, 87.

trickster, it always remained possible for women to recognize that something was not right because the devil could never achieve a perfect disguise. He, for instance, had the hoof of a goat, felt cold, smelled of sulphur, or lacked a shadow. However, due to their moral weaknesses and sexual appetite, women often did not pay careful attention and did not realize who they were dealing with. They accepted the stranger's gifts, had intercourse with him, and made the fatal mistake of renouncing God. Disillusionment immediately followed. The intercourse was not as pleasant as expected, as the stranger pushed them down like a dog, and the sulphur smell, goat-like hoof, and coldness became apparent. The genital organ turned out to be long and hard, while the sperm and penis were as cold as ice, causing horrible pains.[79] As soon as the stranger's identity had become apparent, his promises were retracted: the coins turned into dung or leaves and the jewellery into trash or broken pieces. But at that point, the disgraceful decision could not be reversed. Like in a marriage, Satan had effectively made these women "his." They had become a witch and were now a part of the devil's hateful fight against Christianity.

While women were far more likely to enter the pact, Satan also entangled men in his nets.[80] For instance, when men were drunk, suddenly, a young and handsome woman appeared who tried to seduce them and asked them to renounce God. If careful enough, men could observe that this woman's hands were rough and hairy, demonstrating his true masculine nature. In her research of German trial records, Lyndal Roper noticed that male witches were more often provided with multiple partners in contrast to female witches, who usually had one diabolical lover.[81] But male initiation into witchcraft could, for example, also happen when Satan promised to take revenge on enemies or offered them knowledge in exchange for their souls, as happened in the legendary case of the German doctor Faustus.[82] Narratives about men becoming witches were somewhat more diverse and unstable and were indeed no match for the story about female initiation into witchcraft through diabolical intercourse. The *Malleus Maleficarum* typically brought in many antique and biblical quotes to explain it. Had not the early church father Saint John Chrysostom already said, "What else is a woman but the enemy of friendship, an

[79] Boguet, *Discours exécrable des sorciers*, 57–58.

[80] Briggs, *Witches & Neighbours*, 350; Goodare, *The European Witch-Hunt*, 301.

[81] Roper, *Witch Craze*, 90/1.

[82] Füssel and Kreutzer, *Historia von D. Johann Fausten*.

inescapable punishment, a necessary evil, a natural temptation, a desirable disaster, a danger in the home, a delightful detriment, an evil of nature, painted with nice color?"[83]

When the pact was sealed, and the intercourse had taken place, witches would then regularly attend a so-called witches' sabbath, also referred to as "the synagogue," "nocturnal assemblies," "schools," or "the dance." This constitutes the cumulative concept's third essential element.[84] The gathering, named after what Christians saw as the Jewish assemblage, was often imagined as a dreadful event where witches of all kinds, young, old, rich, and poor, assembled to meet the devil and plot against Christianity. These sabbaths occurred regularly at night, often close to the edges of the cultivated world or in even more isolated places like woods, moors, or hilltops. The assembly was often thought to be chaired by the devil, who sat on a throne with the horrific appearance of a big goat, a dog, or a toad, while the witches paid homage to him. All the moral principles of Christianity were reversed: the witches worshipped Satan while they were standing upside down, and instead of kissing his lips, they licked his anus. Demons joined in, and the sacred symbols of Christianity were mocked. The cross was trampled on, the host desecrated, and the Holy Bible spat on. After the blasphemy, the time was ripe for intense feasting. The flesh of Christian children became the object of frenzied cannibalistic rituals, and demons and witches were sitting at long tables with plenty of food and drinks, which looked nice but were tasteless and unsatisfying.[85]

The exact events of sabbaths were a topic of much scholarly consideration, for instance, in the *Disquisitiones Magicae* (Investigations into Magic) of the Jesuit scholar Martin Delrio from 1600.[86] This renowned intellectual depicted the witches' sabbath in great colour and detail and maintained that when the witches and demons had finished their meals, the following happened:

> Then they begin to dance. Sometimes they hold lighted candles in their hands, with which they worship the evil spirit, and exchange kisses in his presence. They sing very obscure songs in his honour, or jump up and down, to a drum or a pipe which is played by someone sitting in the fork of

[83] Institoris, *The Hammer of Witches*, 162, 170.

[84] Clark, *Thinking with Demons*, 3–94; Roper, *Witch Craze*, 104–24; Levack, *The Witch-Hunt in Early Modern Europe*, 37–41.

[85] Rémy, *Demonolatry*, 57–59.

[86] Rio and Maxwell-Stuart, *Investigations into Magic*, 83–98.

> a tree. They behave ridiculously in every way, and in every way contrary to accepted custom. Then their demon-lovers copulate with them in the most repulsive fashion.[87]

The lovemaking was one of the most shocking elements. Many witches were naked, and intercourse happened between witches and demons, between witches and other witches, and in heterosexual, homosexual, and incestuous ways.

The fact that witches' sabbaths were often held at relatively remote places created a problem for the diabolical conspiracy, though. This age was without cars, bikes, or street lighting, so attending sabbaths by traveling back and forth within a single night was physically challenging. However, the devil offered the solution of the "nightly flight," which constitutes our fourth essential element.[88] As Martin Delrio wrote about the witches' voyages to the sabbath: "If they live nearby, they go on foot. If they live further away [...], they are transported to meetings."[89] The means of aerial transport varied greatly, as witches could use items like broomsticks, pitchforks, or pieces of wood. These tools were turned into vehicles for flight by smearing them with magical ointments, partly made of the fat of dead babies. The voyages could happen alone and in cohorts resembling the Wild Hunt. Witchcraft thus formed a collective and even more powerful and dangerous effort of people who consorted with each other over long distances.

If the diabolical pact, the intercourse, the sabbath, and nightly flight were not awful enough, the concept contained a fifth element that turned demonic witchcraft into a threat in people's daily lives: the devil gave his witches the skill of performing "harmful magic."[90] A theological subtlety here, at least to the educated, was that witches did not possess magical power themselves. Their abilities always depended on the cooperation of demons, which, in turn, relied on divine permission.[91] Yet, this nuance did not make its horrors any less frightening. They were known to cause

[87] Rio and Maxwell-Stuart, 93.

[88] Levack, *The Witch-Hunt in Early Modern Europe*, 41–45; Hutton, *The Witch*, 204–6; Dillinger, *Hexen und Magie*, 44–62.

[89] Machielsen, *Martin Delrio*, 94.

[90] Levack, *The Witch-Hunt in Early Modern Europe*, 3–12; Dillinger, *Hexen und Magie*, 19–24.

[91] Förner, *Dämonenglaube und Zauberei im Jahre 1625*, 82; Bodin, *De la démonomanie des sorciers*, 278.

sickness in people as well as in cattle by putting poisons in food and drinks or by breathing or blowing on them. They threw herbs or pieces of straw on the ground, which caused terrible pain when people or animals crossed them. Witches drained people's resources, for instance, by flying into cellars to drink barrels of wine or by creating weather conditions, like frosts and hailstorms, that destroyed harvests. A particular focus of the witches' malign intents was children.[92] They dropped poison on them, suffocated them with the mattress, or snatched children from the cradle to cut them in pieces or to suck their blood.

Another target was male sexuality. One of their notorious magical activities was the tying of knots, which caused impotence among men.[93] A woodcut from 1544, made by the German painter Hans Baldung Grien, also vividly illustrates how the witchcraft concept could tap into male anxieties about women.[94] It shows how a groom is lying stretched out on the floor of a stable—a visible crotch emphasizes his masculinity—after a witch has apparently harmed him with a magical torch. The hag is visible in the upper right corner of the picture, with wild hair and breasts hanging out of her clothes. For men at the time, it must have been a frightening picture, showing male powerlessness against the treacherous forces of female witchcraft.

Diabolical witchcraft was extremely dangerous but also difficult to combat. The identity of witches remained concealed, and they pretended to be good neighbours.[95] An aunt, the baker, or a friendly neighbour could be a witch for months or years without anyone noticing anything. The magic, flights, and sabbaths typically happened secretly, and the devil and his witches tried to ensure that nobody saw such things happening.[96] It was pointless to check whether alleged witches were sleeping in their beds at night because demons created an illusion that made it look as if the witches decently slept at home. Even their husbands and wives could be clueless about the feasting at night.[97] There were some occasional reports

[92] Rio and Maxwell-Stuart, *Investigations into Magic*, 119–20; Binsfeld, *Tractat Von Bekantnuß der Zauberer unnd Hexen*, 55.

[93] Wilson, *The Magical Universe*, 140; Bodin, *De la démonomanie des sorciers*, 182–83, 421.

[94] Zika, *The Appearance of Witchcraft*, 32–34.

[95] Schultheis, *Eine Außführliche Instruction*, 308.

[96] Anonymous of Arras, "A History of the State, Case, and Condition of the Waldensian Heretics," 28.

[97] Rémy, *Demonolatry*, 43–44.

about people who had seen witches flying or who had gotten a glimpse of their gatherings, but these were exceptions.[98]

Hence, if the witches' activities and identities were concealed, how would one find out who the witches were? A first indication of someone being a witch could be a person's bad name within a local community. A reputation for maleficent magic could be an indication, but other dimensions of people's daily behaviour could also offer clues.[99] For instance, witches often tended to harm people with whom they had lingering, albeit somewhat unexpressed enmities. So, were there perhaps individuals with whom others did not get along very well?[100] Why had a specific person suddenly appeared in someone's house just before the cow fell ill, and had this person not lately been heard talking with someone while they were supposed to be alone? In other words, was there contact with demons? Thus, bad reputations and other suspicious clues could be a starting point for an investigation.[101]

A few hurdles still needed to be cleared before a witch's identity could be established. Many parts of Europe at the time applied Roman Law, which required certain degrees of evidence, and rumours alone were insufficient. This made the hidden crime of witchcraft notoriously difficult to prove. However, there was an opening. Roman Law gave much credence to the confessions of criminals, so what if one could make suspected witches confess to their deeds, for instance, through painful questioning?[102] The use of torture during judicial procedures was not unusual at the time, and authorities often had a vast arsenal of devices at their disposal, like the rack, the ladder, thumb screws, or leg screws. Yet, witch-hunters still faced the problem that the practice of torture was legally restricted. Law books prescribed that torture could not be applied more than three times, that leading questioning was prohibited, and that prisoners had the right to retract their confessions later on. But crucially, witchcraft gradually became known as a "crimen exceptum"—an exceptional crime that required exceptional means of investigation.[103] This implied that normal restrictions on torture could be discarded. Painful questioning also created the

[98] Stephens, *Demon Lovers*, 135.

[99] Dillinger, *Böse Leute*, 186–91; Briggs, *Witches & Neighbours*, 42.

[100] Briggs, *Witches & Neighbours*, 94, 138.

[101] Rio and Maxwell-Stuart, *Investigations into Magic*, 191.

[102] Roper, *Witch Craze*, 44–66; Rummel and Voltmer, *Hexen und Hexenverfolgung*, 48–51; Levack, *The Witch-Hunt in Early Modern Europe*, 68–81.

[103] Boguet, *Discours exécrable des sorciers*, 405–6.

possibility of making witches confess who else they had seen at the sabbath. In an extensive book on witchcraft, the renowned French jurist Jean Bodin optimistically wrote that "only one is needed to denounce an infinite number of them."[104]

There were also some additional forms of evidence.[105] Persecutors could ascertain whether the accused had suspicious witchcraft equipment, like unguents and wax figures. There was also the belief that suspects could have a "devil's mark" somewhere on their skin, like a blemish, given by the devil after the witches had entered into the pact. Somewhat more unusual and rejected by many persecutors was the so-called water ordeal, in which the floating of suspects in water was seen as proof of witchcraft.[106] Last but not least, one category of witnesses was potentially helpful during an investigation: children.[107] Especially in the later period, it was highlighted that children could be witches too and were taken to the sabbaths by their witch parents. This created persecutory opportunities because children were less reluctant to tell inquisitors what they wanted to hear.

The cumulative concept of witchcraft gained more and more credence in the course of the fifteenth and sixteenth centuries. It spread through scholarly works, pamphlets, sermons, newssheets, conversations, and visual images.[108] The concept continually faced resistance but increasingly developed abilities to evade those criticisms. Critics might wonder whether the use of torture stimulated false confessions and denouncements, but here, witch-hunters retorted that God disallowed that.[109] If critics pointed out that the proof for witchcraft during an investigation was shaky, proponents could respond that demonic forces were continually removing all the evidence, implying that procedures needed to be continued.[110] *Any* subsequent event during a trial could then be interpreted as further proof. If physical torment made suspects confess quickly, it showed their guilt. But if they did not confess soon, it probably resulted from demons who

[104] Bodin, *De la démonomanie des sorciers*, 355.

[105] Rummel and Voltmer, *Hexen und Hexenverfolgung*, 45–48; Goodare, *The European Witch-Hunt*, 197–202.

[106] Schultheis, *Eine Außführliche Instruction*, 109–10.

[107] Dillinger, *Kinder im Hexenprozess.*

[108] Levack, *The Witch-Hunt in Early Modern Europe*, 47–56.

[109] Binsfeld, *Tractat Von Bekantnuß der Zauberer unnd Hexen*, 142–47; Schultheis, *Eine Außführliche Instruction*, 57.

[110] Roper, *Witch Craze*, 76.

helped suspects endure torture, which also indicated guilt.[111] Probably most suspect of all were the people who criticized witch-hunts and were unwilling to cooperate. These individuals were likely witches themselves.[112]

Moreover, the cumulative concept increasingly gained the weight of authority. There had been the papal bull of 1484, and some of the period's best minds endorsed the concept. Martin Delrio is somewhat forgotten today, but contemporaries considered him a genius—"the miracle of our age," Justus Lipsius called him.[113] Many eminent rulers condoned witch trials, and the increasing number of persecutions seemed to substantiate that the network was real. People were also warned not to think lightly of *not* acting against this sect.[114] Diabolical witchcraft combined some of the worst crimes imaginable, like heresy, blasphemy, murder, unnatural sex, and hatred of God. This was a super-crime. The German cleric Peter Binsfeld rhetorically asked his readers: "Should we let this horrendous crime continue unpunished? Will we condone that their attacks and deeds against Holy Grace and the general good remain hidden?" Binsfeld presaged that God would not think lightly of that.[115]

It is worth pointing out that we should not interpret this idea of a "cumulative concept of witchcraft" in an overly rigid manner, as there was significant variation in what people believed. Did witches have the ability to metamorphose? Was the water ordeal any good? Did flights to the Sabbath happen literally, or was this only an illusion created by demons?[116] Experts disagreed. Or take the witches' sabbath.[117] To the literate classes, this often mattered greatly because they were primarily focused on the demonological aspects of witchcraft. On the other hand, many peasants' notions of a sabbath do not seem to have mattered that much, at least not initially. For them, maleficent magic within their communities was the key issue.[118] In addition, there was much diversity as to what people thought was happening at the sabbath. Some perceived it as a highly elaborate inversion of the Holy Mass, while others tended to think of an ordinary

[111] Krause, *Witchcraft, Demonology, and Confession in Early Modern France*, 26.

[112] Lancre, *On the Inconstancy of Witches*, 7; Meder, *Acht Hexenpredigten*, 15.

[113] Machielsen, *Martin Delrio*; Trevor-Roper, *The European Witch-Craze*, 80.

[114] Schultheis, *Eine Außführliche Instruction*, 502.

[115] Binsfeld, *Tractat Von Bekantnuß der Zauberer unnd Hexen*, 24–27.

[116] Clark, *Thinking with Demons*, 91; Levack, *The Witch-Hunt in Early Modern Europe*, 45/6.

[117] Blécourt, "Sabbath Stories"; Dillinger, *Hexen und Magie*, 62–72.

[118] Goodare, *The European Witch-Hunt*, 151.

village festival with some grotesque elements. Sabbath beliefs were also characterized by much regional and temporal variation. The notion was central to late sixteenth- and early seventeenth-century imaginations about witchcraft but far less significantly so in earlier ones, and in England and Holland, it never gained much traction.[119]

It is essential to keep in mind that there was no such thing as "the" cumulative concept of witchcraft. Concepts were diverse and continually changed. Some experts today are thus wary about using the term "cumulative concept" at all.[120] Still, there are good reasons for treating the belief in diabolical witchcraft as a somewhat unified phenomenon. First of all, because this is no anachronism. Demonologists at the time emphasized that witches did different things in different countries but that they were still a part of the same diabolical sect.[121] Secondly, during a particular period, the preceding ideas about witchcraft often clustered and retained a certain degree of coherence and stability. When combined, they could also have quite similar effects. This amalgam of ideas developed a striking capacity to trigger escalating witch panics. We now move to the German prince-bishopric of Bamberg in 1626 to see how this could work.

5.4 The Dynamics of Witch-Hunting

In the spring of 1626, the Catholic prince-bishopric of Bamberg seemed to be facing a relatively favourable future.[122] The small principality was located in the region of Franconia, in what is now southern Germany, and was part of the old and heterogeneous political body of the Holy Roman Empire. Like much of the empire, the prince-bishopric had recently experienced economic restraints, extreme weather conditions, and religious and political tensions. The region was fractured between Catholic and Protestant principalities, and militant versions were on the rise on both sides. The assertive Counter-Reformation movement had alerted the Protestants. In 1618, it was one of the triggers for a Protestant revolt against the Catholic emperor in the nearby kingdom of Bohemia, marking the beginning of the Thirty Years' War. As a devoutly Catholic territory,

[119] Thomas, *Religion and the Decline of Magic*, 526–29; Waardt, *Toverij en samenleving*.

[120] Briggs, "Many Reasons Why," 54; Broedel, "Fifteenth-Century Witch Beliefs," 33; Blécourt, "Sabbath Stories."

[121] Bodin, *De la démonomanie des sorciers*, 79, 385; Rio and Maxwell-Stuart, *Investigations into Magic*, 232.

[122] Gehm, *Die Hexenverfolgung im Hochstift Bamberg*, 90–99.

Bamberg had already expelled its own Protestants, and when the war broke out, it supported the emperor with troops.[123] Later, armies traversed Bamberg's lands. In the spring of 1626, however, the prospects began to look more fortunate again. Catholic armies progressed against their enemies, infringing on the dominantly Protestant north. The victories triggered an atmosphere of Catholic euphoria. On top of that, the winter of 1626 had been mild, so the corn was already flowering in May, promising rich harvests.

But then, on the night of 27 May, tragedy struck. The weather had been terrible throughout southern Germany, with hailstones the size of walnuts and a frost so heavy that it seemed mid-winter. The effects in Bamberg were dramatic. The fruits on the trees, the corn in the fields, and the wine grapes in the hills and mountains were frozen.[124] People in the region were upset, as the frost had severely damaged the harvests. Soon, questions were asked, and rumours spread. This kind of weather at this time of year was strange and suspicious, so had it not been caused by internal enemies? Was this not the work of evil witches? In the small city of Zeil, located close to Bamberg, a farmer and city official named Johann Langhans reported in his diary that "there were many pleas and requests from the populace, asking why we were standing by while the sorcerers and witches were spoiling the fruits, so his princely highness did nothing less than start the punishment of this evil."[125]

That so many people in Bamberg thought of witches as the cause of their troubles did not come out of the blue. Fears about witchcraft had become rampant in many parts of Germany during the preceding decades.[126] Newssheets, sermons, gossip, and scholarly tracts warned about the terrible harm that witches might cause, and throughout the empire, people demanded and petitioned their authorities to act. From around 1590, a new and ominous kind of witch persecution had appeared—William Monter calls it the "super-hunt"—which could kill up to hundreds of people.[127] For quite some time, the prince-bishopric of Bamberg had mainly remained unaffected because, amongst other things, its ruling bishop until 1609 was more interested in power and pleasure than religious

[123] Gehm, 94–100.

[124] Jung, "Die Bamberger Hexenverfolgungen," 73–74; Gehm, *Die Hexenverfolgung im Hochstift Bamberg*, 100–102.

[125] Stickler, *Eine Stadt im Hexenfieber*, 74–5.

[126] Behringer, *Witches and Witch-Hunts*, 83–111.

[127] Monter, "Witch Trials in Continental Europe 1560–1660," 23.

zealotry. Yet, his successors were influenced by the religious fervour of the Counter-Reformation, possessing an ardent desire to turn their prince-bishopric into a godly state without drunkenness, cursing, blaspheming, dancing, and also without witches.[128]

The first significant trials in Bamberg occurred in 1612 in the small town of Kronach.[129] An older woman named Lena Pantzerin had a reputation for magical powers within her community. After her daughter and son-in-law officially accused her of witchcraft, the authorities started a trial against her. Pantzerin was arrested, tortured, and confessed to being a witch. She also mentioned some names of accomplices, which triggered a small chain of accusations that led to the arrest and torturing of fifteen more people, including the daughter of Lena Pantzerin, who had initiated the trial. At least six people were killed. These events were still relatively moderate compared to trials that began in 1616. Around this time, the region experienced a dry summer, followed by an unusually wet autumn, which damaged many crops. In the small city of Zeil, a persecution of the allegedly responsible witches began. Now the snowballing of accusations led to many more arrests as the trials spread from Zeil to other towns and villages of the prince-bishopric. It killed probably around two hundred people. The persecution only ended in 1619, after the authorities of Bamberg had become divided about it.[130] A faction influenced by the Counter-Reformation wanted the hunt to continue. Still, they were resisted by another faction with the powerful and wealthy bishop's chancellor, Georg Haan, as its most important representative. He cut off the trials' money supply. The end of the persecution left the opposing faction with resentment against Haan and a fear that the witches' network had not been entirely wiped out.

The ultimate representative of the zealous mood in Bamberg was Friedrich Förner, the suffragan bishop.[131] Förner was born in the region and was later educated in Rome, where he became an uncompromising proponent of Counter-Reformation ideas. After his return, he grew into a charismatic preacher of Bamberg's imposing cathedral. Many sermons that he gave during the years before 1626 have been published, and they

[128] Walinski-Kiehl, "Witch-Hunting and State Building," 251; Dippold, "Die 'Hexenpolitik' Der Bamberg Fürstbischöfe," 48.

[129] Gehm, *Die Hexenverfolgung im Hochstift Bamberg*, 47–56.

[130] Gehm, 41–89; Dippold, "Die 'Hexenpolitik' Der Bamberg Fürstbischöfe," 48–51.

[131] Gehm, *Die Hexenverfolgung im Hochstift Bamberg*, 42–47, 97–100; Jung, "Die Bamberger Hexenverfolgungen," 70–85.

provide a frightening picture that he presented to his congregation.[132] Förner warned that Satan, filled with hatred of humankind, was encircling the flock like a roaring lion, seeking whom to devour. By becoming Protestants, many people had already been entangled in his nets, but to the devil's dismay, Bamberg had expelled its Protestants, robbing him of one of his essential tools. With his sharp eyes, Satan was now looking for Catholics whom he could remove from the true faith by pulling them into his secret arch-heresy of evil magic and poisoning.

Förner described evil sorcery as a disease, with rootlets penetrating Catholic society more deeply every day. Through their sins, people offered the devil an opening, for instance, when they sought the protection of amulets, consulted older women for magical cures, or were lustful, gluttonous, greedy, idle, or overly curious. Women were particularly susceptible to lust, as "all-female witches are the slaves of Venus."[133] Förner referred to children as being particularly vulnerable to curiosity. He described a recent case of a boy who was fascinated by the story of Dr. Faustus making a pact with the Devil, but then the boy was himself seduced by the devil into witchcraft.[134] He urged his audience to seek the protection of the Holy Faith and all its practices and institutions. But this was not yet enough, as diabolical evil would never stop by itself. The preacher reminded the divinely ordained authorities of their obligation to:

> *Accuse, scold, tear out, destroy, chastise, and eliminate* this plague of the soul; destroy this fatal monstrosity within our community, tear it out by the root, tear off its poison-soaked limbs within the republic (…), to ensure that *the healthy part will not be impaired* and *the sick cattle will not infect the entire sheepfold.*[135]

Luckily for Förner, the new prince-bishop of Bamberg since 1623 was Johann Georg Fuchs von Dornheim, who ardently believed in the danger of diabolical witchcraft.[136] So when the populace in 1626 demanded their authorities act, they found a sympathetic ear. The first serious allegations

[132] Förner, *Dämonenglaube und Zauberei im Jahre 1625.*
[133] Förner, 247–48, 306–7.
[134] Förner, 247–48, 306–7.
[135] Förner, 131–2.
[136] Gehm, *Die Hexenverfolgung im Hochstift Bamberg*, 90–102.

occurred again in the small city of Zeil.[137] A boy named Paul Rüghamer allegedly told stories about visits to a witches' dance, which alerted the local bailiff. The boy was arrested and began to tell staggering stories about visits to a sabbath and the specific adults from Zeil whom he had seen there. The bailiff reported the case to Bamberg, and the prince-bishop decided to form a special *Hexenkommission*, a witch commission, to lead an investigation in Zeil. The committee installed itself in one of the local towers, and inquiries began.

The commission arrested some of the people the boy had mentioned. The suspects were subjected to a standardized investigatory protocol containing questions such as: "Did you dance with the devil, and what other witches were present?" and "How often did you bewitch livestock, cause tempests, hailstorms, and night frosts?"[138] The suspects were unwilling to confess, so the investigators took them to the torture chamber in the tower, showing them devices like thumb screws, the wooden horse, and the strappado. When the sight of these tools still failed to make them talk, the suspects' bodies were shaven, and painful questioning was started. Not surprisingly, confessions soon followed. One woman crucially admitted that she had taken part in a diabolical conspiracy that had created the recent frost.[139] To the witch commission, this confirmed they were onto the network. More confessions followed, more names of accomplices were mentioned, and the circle of suspects and arrests widened. By January 1627, the commission had imprisoned thirty people, of whom twenty confessed and were sentenced to death. The convicts were first decapitated and then burned in public at the stake for the sake of destruction and purification.[140]

In the spring of 1627, the city of Bamberg was also confronted with the voluntary confessions of a boy named Hans Morhaupt, the son of a mayor. The witch commission took him into custody for interrogation.[141] At first, the boy seemed intimidated, but after some time, he apparently began to enjoy all the attention that he received. Without any torture, he told astounding stories about a diabolical conspiracy in the city. For him, it had

[137] Dippold, "Die 'Hexenpolitik' Der Bamberg Fürstbischöfe," 52–53; Gehm, *Die Hexenverfolgung im Hochstift Bamberg*, 100–106.

[138] Stickler, *Eine Stadt im Hexenfieber*, 46.

[139] Walinski-Kiehl, "Witch-Hunting and State Building," 258–59.

[140] Dülmen, *Theater des Schreckens*, 125.

[141] Walinski-Kiehl, "The Devil's Children"; Jung, "Die Bamberger Hexenverfolgungen," 81–82.

all started at his Jesuit school, where he read a book about Dr. Faustus, which his teachers took away from him. To retrieve the book, Hans made a pact with the devil, which introduced him to a marvellous world of witchcraft. He entered a sexual relationship with a female demon named "Kandel." With his mother and her maid, he flew on a pitchfork through a window in the attic to a witches' dance, where he met people from the prominent circle of his family. Whether Hans Morhaupt understood the consequences of his words is questionable, as it resulted in his death at the stake, along with that of much of his family.

The new trials in the city created a novel sense of anger and panic because the problem of witchcraft seemed worse than anticipated, even infiltrating the highest classes.[142] The persecution was continued with the support of much of the populace, qualified lawyers, clerics such as Förner, and the prince-bishop eager to rid his principality of this terrible "witch plague."[143] Meanwhile, the weather remained awful. This indicated that the network of witches had not been defeated yet. The commission worked relentlessly. One of its members, Ernst Vasoldt, was obsessed with sabbath attendance and even managed to extract 126 names of accomplices from one single suspect.[144] More people were thrown into the dungeons and robbed of daylight, sufficient food, and contact with others, while the legal restrictions on the use of torture were violated. With all the new suspects in jail, even the extended prisons in Zeil were no longer sufficient. In the summer of 1627, a sizeable new witch prison was built in Bamberg.[145]

With the new facilities at its disposal, the witch commission found a new target in Georg Haan. In 1619, Haan had been able to stop the hunt, but this time, things took an unfortunate turn for the chancellor. First, his wife and daughter were arrested and killed, followed by his arrest and that of his son. Haan's earlier resistance to the trials was now explicitly used against him. Eventually, the whole family was burned at the stake, including many of their servants.[146] With the most powerful opponent of trials crushed, the chain of accusations further exploded, with hundreds of new

[142] Gehm, *Die Hexenverfolgung im Hochstift Bamberg*, 107–70; Jung, "Die Bamberger Hexenverfolgungen," 74.

[143] Dippold, "Die 'Hexenpolitik' Der Bamberg Fürstbischöfe," 44.

[144] Walinski-Kiehl, "Witch-Hunting and State Building," 256.

[145] Gehm, *Die Hexenverfolgung im Hochstift Bamberg*, 119–21.

[146] Renczes, *Wie löscht man eine Familie aus?*

arrests and burnings.[147] Hardly anyone was safe from the investigations—women and men, seniors and children, as well as the rich and the poor, were rounded up. Anyone who entered the witch prison was almost certainly doomed. A significant part of the population in Zeil was wiped out, including nearly the entire city council. We can still trace the course of events in Johann Langhans's diary, who had to report about family members being killed.[148] During the trials of 1616–1619, he had already lost two aunts, who, as Langhans commented, had "turned away from God our saviour and creator, forsaken the Lord, and given themselves to the Devil."[149] This time, it became much worse, as an uncle was convicted, a stepsister, then his wife, and his daughter, and in February 1628, the diary ended because Langhans was accused and burned as a witch himself. The effects in the city of Bamberg were not as drastic as in Zeil, but especially for the elite, the hunt was turning into a carnage.

Amongst the victims was one of the mayors of Bamberg, Johannes Junius. There are no clues that Junius was an unpopular figure or that he cultivated enemies. Still, after his wife was killed in early 1628, he was imprisoned in June of that year.[150] From his jail, Junius wrote a letter to his daughter, which he tried to get smuggled out. His guards discovered the letter and added it to the trial records. It offers us a first-hand account of life in the witch prison:

> Many hundred thousand good nights, dearly beloved daughter Veronica. Innocent have I come into prison, innocent have I been tortured, innocent must I die. For whoever comes into the witch prison must become a witch or be tortured until he invents something out of his head – God pity me – bethinks him of something. I will tell you how it has gone with me.[151]

Junius told his daughter about his agonies and excused himself for his bad handwriting—his interrogators had used thumbscrews that made the blood run out of his nails, making writing difficult. Eventually, after long resistance, the former mayor confessed almost everything his interrogators

[147] Gehm, *Die Hexenverfolgung im Hochstift Bamberg*, 127–70; Stickler, *Eine Stadt im Hexenfieber*, 74, 142.

[148] Stickler, *Eine Stadt im Hexenfieber*, 61–67.

[149] Stickler, 64.

[150] Walinski-Kiehl, "Males, 'Masculine Honor', and Witch Hunting"; Gehm, *Die Hexenverfolgung im Hochstift Bamberg*, 152–57.

[151] Levack, *The Witchcraft Sourcebook*, 201.

wanted to hear, including the names of thirty accomplices. Junius urged his daughter to leave Bamberg as soon as possible, ending his letter with the words: "Farewell, you will never see your father Johannes Junius again!"[152] Shortly after that, Junius was burned.

From the investigators' perspective, however, they were doing the noble work of uncovering a grotesque network of witches ravaging the area.[153] During the interrogations, the committee learned that the witches were in continuous contact with Satan, had sex with demons, and desecrated holy symbols. At night, the witches flew out to dance together, for example, in a neighbouring forest. The harm that these servants of Satan were inflicting upon the population was startling: they stole milk, spoiled the wells, and sickened livestock.[154] It was further discovered that the conspiracy had caused recent storms and a drought that ruined the hay. The witch commission had been just in time to thwart their new plans for creating a storm that was so powerful that it would destroy the cathedral.

The events in Bamberg were spectacular, and the news about the hunt spread throughout the region. Newssheets told people about the terrible things witches had done and how they had been justly punished.[155] But there were also critical voices. Citizens of Bamberg increasingly began to flee the city, telling their acquaintances elsewhere about the trials' horrors and injustices. When the wealthy families were attacked, they appealed to higher authorities elsewhere. The *Reichskammergericht* (Imperial Chamber Court), located in Speyer, indeed denounced many of the trial procedures.[156] There were also some internal problems. At a certain point, witch commissioner Ernst Vasoldt became drunk in a tavern in Zeil, losing his extensive list of further suspects. The document became known in Zeil and created upheaval in the city.[157] Vasoldt was discharged, but when the number of arrests kept growing, the initial widespread support for trials in Zeil wavered. The committee faced an increasingly hostile mood and, at

[152] Gehm, *Die Hexenverfolgung im Hochstift Bamberg*, 156.

[153] Behringer, *Hexen und Hexenprozesse in Deutschland*, 261–64; Dippold, "Die 'Hexenpolitik' Der Bamberg Fürstbischöfe," 35.

[154] Gehm, *Die Hexenverfolgung im Hochstift Bamberg*, 143; Behringer, *Hexen und Hexenprozesse in Deutschland*, 261–64.

[155] Behringer, *Hexen und Hexenprozesse in Deutschland*, 262–64.

[156] Dippold, "Hexereiprozesse Im Hochstift Bamberg," 57–58; Gehm, *Die Hexenverfolgung im Hochstift Bamberg*, 127–70.

[157] Gehm, *Die Hexenverfolgung im Hochstift Bamberg*, 108–22.

some point, began arresting suspects at night to avoid trouble. Even the people who assisted the persecutors could not protect family members from being killed. It stimulated discussions within the witch commission about the risks of harming the innocent. A further problem was financial costs.[158] The procedures were so expensive that they began to burden civic budgets, and in Zeil, the committee was forced to build a witch oven to save on firewood costs. However, the high number of rich witches executed, made financial problems surmountable: the rich witches' possessions were used for further funding. This was in contravention of judicial rules, but Bamberg ignored those rules, just as it ignored the verdict from Speyer. The risks of self-incrimination also discouraged further internal criticism. By 1630, there was still no prospect of an end.

The trials' eventual slowdown was mainly due to outside forces.[159] Citizens of Bamberg were making ever more appeals to authorities elsewhere about the "bloodthirsty, unchristian, and ruthless" policies of the witch commission, which brought "the city to the brink of ruin" and "the entire honourably citizenry to the ground."[160] In 1629, a woman from the prosperous and well-connected Flock family was arrested and her husband moved heaven and earth to free her.[161] His pleas to the emperor and the *Reichshofrat*—another judicial body of the empire—delivered two further condemnations of Bamberg's policies. At the Diet of Regensburg in 1630, officials from the prince-bishopric were openly criticized.[162] As a defence, Bamberg relied on the concept of *crimen exceptum*, and the prince-bishop replied that he had "only lawfully conducted the trials for the sake of promoting and propagating the glory of God and for the salvation of many seduced souls, as was also done by other princes and electoral princes in our Holy Roman Empire."[163] This was to no avail. In 1631, the *Reichshofrat* issued a binding imperial mandate that ordered the prince-bishop to stop the trials immediately.

The end of the trials was due to another development, however. The Thirty Years' War odds had turned against the Catholic party, and the war

[158] Dippold, "Die 'Hexenpolitik' Der Bamberg Fürstbischöfe," 44–58.

[159] Gehm, *Die Hexenverfolgung im Hochstift Bamberg*, 171–228.

[160] Behringer, *Hexen und Hexenprozesse in Deutschland*, 389.; Bradford Smith, "The Persecution of Witches," 68.

[161] Dippold, "Die 'Hexenpolitik' Der Bamberg Fürstbischöfe," 58.

[162] Gehm, *Die Hexenverfolgung im Hochstift Bamberg*, 171–210.

[163] Gehm, 209; Behringer, *Hexen und Hexenprozesse in Deutschland*, 390.

moved southwards again. In 1632, a Protestant Swedish army approached the city, and to its chagrin, Bamberg found out that it could no longer defend itself.[164] The public budgets had largely been spent on witch trials, so not much money was left for troops and fortifications. Johann Fuchs von Dornheim and his allies fled the city. When the Swedes conquered Bamberg, they encountered communities that the preceding witch-hunt had ravaged.[165] The town of Zeil was depopulated to a large extent, and hardly any family had remained unharmed. In Bamberg, the destruction of much of the upper class had damaged the economy. Many houses in Bamberg's most prestigious street, the Lange Gasse, were emptied, and no one was interested in buying them. Around an approximated six hundred people in the prince-bishopric had been killed, of whom about seventy percent were women.[166] The Nordic rule did not last long, as the Catholics soon recaptured the prince-bishopric, but not before the Swedes pillaged the lands even further. Yet, after all these horrors, Bamberg did seem to have learned a lesson: apart from a few isolated burnings in the 1670s, the prince-bishopric would never experience a witch-hunt again.[167]

What happened in Bamberg was unusual. In many cases, witch trials involved the persecution of one or just a few people, and even those trials were only the tip of an iceberg of local angry exchanges and suspicions of witchcraft that did not result in a criminal investigation. This nuance made some experts observe that focusing on extensive persecution is somewhat misleading.[168] Big hunts often get much attention because they are spectacular, but they are not representative of the experience of witchcraft in early modern Europe. Still, the Bamberg hunt was far from unique.[169] It had been preceded by somewhat similar hunts elsewhere, and persecution in nearby Würzburg killed around nine hundred people. Concurrent trials in the Electorate of Cologne even killed about two thousand people and

[164] Callow, *Embracing the Darkness*; Gehm, *Die Hexenverfolgung im Hochstift Bamberg*, 224–28.

[165] Gehm, *Die Hexenverfolgung im Hochstift Bamberg*, 139–48, 170–74.

[166] Gehm, 92–4.

[167] Gehm, 224–31; Dippold, "Die 'Hexenpolitik' Der Bamberg Fürstbischöfe," 57–60.

[168] Briggs, *Witches & Neighbours*, 7; Midelfort, "Witch Craze?"; Levack, *The Witch-Hunt in Early Modern Europe*, 171–72.

[169] Levack, *The Witch-Hunt in Early Modern Europe*, 172–75; Goodare, *The European Witch-Hunt*, 225–66.

possibly even more.[170] While chain-reaction hunts were only a small percentage of the total number of trials, they still seem responsible for a majority of the total number of executions.[171] The Bamberg trials also possessed many of witch-hunting's typical traits. Hence, Bamberg offers an exemplary case study of the cumulative concept of witchcraft's inflammatory potential.

These trials illustrate how the cumulative concept could make unusual misfortune be interpreted as an act of diabolical witchcraft.[172] The frost of May 1626 was not just unusual—it resulted from the maleficent magic of satanic allies. The trials of Bamberg further show how two categories of potential suspects could get a trial going. First, individuals, like Lena Pantzerin from Kronach, who had reputations for harmful magic within their local communities.[173] Persecutions often began when local aggression against such people had reached intolerable levels. In some cases, communities took the initiative to kill such "village witches" themselves, but since authorities discouraged that, the more usual mechanism was that these people were reported to the local government. A trial was started when the authorities granted such requests. The second category was children.[174] More so than adults, children were likely to offer their interrogators imaginative testimonies about witchcraft, making the use of torture unnecessary.

When the first suspects were identified, the idea of witchcraft as a "crimen exceptum" provided a crucial next step. Bamberg exemplifies how excessive torture enforced confessions. This was no exception. The historian Gerhard Schormann writes that "torture may not have been the cause of the witch trials, but it was an essential prerequisite for their massive implementation."[175] Without it, most of the seemingly consistent confessions that kept the trials going on this scale would have been unthinkable.[176] The questions that interrogators in Bamberg asked also typify the

[170] Behringer, *Witches and Witch-Hunts*, 130; Grawlich, "Der Hexenkommisar Heinrich von Schultheiß," 303–4.

[171] Goodare, *The European Witch-Hunt*, 118, 250.

[172] Rummel and Voltmer, *Hexen und Hexenverfolgung*, 87–88.

[173] Briggs, *Witches & Neighbours*, 95; Goodare, *The European Witch-Hunt*, 103–18.

[174] Walinski-Kiehl, "The Devil's Children"; Behringer, "Kinderhexenprozesse."

[175] Schormann, *Hexenprozesse in Deutschland*, 43, 52.

[176] Rummel and Voltmer, *Hexen und Hexenverfolgung*, 44–51; Dillinger, *Hexen und Magie*, 80–86.

sabbath's importance.[177] Its key implication was that captured witches could tell who the other witches were. Julian Goodare rightly observes that chain-reaction panic became possible only after elites started believing in conspiracies.[178] The snowballing of accusations could also lead to the killing of people who were not associated with harmful magic at all and who, in many cases, even had respectable positions within their communities, like Mayor Johannes Junius. Moreover, Bamberg's interrogations illustrate why the nightly flight was significant: it enabled witches to meet each other at various places in the middle of the night.[179] Furthermore, the belief had the potential to spread accusations over long distances, as witches were also supposed to know accomplices who lived far away.

Friedrich Förner's references to women as dangerous magical practitioners and "slaves of Venus" highlight another critical aspect: the predominant focus on women. Overall, the average percentage of female victims was even higher than in Bamberg because estimations assume that the average percentage lay somewhere between seventy-five and eighty percent.[180] The local differences were significant, though, as some regions almost exclusively executed women, while other areas had much higher percentages of male victims. Finally, the hunt in Bamberg was characteristic of the time and place in which it occurred. The overwhelming majority of the executed died between 1560 and 1660, and probably more than half of them had lived within the boundaries of the Holy Roman Empire, speaking some dialect of German.[181] Many burnings also occurred in parts of Switzerland, Poland, and the eastern borderlands of France and the Southern Netherlands. More peripheral pockets of intense hunting were found in the Basque Country, Scandinavia, and Scotland. In contrast, countries like England, Italy, France, the Northern Netherlands, and Spain were characterized by more modest levels of persecution.

[177] Cohn, *Europe's Inner Demons*, 252; Levack, *The Witch-Hunt in Early Modern Europe*, 74–81.

[178] Goodare, *The European Witch-Hunt*, 225–63.

[179] Cohn, *Europe's Inner Demons*, 228–29, 252; Levack, *The Witch-Hunt in Early Modern Europe*, 41–42.

[180] Levack, *The Witch-Hunt in Early Modern Europe*, 128–35; Rummel and Voltmer, *Hexen und Hexenverfolgung*, 79–80; Rowlands, "Witchcraft and Gender in Early Modern Europe," 449.

[181] Monter, "Witch Trials in Continental Europe 1560–1660," 14–16; Rummel and Voltmer, *Hexen und Hexenverfolgung*, 80–83.

Bamberg thus gruesomely illustrates how the cumulative concept could trigger witch trials that were difficult to stop once set in motion. The underlying mechanisms of witch-hunting and how cumulative notions produced such lethal dynamics have been analysed by historians in detail. However, while their observations help us explain a lot about witch-hunting, they still leave much unsolved. How could a phenomenon like this come into being? Why did it persist for such a long time? We will now integrate the case study into this book's theoretical concerns. What may this case study tell us about the origins and purposes of socio-cultural functions?

CHAPTER 6

Explanations

6.1 An Explanatory Problem

What was the function of the witch-hunts? What was their purpose? If we would pose these questions to the historical actors who created and fostered the trials, they would probably be surprised. After all, the answer would have been evident to them: evil minions of the devil were involved in conspiracies against Christianity and launching magical attacks on the population. Witch-hunting reduced witchcraft's danger, and this intended and recognized effect constituted its functionality. It is only to those who doubt the guilt of convicted witches or to those who doubt the existence of diabolical witchcraft altogether that questions about the hunts' functionality become a more challenging and theoretically exciting problem. If witch-hunts did *not* have the effect of stopping demonic activities, what else were they for? Were they shrewdly used for alternative purposes, or did they have concealed functional causes and effects that the witch-hunters themselves were unaware of? Or is the question about their functionality wrongly asked? The next section will look at various answers. But first, we need to address the issue of diabolical witchcraft's reality because it is the repudiation of its reality that gives rise to these questions in the first place. This is also more of an empirical and theoretical puzzle than one might first assume.

S. Hofhuis, *The Evolutionary History of Witch-hunting*, https://doi.org/10.1007/978-3-031-66936-1_6

Doubts about the reality of the crime of diabolical witchcraft are as old as the persecutions themselves. Proponents of witch-hunting always faced resistance, and Julian Goodare even states that some scepticism remained "the default position" amongst European judges.[1] But in many cases, this did not imply that people doubted the possibility of diabolical witchcraft, let alone the central tenets of demonology. Mainstream scepticism at the time especially concerned the guilt of the actual people who were burned as witches at the stake.[2] In this sense, some scepticism can even be found amongst zealous advocates of witch-hunting, like Peter Binsfeld, who expressed worries about inexperienced persecutors who would proceed too quickly and ran a risk of torturing and imprisoning older women who had done nothing wrong.[3] There was no sharp distinction between "believers" and "sceptics."[4]

However, there was a contemporary strand of critique that had more pronounced misgivings about the reality of diabolical witchcraft and the validity of trials. Today's scholarship calls it "the sceptical tradition."[5] Amongst its most well-known exponents are the Dutch-German physician Johann Weyer, who in 1563 wrote an extensive critique named *De praestigiis daemonum* (On the Tricks of Demons), and the German Jesuit Friedrich Spee, who in 1631 published his *Cautio criminalis* (Precautions for Prosecutors).[6] Nineteenth and twentieth-century historiography has sometimes heralded such authors as precursors of a modern scientific outlook.[7] Indeed, their works contain critiques of the beliefs of their contemporaries that a modern rationalistic audience will be able to appreciate.

Johann Weyer mocked claims about the alleged powers of witches in ways that are still amusing today. He described how witches were supposed to ruin crops and pollute waters and added that "these old women called *Lamiae,* by a single nod, will not only accomplish these things, but also smite armies, and shatter cities and regions, so that the enemy will

[1] Goodare, *The European Witch-Hunt*, 354.

[2] Goodare, 79–82.

[3] Binsfeld, *Tractat Von Bekantnuß der Zauberer unnd Hexen* Vorrede.

[4] Schwerhoff, "Rationalität im Wahn"; Clark, *Thinking with Demons*, 173–213.

[5] Stephens, "The Skeptical Tradition"; Lehmann and Ulbricht, *Vom Unfug des Hexen-Processes.*

[6] Weyer, *Witches, Devils, and Doctors in the Renaissance*; Spee von Langenfeld, *Cautio criminalis.*

[7] White, *A History of the Warfare of Science with Theology in Christendom*, 350–63.

consider themselves fortunate, if they are permitted to surrender."[8] He then ironically wondered why generals do not use these marvellous powers in wars: "What need then for expensive artillery? For quantities of that diabolical powder? For heaps of cannonballs and bullets? (...) Indeed, we seem to have been going about blindly amidst the full light of day – we who have so stupidly neglected this possibility!"[9] Weyer acknowledged that using diabolical forces in such ways would be utterly unchristian. Still, many Christian princes did all sorts of disgraceful things, so why did they leave this option aside? Friedrich Spee took a similarly critical stance on his contemporaries' beliefs. As a priest, he had been responsible for the pastoral needs of convicted witches, and he was horrified by what he saw. His *Cautio criminalis* combined emotional observations about the suffering of the convicts with a sharp critique of the inconsistencies of trials. Spee, for instance, noted that persecutions were built upon testimonies from alleged witches that were unreliable by definition: either the convicts were real witches, which would make their testimonies untrustworthy because the devil's allies are unlikely to tell us the truth, or they were not real witches, which would make them equally deceptive, as their witness statements had to be false.[10]

But we should be careful again and not read such contemporary sceptics as modern rationalist companions in a different age. In surprising ways, these authors shared much of their witch-hunting contemporaries' worldview and polemical habits.[11] Weyer and Spee were profoundly Christian and did not at all deny the active presence of demons in the world. On the contrary, they used demonic forces as an explanation for why witch-hunts occurred. Weyer maintained that demons created illusions in people's minds about witchcraft, amongst persecutors as well as amongst victims, and the ultimate diabolical goal of that was to trigger the human suffering of the trials.[12] According to Spee, many of the inquisitors responsible for witch-hunting were likely to be witches themselves![13] Eric Midelfort rightly commented that the contemporary debate was "not an

[8] Weyer, *Witches, Devils, and Doctors in the Renaissance*, 218.
[9] Weyer, 218.
[10] Spee von Langenfeld, *Cautio criminalis*, 380.
[11] Schwerhoff, "Rationalität im Wahn."
[12] Weyer, *Witches, Devils, and Doctors in the Renaissance*.
[13] Spee von Langenfeld, *Cautio criminalis*, 242, 253.

issue between enlightened rationalists and fanatic obscurantists. The basic argument was really between two groups of pious men."[14]

Both "skeptics" also devoted attention to the overrepresentation of women amongst the victims in ways that were characteristic of their times. They wrote with compassion about the torments of female suspects but simultaneously used stereotypes about women that were eerily similar to those of fervent witch-hunters.[15] Weyer stated that the devil shrewdly seduced people into believing that they were real witches, and due to their overall weakness in character, women were his most likely victims: "That crafty schemer the Devil thus influences the female sex, that sex which by reason of temperament is inconsistent, credulous, wicked, uncontrolled in spirit."[16] Spee pointed out that the testimonies about witchcraft were mainly derived from females who were tortured, which allegedly made these accounts twice as unreliable. After all, torturing men already stimulated false confessions, so what to expect from women? The convicted suspects tended to be "insane, incomprehensible, reckless, talkative, jealous, deceitful, lying" women who made judicial procedures run rampant.[17]

Was there no more fundamental scepticism about things like the reality of demons, magic, or even divine providence itself? This is a much broader question about late Medieval and early modern European history, rife with interpretative difficulties.[18] There are examples of apparent scepticism, but what complicates things is that people at the time could not speak freely about such matters; outright sceptics were likely to be put to death. Thus, one can easily read too little disbelief in early modern documents or too much, as speculations about hidden messages can run more freely.[19] A case in point is the sixteenth-century French sceptic Michel de Montaigne, author of a renowned collection of essays. In an essay "On Cripples," Montaigne made some highly critical remarks on witch trials. He considered the evidence used by persecutors thoroughly insufficient, adding that "it is putting a very high price on one's conjectures to have a man roasted

[14] Midelfort, *Witch Hunting in Southwestern Germany*, 65.

[15] Clark, *Thinking with Demons*, 118.

[16] Weyer, *Witches, Devils, and Doctors in the Renaissance*, 181.

[17] Spee von Langenfeld, *Cautio criminalis*, 226, 278.

[18] See, for instance: Greenblatt, *The Swerve*; Closson, *L'imaginaire démoniaque en France (1550-1650)*, 50–62; Febvre, *Le problème de l'incroyance au 16e siècle.*

[19] Lehmann and Ulbricht, "Motive und Argumente von Gegnern der Hexenverfolgung von Weyer bis Spee," 10.

alive because of them."[20] But to understand such remarks as indicative of a more systematic, critical stance on the existence of witchcraft, demonic powers, or religion's authority in society is questionable.[21]

Arguably, a better place to look for clues of outright scepticism is not so much in the work of the "sceptics" but in the works of the witch-hunters and demonologists. In his book *Daemonolatreiae libri tres* (Demonolatry) from 1595, the Lorraine witch-hunter Nicolas Rémy spoke with horror about the epicurean school, which held that "no such things as spirits and Demons exist in the whole realm of nature."[22] Such remarks were not unusual, indicating that the contestation of the spirit world was conceivable.[23] The literary scholar Walter Stephens has recently delved into this topic and proposes that the excessive interest in witchcraft theories between 1400 and 1700 should primarily be understood as a response to such scepticism—the perceived threat concerned not only witchcraft but also central tenets of Christianity.[24] The excessive curiosity about contact between demons and witches provided people with a means to gain knowledge about the spirit world at a time when doubt had been cast on that world's mere existence. Walter Stephens: "The attitude of witchcraft theorists towards their theories was not belief but rather *resistance to skepticism*, a desperate attempt to maintain belief, and it betrays an uncommonly desperate *need to believe*."[25] Several key experts, however, think that Stephens has exaggerated the point.[26] True, witchcraft theorists were often on the defensive, but this does not mean that it was an inner atheist that they wanted to convince, but, rather, their thoroughly Christian contemporaries who doubted their more extravagant claims about witchcraft.

It was only in the seventeenth century that fundamental doubts about demonology and concepts of witchcraft became more common and explicit, becoming unambiguous in the eighteenth century. The spirit of the Enlightenment in the latter century made significant parts of the

[20] Montaigne, *Les Essais*, 1079.

[21] Brian Ribeiro, "Montaigne on Witches and the Authority of Religion in the Public Sphere."

[22] Rémy, *Demonolatry*, 79.

[23] Goodare, *The European Witch-Hunt*, 82; See, for instance in: Bodin, *De la démonomanie des sorciers*, 76, 215.

[24] Stephens, *Demon Lovers*.

[25] Stephens, 27.

[26] Behringer, "Walter Stephens. Demon Lovers"; Machielsen, "Thinking with Montaigne."

educated classes think of witch beliefs as outdated or even laughable.[27] Still, we need to be careful and not place the complete denial of witchcraft too early or too pervasively in time.[28] Throughout much of the eighteenth century, cumulative notions continued to receive influential support, and witch trials still occasionally occurred. Only in the late eighteenth century did elites almost entirely distance themselves from witchcraft beliefs.

This spirit of the Enlightenment, embracing modern science and rationalism, continued in subsequent centuries and gave rise to a crucial school of thought in witchcraft historiography, "the liberal tradition."[29] The non-existence of diabolical witchcraft was its focal point of attention. Exponents often looked back in anguish at what their ancestors had done and believed. Reflecting on the large trials of the early seventeenth century, the Scottish author Charles Mackay 1841 found it "almost impossible to believe that mankind could ever have been so maddened and deluded."[30] The liberal tradition was not only about history but also had contemporary aims in mind. Nineteenth- and twentieth-century liberalism contended with resilient forms of conservative religion, Roman Catholicism in particular, and the history of witch-hunts provided them with a useful rhetorical weapon. By exposing the Catholic fanaticism, cruelty, and superstition of the "Witch Crazes," they criticized what they saw as the religious superstition and stupidity of their age.

During the same period, witchcraft scholarship gave rise to a powerful counter-trend, "the romantic tradition."[31] The romantics also took a contemptuous stance towards the witch-hunts, but what made them different was their take on underlying realities. The romantics also considered diabolical witchcraft an illusion but contended that persecutors were after something else that was real. According to the nineteenth-century philologist Jacob Grimm, renowned for his fairy tales, many "witches" had been wise women who preserved precious elements of German folk culture. Witch-hunts should thus be understood as an attempt by religious and

[27] Goodare, *The European Witch-Hunt*, 335–44; Closson, *L'imaginaire démoniaque en France (1550-1650)*, 53–62.

[28] Behringer, "Späte Hexenprozesse - ein Pfahl im Fleisch der Aufklärung."

[29] Tuczay, "The Nineteenth Century," 56–60; Goodare, *The European Witch-Hunt*, 363–65.

[30] Mackay, *Memoirs of Extraordinary Popular Delusions and the Madness of Crowds*, 158.

[31] Tuczay, "The Nineteenth Century," 52–56; Goodare, *The European Witch-Hunt*, 366–69.

political elites to get rid of those women.[32] The anthropologist and Egyptologist Margaret Murray developed an influential twentieth-century version of the romantic viewpoint. She maintained that witch-hunts were a Christian fight against a lingering Stone Age fertility cult. This cult had worshipped a horned god, rebranded by the elites as the devil.[33]

The liberal and romantic perspectives are essential to witchcraft historiography, as they dominated it well into the twentieth century and continue to shape many popular notions about witch-hunts today. Yet, from the 1960s and 1970s onwards, professional historical scholars have distanced themselves from both traditions.[34] The study of witch beliefs and trials achieved higher levels of empirical precision and interpretative sophistication, which made both schools of thought run into trouble. The key problem with the romantic approach is that meticulous examination of historical documents failed to provide convincing clues for real dissident networks.[35] All contemporary evidence for their existence turned out to be second-hand, thoroughly contaminated by leading questioning and torture, or unreliable for other reasons. Not to mention other things the witches were supposed to do, like weather magic or nightly flights, which are impossible deeds.[36]

However, dismissing real networks did not make current witchcraft scholarship endorse the liberal viewpoint. The tendency to interpret the behaviour and convictions of historical actors as irrational and ridiculous no longer seemed helpful.[37] Describing historical behaviour as "crazed" or "deluded" does not offer much of an explanation and may actually misapprehend how the historical actors involved saw things. Current scholarship thus tends to treat early modern witch beliefs as a complex paradigm in its own right that possessed a certain degree of internal consistency that could be difficult to contest at the time. Only extravagant ideas about the sabbath or extensive persecutions were easier to dismiss within the available frameworks. This stance, which we could summarize as "understanding denial," was explored earlier in anthropology. In 1937, the British anthropologist Edward Evans-Pritchard published his book *Witchcraft, Oracles,*

[32] Grimm, *Deutsche mythologie*, I:836–1058.
[33] Murray, *The Witch-Cult in Western Europe.*
[34] Midelfort, *Witch Hunting in Southwestern Germany*; Cohn, *Europe's Inner Demons.*
[35] Goodare, *The European Witch-Hunt*, 366–69; Dillinger, *Hexen und Magie*, 113–36.
[36] Levack, *The Witch-Hunt in Early Modern Europe*, 12–18.
[37] Schwerhoff, "Rationalität im Wahn"; Gibson, "Thinking Witchcraft."

and Magic among the Azande, which made a profound attempt to understand the beliefs of the African Azande people on their own terms. Evans-Pritchard stated that "witches, as Azande conceive them, cannot exist," but at the same time, he treated their convictions as an "intellectually coherent system."[38]

The debate about what was real about witchcraft is not over yet, however, as this position of "understanding denial" fails to satisfy all current experts. It is contested from two angles. The first maintains that the debate about witchcraft's reality is redundant, as the distinction between "real" and "non-real" is irrelevant. This viewpoint comes forward in the influential book *Thinking with Demons* by the English historian Stuart Clark.[39] In Clark's view, much historical scholarship suffers from a "referential fallacy" that mistakably assumes that our language can correspond to a reality that exists independently from ourselves.[40] Clark considers it a form of "naive positivism," "a piece of mystification," and thus wants to unhook witchcraft scholarship from its realist moorings.[41] Clark writes: "Whether witchcraft beliefs did *in fact* correspond with reality becomes, therefore, a question not worth asking." It is a "non-issue."[42] Instead, historians are obliged to "take up a relativist position" that moves them "away from causes and towards meaning."[43] In other words, Clark's work vividly shows that postmodern relativism has appeared in witchcraft scholarship.

How to react to this relativist viewpoint? It is important to note that it depends on the research question whether it is relevant to know to what extent beliefs were illusory or not. If the sole aim is to find out what people believed about witchcraft and what meaning it had to them, the questions about witchcraft's existence are not necessarily relevant. Arguably, it can be a helpful heuristic device to leave those questions aside, and indeed, Clark delivered the most profound account of the demonological worldview. But Clark is after bigger game, as he renders the question about witchcraft's reality irrelevant to witchcraft research *in general*. At this point, he enters the domain of postmodernist self-contradiction and

[38] Evans-Pritchard, *Witchcraft, Oracles and Magic among the Azande*, 63, 475.

[39] Clark, *Thinking with Demons*; For other prominent examples, see for instance: Purkiss, *The Witch in History*, 59–90; Stephens, *Demon Lovers*, 10.

[40] Clark, *Thinking with Demons*, 6.

[41] Clark, 317.

[42] Clark, 6–7.

[43] Clark, 396, 402.

confusion.[44] Clark accuses others of the "referential fallacy," but interestingly, his book is full of references to reality! These references include countries, rulers, or courts, and his footnotes lead us to publishers, book titles, and page numbers. His underlying assumption is that those things exist out there and can be referred to and checked by others accordingly. Furthermore, Clark accuses colleagues of a scientific "triumphalist reading of the past," which supposedly miscomprehends early modern Europe's intellectual history.[45] However, if Clark can determine that colleagues are wrong about something that exists independently from themselves, then why is it impossible for others to determine that early modern demonologists were wrong about things like nightly flights on broomsticks, or intercourse with the devil? In a typically postmodernist fashion, Clark maintains that others cannot refer to reality out there, but for reasons that have yet to be clarified, he does have this ability himself.

Much of Clark's argument is built upon a false opposition of *real causes* versus *meaning*. As the work of Evans-Pritchard illustrates, the one does not exclude the other. Moreover, from a broader academic perspective, knowing what was really going on *is* relevant. If there was a lingering pagan cult out there that Christians wanted to destroy, then Margaret Murray's good name needs to be restored. Or if the beliefs of demonologists like Kramer, or Binsfeld were correct, that would open up profoundly fascinating research agendas. How exactly were broomsticks turned into tools for aerial journeys? What are the implications for the law of gravity, and are flights on broomsticks still possible today? It requires a remarkable lack of scientific curiosity to render questions about the reality of witchcraft a "non-issue." It is also quite an affront to the people tortured and burned as witches to pretend that the question of their guilt is irrelevant.

Then, a second angle of critique on the "understanding denial" position deserves consideration, which aims to put the question of witchcraft's reality back on the agenda. This position is mainly represented by the American historian Edward Bever, who seeks to revive the notion of powerful underlying realities. Strikingly, he does so in a sense close to what early modern contemporaries thought was going on.[46] According to

[44] For critiques of the relativist viewpoint, see: Goodare, *The European Witch-Hunt*, 3–6; Hutton, *The Witch*, 40–41.

[45] Clark, *Thinking with Demons*, 317.

[46] Bever, *The Realities of Witchcraft*; For another example, see: Henningsen, "The Witches' Flying and the Spanish Inquisitors, or How to Explain (Away) the Impossible."

Bever, current scholarship grossly underestimates what was real about witchcraft. "As with every crime, of course, not every allegation was valid, but the belief that *maleficum* might be perpetrated and fear that it might cause harm were not irrational."[47] Much of Bever's argument is not entirely new for historians. It has already been recognized that people attempted harmful magic and that psychosomatic mechanisms could make such attempts have real damaging consequences.[48] Bever's work's merit is that he delves much deeper into this line of reasoning and abundantly refers to recent scientific findings, mostly regarding neuroscience and subliminal communication.

Bever's argument contains some quite surprising twists, though. He upholds that investigations into the "paranormal" have delivered "a substantial body of evidence suggesting that there are natural processes which are not adequately explained by current scientific understanding."[49] Bever is keenly interested in apparently unexplained phenomena like ball lightning, poltergeists, "urns crashing on the ground for no apparent reason; cloth torn up before peoples' eyes."[50] This would expand the possibilities of witchcraft's reality. Bever argues that many of the accusations that he encountered in southwestern German trial records concern cases of actual harm done. This also bears upon his functional analysis of persecutions. He proposes that witch trials were quite effective in what they wanted to achieve: a reduction of harmful magical practices. The persecution's decline would, in part, thus be a consequence of its own success.[51]

Much of Bever's argument is thought-provoking, but this introduction of the paranormal into witchcraft scholarship is challenging to adopt. The reality of such phenomena is not accepted by mainstream science today, so historians cannot rely on that. At some point, Bever also acknowledges that the existence of the paranormal is contested and that early modern documents are unlikely to settle the debate. Additionally, it needs to be stressed that Bever's statements are primarily about village witchcraft. His claims remain far more modest when it concerns collective diabolical witchcraft. Regarding the cumulative notions of witchcraft, even Bever

[47] Bever, *The Realities of Witchcraft*, 63.

[48] See, for instance: Thomas, *Religion and the Decline of Magic*, 249; Briggs, *Witches & Neighbours*, 64.

[49] Bever, *The Realities of Witchcraft*, 39.

[50] Bever, 329–34, 438–39.

[51] Bever, "Witchcraft, Female Aggression, and Power in the Early Modern Community," 974.

speaks of "an illusory whole."[52] This brings us back to the question we posed above. If witch-hunts did not stop a sect of magic-using devil worshippers, then what should a functional analysis of these beliefs and trials look like? As we shall see, this is a theoretical and empirical puzzle that is remarkably difficult to solve.

6.2 Explanatory Attempts

"Too well designed," wrote the renowned cultural anthropologist Marvin Harris about the European witch persecutions.[53] A presumed danger that was omnipresent but difficult to detect, belief in nightly flights to collective meetings, and leading questioning through the relentless use of torture: to Harris, it all had the appearance of something purposeful. Since witch trials did not fulfil the function of eradicating real networks of witches, this adaptedness demonstrated that there was an alternative function involved. The system of witch-hunting had practical uses other than the stated goals of the witch-hunters. It was a means to a hidden end.[54] Harris was not the only one to think along these lines. From a wide range of theoretical perspectives, modern scholars have been searching for the ulterior aims of the witch trials and their concealed beneficiaries. What is relevant for our purposes is that the three essential perspectives on the *cui bono* question discussed above made their appearance. Scholars have tried to understand witch-hunts as functional for social collectives, maintaining power structures, or individual needs.

A perspective that interpreted witch trials as a means to exert power was the aforementioned liberal tradition. During the nineteenth and early twentieth centuries, these scholars were inclined to explain witch trials as a top-down tool used by the Catholic church to exert power over the ordinary people.[55] This makes the liberal viewpoint ambiguous—on the one hand, witch-hunts were crazed, but on the other hand, the evil clergy used them shrewdly. Joseph Hansen stressed the clergy's cleverness by pointing out that cumulative notions of the witches' sabbath and nightly flight were "the real source of the mass persecution," making trials spread

[52] Bever, *The Realities of Witchcraft*, 414.
[53] Harris, *Cows, Pigs, Wars, & Witches*, 236.
[54] Harris, 207–40.
[55] Soldan and Heppe, *Geschichte der Hexenprozesse*.

over long distances.[56] Such cumulative notions had been "interwoven" by the late Medieval Inquisition in the Alps before they were applied elsewhere. "With all the means of educational and religious policing that it had at its disposal, [the Medieval church] created and upheld a strong delusion that, despite all its internal inconsistencies, attained a significant propensity for propagation and persistence."[57]

This interpretation of witch-hunting as a top-down tool of religious elites has outlived the liberal tradition. The English historian Hugh Trevor-Roper argued that trials functioned as an instrument of the late Medieval church, missionaries in particular, who wanted to "fabricate a second set of heresies." Witch-hunting had been "perfected in the course of a local struggle" in the Western Alps and was then extended elsewhere.[58] From this point of view, the intensification of witch-hunting in the 1560s seemed to make sense, as it was closely related to the escalation of the religious wars. In ongoing struggles between Catholics and Protestants, the hunt for witches provided a helpful tool for discrediting and combating religious opponents. An instrumental link between witch-hunting and the conflicts of the Reformation has been reaffirmed more recently by Peter Leeson and Jacob Russ.[59] These two economists analysed substantial datasets on the timing of witch trials. The correlation between witch-hunting and religious warfare came out as the strongest one, in contract to bad weather, negative income shocks, weak states. Leeson and Russ then projected that witch trials "reflected non-price competition between Catholic and Protestant churches for religious market share in confessionally contested parts of Christendom."[60] Populations were often undecided about what side to choose in the religious wars, and simultaneously afraid of satanic witchcraft. This provided opportunities for religious competitors to make themselves popular. Drawing inspiration from rational choice theory and Michel Foucault, these economists explain witch-hunts as a way prosecutors could advertise their religious denomination's superior powers. It provided an effective cost-benefit ratio for winning over converts.

[56] Hansen, *Zauberwahn, Inquisition und Hexenproceß im Mittelalter*, 510.
[57] Hansen, 388–89, 535.
[58] Trevor-Roper, *The European Witch-Craze*, 113, 115.
[59] Leeson and Russ, "Witch Trials."
[60] Leeson and Russ, 2066.

Marvin Harris proposed a different top-down power campaign.[61] In his view, impacted by Marxism, the witch-hunt system represented a material power struggle between the ruling classes and the oppressed. Poor people were unsatisfied with their lot and potentially rebellious, so the governing classes created and sustained the scapegoat of the witch. By the time Kramer and Sprenger had finished their *Malleus Maleficarum*, the witch-hunting system had become "ready to be applied throughout Europe for the next two hundred years, with devastating results."[62] One of the ingenious effects of the trials was their ability to cause divisions among ordinary people. The aggression of the people could be targeted against alleged witches instead of the ruling classes, who reaffirmed their position of power.

A poststructuralist orientation can be found in the work of the prominent French historian Robert Muchembled, who thought along the lines of Michel Foucault.[63] Echoing the romantic tradition, Muchembled upheld that persecutions were part of a cultural power struggle between traditional peasant society and the upcoming world of the modern city. Older women held a central and respected position in traditional peasant culture, and city elites created the idea of the witch to reverse this image. It told peasants that the women they had always respected so much were evil witches. By killing these women, the city elites destroyed something crucial to the peasant world. The repression of witchcraft thus formed a visible part of a more or less conscious campaign of disciplining traditional rural society; "it constituted a form of pedagogy."[64]

The gender dimension of witch trials and the substantial overrepresentation of women amongst the victims have been highlighted most particularly by feminist historians.[65] Anne Barstow interpreted witch-hunts as a powerful symbolic tool that men used to mark the subordinate position of women. Male interrogators, judges, inquisitors, and executioners used the trials as a public means by which to demonstrate how lustful and dangerous women could be. This underlined the need to keep them confined. The persecutions made the role of women in European public and working

[61] Harris, *Cows, Pigs, Wars, & Witches*, 207–40.

[62] Harris, 217.

[63] Muchembled, *Sorcières*.

[64] Muchembled, 23–24, 69.

[65] See, for instance: Ehrenreich and English, *Witches, Midwives and Nurses*; Honegger, *Die Hexen der Neuzeit*.

life all the more circumscribed, and thus, "as a didactic device, the ritual execution of witches succeeded superbly."[66] In a more moderate version of the argument, Elspeth Whitney stated that the period of the witch trials gave rise to an increasingly hierarchal and rigid view of women's place in society. The trials contained elements of "sexual terrorism" against women, which "suggest that the hunts were a more or less direct projection of sexual anxiety."[67] The misogynous rants of Kramer and Sprenger in the *Malleus Maleficarum* also represented a broader tendency of using the trials as a means to discredit the power of female healers. Whitney complains that modern historians fail to take sufficient notice of such dimensions, illustrating how the misogynous "shadow of the hunts still lies over even its modern observers."[68]

Another perspective applied anthropological and sociological functionalism to European witch trials, which interpreted non-European witch-hunting as functional.[69] Emblematic was Clyde Kluckhohn's analysis of witch killings amongst the Navaho Indians in the southwest of the USA. Kluckhohn presumed that "no cultural forms survive unless they constitute responses which are adjustive or adaptive, in some sense, for the members of the society or the society considered as a perduring unit."[70] Subsequently, he presented a plethora of manifest and latent functions that Navaho witch killings fulfilled. Navaho societies experienced significant strains and stresses as a result of contact with white Americans, such as defeats, growing inequality, ecological problems, and the prohibition of tribal warfare. These created internal fears and hostilities that, if not expressed in a channelled outlet for displaced aggression, would make neuroticism and conflict endemic in society. The murdering of witches thus offered an institutionalized means of relieving anger and anxiety.[71]

In the early 1970s, two English historians applied such insights to the witch trials of early modern England. In his seminal *Religion and the Decline of Magic,* Keith Thomas wrote that witch beliefs "discharged a function in early modern England similar to that which they perform in many primitive societies today."[72] Thomas highlighted how

[66] Barstow, *Witchcraze*, 156.
[67] Whitney, "The Witch 'She'/the Historian 'He,'" 97.
[68] Whitney, 93.
[69] Gluckman, *Custom and Conflict in Africa*; Marwick, *Sorcery in Its Social Setting*.
[70] Kluckhohn, *Navaho Witchcraft*, 79.
[71] Kluckhohn, 89.
[72] Thomas, *Religion and the Decline of Magic*, 675.

sixteenth-century English society also experienced severe stress. The rise of capitalism had put much pressure on the traditional mechanisms of solidarity and social control within villages. Simultaneously, the Reformation broke down protective armours of ecclesiastical magic that had offered relief in the face of danger. In those circumstances, witchcraft accusations "served both as an explanation of misfortune and as an expected means of redress."[73] Thomas was particularly interested in the social contexts in which accusations arose. For instance, poor people often asked their neighbours for help, which, as traditional customs prescribed, should have been provided to them. But wealthier neighbours were increasingly unwilling to respond to such demands, which often made beggars go away cursing. To overcome a sense of guilt, wealthier people sought relief by convincing themselves that the beggars were evil witches. Accusations helped to dissolve social relations that had become redundant.[74] The historian Alan Macfarlane similarly concluded on trials in Essex: "Witchcraft prosecution may be seen as a means of effecting deep social change; a change from a 'neighbourly,' highly integrated and mutually interdependent village society, to a more individualistic one."[75]

Sociologist Nachman Ben-Yehuda explained the entire European witch-hunt from a functionalist angle. Europe suffered from severe stresses, confusion, and anomie, which called for reestablishment and a redefinition of the old moral boundaries of society. Witch trials won extensive support because they would help "cleanse the world from all effects of social change and anomie." As Ben-Yehuda put it: "In fact, the witch craze was a fictitious deviance, created for those purposes."[76] Without explicitly endorsing functionalism, Walter Stephens also brought forward that witch persecutions came into being to overcome anxieties. "Persecution of witches had tremendous positive value for everyone but the witches."[77] At the village level, Stephens endorses the Thomas-Macfarlane model, and at the elite level, he makes his aforementioned argument about witch theories and trials helping people overcome mounting doubts about the existence of the spirit world.[78]

[73] Thomas, 652.
[74] Thomas, 677.
[75] Macfarlane, *Witchcraft in Tudor and Stuart England*, 197.
[76] Ben-Yehuda, "The European Witch Craze of the 14th to 17th Centuries," 14, 22.
[77] Stephens, *Demon Lovers*, 7, 14, 104.
[78] Stephens, 26.

Other scholars have looked for more mundane individualistic aims. A common suspicion is that trials were created for financial gain.[79] What the persecutors were after, according to the social scientist Elliot Currie, was the money of the convicts. Mechanisms for profit were "built into the structure of the trials." It was "a large and complex business" that created the "livelihoods of a sizable number of people."[80] That is not to mention other explanations that assumed that witch-hunting was a means of discouraging birth control, enforcing state formation, or indulging in sadistic impulses.[81]

Hence, we can observe that modern scholars have looked for the hidden aims and hidden beneficiaries of witch-hunts in many directions. Yet, no matter how diverse their explanations, they all share that many witchcraft experts do not find them overall convincing. To their critical arguments we now turn.

6.3 Explanatory Failures?

"Les méchants faits détruisent les belles théories"—"nasty facts destroy lovely theories," is how the French historian Marc Bloch warned his colleagues.[82] The field of witchcraft scholarship provides a beautiful illustration of Bloch's dictum, as the facts of the witch trials have been unmerciful to the theories mentioned. When confronted with the complexities of history, these explanatory models did not work or worked only to a limited extent. These theories face three essential problems.

The first is that primary sources have failed to reveal explicit evidence for a hidden agenda or underlying function.[83] What we encounter in historical documents of supporters of witch trials is people telling each other how dangerous diabolical witchcraft could be and how to stop it. They look genuinely afraid. What they did not talk about was subjugating women, oppressing the poor, let alone relaxing social systems. This is not because such topics are absent in the sources. Advocates of trials, for instance, indeed expressed concerns about the well-being of social

[79] Robbins, *The Encyclopedia of Witchcraft and Demonology*, 273, 479–84.

[80] Currie, "Crimes without Criminals," 21–22, 32.

[81] Heinsohn and Steiger, *Die Vernichtung der weisen Frauen*; Lehmann, "Hexenverfolgungen und Hexenprozesse im alten Reich"; Trethowan, "The Demonopathology of Impotence."

[82] Howell and Prevenier, *From Reliable Sources*, 96.

[83] Dillinger, *Böse Leute*, 230; Dillinger, *Hexen und Magie*, 127.

communities and underlined the need for social control.[84] However, they did not address this in an instrumentalist or functionalist sense—diabolical witchcraft as *fictitious* deviance that was used indirectly for those ends—but more straightforwardly: diabolical witchcraft was a *real* threat to communities, and, therefore, social control was needed.

At first sight, it may seem that a stronger case can be made for oppressing women as an underlying aim—think of the rampant misogyny in the *Malleus Maleficarum*. A more careful interpretation of the sources suggests a different approach, however.[85] The *Malleus* contains much misogyny indeed, but that made it somewhat of an outlier amongst demonological works. Many other vital texts put less emphasis on the topic of female overrepresentation. As far as they discussed the subject, it happened in passing and in somewhat formulaic forms before they moved on to things that they thought were of greater importance, such as the (im)possibilities of magic or diabolical evil. Furthermore, the negative stereotypes about women were, as Stuart Clark put it, "entirely representative of their age and culture."[86] They can be found abundantly in contemporary sources that were unrelated to witchcraft, and we already saw that opponents used very similar stereotypes *against* witch-hunting. Claudia Opitz did retort, however, that Stuart Clark plays down the misogyny of an author like Jean Bodin. Indeed, there is some ground for this counter-argument: especially in an appendix of his book, Bodin spilled much ink on why witches were primarily females, relying on a vast arsenal of stereotypes.[87] Yet, still, it was hardly the central topic of his work, and Bodin also used one misogynous platitude about "the imbecility or fragility of the female sex" to argue that female witches should be punished *less* severely than male ones.[88]

Hence, if the persecutors of witches did not discuss the alleged underlying purposes or functions of the trials in print, how did such hidden purposes come about? The second problem is that most of the scholars who developed such models are remarkably vague on this point. Muchembled stated that witch trials were used "more or less consciously" to oppress

[84] Scarre, *Witchcraft and Magic*, 40–47.

[85] Clark, *Thinking with Demons*, 106–31; Stephens, *Demon Lovers*, 32–57; Schwerhoff, "Hexerei, Geslecht und Regionalgeschichte," 331.

[86] Clark, *Thinking with Demons*, 115.

[87] Opitz-Belakhal, *Das Universum des Jean Bodin*, 147–48; Bodin, *De la démonomanie des sorciers*, 261–67, 448–56.

[88] Bodin, *De la démonomanie des sorciers*, 419.

traditional peasant culture, but he does not elaborate.[89] Leeson and Russ offered statistical clues for a correlation between witch trials and religious warfare. Still, correlation does not imply causation, and their claim that trials were used deliberately to win over converts lacks evidence. Marvin Harris called the system of witch-hunts "well designed," and his book argues that involved actors themselves did not intend or understand such alleged hidden functions. But an alternative explanation of how such "designs" came into being was not forthcoming. Keith Thomas's theories about social and psychological functions suffer from a similar problem.[90]

The absence of explicit evidence in the sources does not, as such, refute these explanatory models. If the witch persecutions were deliberately used for a hidden purpose, it is understandable that the involved actors did not openly discuss this in print.[91] Asking for such evidence creates a high burden of proof. Perhaps there *was* a hidden agenda, but the crucial documents were lost, or people remained deliberately silent about their true aims. Or maybe they were only subconsciously driven forward by such motives or something between these options. Or perhaps the alleged hidden agendas or functions came into being through mechanisms that current witchcraft scholarship still needs to unveil. However, if one of those scenarios were true, the hidden agendas or functions should have become apparent in the course of the trials. But problem number three is that this is not borne out either. Historians have meticulously examined under what conditions witch trials occurred, who initiated them, who became the victims, and what the overall effects tended to be like. It did not happen along the lines that the theories mentioned above would lead us to expect.[92]

The issue with a top-down Roman Catholic campaign is that the role of clerics in the development and propagation of persecutions was far more limited than claimed.[93] There are certainly examples of witch-hunts that crucially depended on confessional Catholic support, for instance, in Counter-Reformation Germany. Still, in a majority of cases, persecutions were carried out by local secular authorities. Very often, trial records also hardly demonstrate a significant involvement of clergy, and there is a

[89] Muchembled, *Sorcières*, 23–24.

[90] Geertz, "An Anthropology of Religion and Magic, I," 77.

[91] Dillinger, *Hexen und Magie*, 127.

[92] See, for instance: Midelfort, *Witch Hunting in Southwestern Germany*; Behringer, *Hexenverfolgung in Bayern*; Dillinger, *Böse Leute*; Briggs, *Witches & Neighbours*.

[93] Dillinger, *Hexen und Magie*, 91–94.

substantial number of cases where clergy were victims.[94] The English historian Robin Briggs writes: "It is evident that had the clergy across Europe been committed enthusiasts for persecution, they could have mobilized local feeling to devastating effect."[95] There were even counter tendencies. The number of trials remained modest in places where central Catholic authorities were relatively strong, such as Spain or the Papal State. The Spanish Inquisition, for instance, was an essential voice of scepticism.[96] There was also a significant involvement of Protestants; parts of Calvinist Scotland and Lutheran Germany experienced severe witch trials, and demonologists on both sides of the religious divide were eager to adopt each other's ideas.[97]

The theory that witch-hunts had an ulterior instrumental function during the wars of religion is also problematic.[98] Witch persecutions preceded those wars, and the correlation later on is far from precise. During the first phase of the Reformation, persecutions decreased, and the most intense episodes of witch-hunting often occurred during intervals of peace. Before the Thirty Year's War (1618–1648), witch trials were already sharply on the rise, they decreased during the late period of the war and resurged somewhat when the fighting was over.[99] Persecutions also hardly focused on religious opponents—it was mostly Catholics versus Catholics and Protestants versus Protestants. The theory that witch-hunting had a positive cost-benefit effect on winning over converts is also questionable. Leeson and Russ provide no evidence for their claim that this motivated rulers. Admittedly, there do seem to have been some cases where the desire for religious following in the competitive environment of the Reformation made rulers condone witch-hunts, but it does not work as a generalized explanation.[100]

Experts have also abandoned the secular top-down aims proposed by Muchembled and Harris. Some trials were primarily instigated from above, but they more usually entailed complex interplays involving various levels

[94] Schwillus, *Kleriker Im Hexenprozess*; Voltmer, "Debating the Devil's Clergy."

[95] Briggs, *Witches & Neighbours*, 199.

[96] Henningsen, *The Witches' Advocate*.

[97] Goodare, *The European Witch-Hunt*, 162.

[98] Monter, "Witch Trials in Continental Europe 1560-1660," 10–11; Levack, *The Witch-Hunt in Early Modern Europe*, 12–16.

[99] Behringer, *Hexen und Hexenprozesse in Deutschland*.

[100] Pohl, *Zauberglaube und Hexenangst im Kurfürstentum Mainz*, 289; Rummel, *Bauern, Herren und Hexen*, 26–27.

of society.[101] Without popular support, elites' efforts would also have remained relatively toothless. In many cases, the initiative came primarily from below. Witch-hunting could also be quite risky for the ruling classes, as the dynamics of a panic were difficult to control.[102] Accusations tended to spread beyond one's territories, endangering precarious relationships between dominions.[103] Or worse, when popular anxieties about witchcraft mounted, elites could be suspected of using double standards. Were the ruling classes not protecting rich witches while the poor ones were being burned? Such popular sentiments could be challenging to resist and, in quite some cases, resulted in members of the elite being burned at the stake.

In contrast to what Harris's model would lead us to expect, relatively strong higher authorities often hindered witch-hunting. Many historians have noted that large and comparatively centralized states, such as France, England, Austria, and Saxony, experienced relatively few trials. In contrast, medium-sized and small territories were more fertile grounds for panics.[104] It is argued that larger states possessed more elaborate judicial bureaucracies, with various checks and balances preventing witch trials from spiralling out of control. In smaller territories, fewer distant supervisory authorities could stop them. This would also be one of the explanations for the relatively high number of trials in politically fragmented Germany.

Then there is the theory of witch-hunting as "women hunting." We already mentioned the lack of direct, convincing evidence in the sources, but the course of trials confronts us with further problems for this theory.[105] What this perspective would lead us to expect is that men instigated persecution against women. However, very often, trials started when women accused other women of witchcraft, and there were ample examples of male judges or juries that subsequently showed restraint. When brought to court, male suspects were not treated more favourably than female suspects, as the conviction rates between the sexes did not differ

[101] Dillinger, *Hexen und Magie*, 95–105; Briggs, *Witches & Neighbours*, 338–41.

[102] Schormann, *Hexenprozesse in Deutschland*, 57; Dillinger, *Hexen und Magie*, 98–105.

[103] Rowlands, *Witchcraft Narratives in Germany*, 63.

[104] Behringer, "Witchcraft Studies in Austria, Germany and Switzerland"; Levack, *The Witch-Hunt in Early Modern Europe*, 88–94.

[105] Briggs, *Witches & Neighbours*, 257–86; Goodare, *The European Witch-Hunt*, 267–316; Moeller, *Dass Willkür über Recht ginge*, 227; Groß, *Hexerei in Minden*, 249–57.

much.[106] It has already been noted that the average percentage of males among the victims lay somewhere around twenty to twenty-five percent, which is still quite substantial and does not fit with the idea that oppressing women was the ultimate aim. Negative stereotypes about female witches and male witches tended to coexist.[107]

Of course, the substantial overrepresentation of women amongst the victims still stands, but witchcraft scholars have developed more subtle explanations for that.[108] Misogyny was unmistakably an essential part of the picture. Still, negative stereotypes about women are no longer treated as a direct cause of persecution but rather as part of a context in which women were far more likely to be associated with witchcraft. We already discussed elite notions about women as morally weak and sexually rapacious creatures, but peasant beliefs also tended to link women to the world of spirits, spells, poisoning, and nightly flights. In addition, women's societal roles possibly made them more vulnerable to accusations. More often than men, women were involved in cooking, healing, and caring for children. If something went wrong in those domains, such as when a child fell ill, it was more obvious to look for a female culprit. When accusations occurred, the weaker position of women in society then gave them less access to means of defending themselves—the unequal distribution of power between the sexes remains a crucial factor. Correspondingly, it has been suggested that women's lack of political and physical power also made them more likely to undertake the kind of covert strategies of attack that would be interpreted as witchcraft by contemporaries, like poisoning and causing spells.[109]

There is also insufficient support for the claim that persecutions were essentially driven forward by greed.[110] A significant proportion of the victims was poor, while witch-hunting was a very time-consuming and expensive activity. When the properties of burned witches were confiscated, they often did not fall into the hands of the persecutors. There are certainly examples of people who gained money from trials, and some made a bit of

[106] Briggs, *Witches & Neighbours*, 261; Pohl, *Zauberglaube und Hexenangst im Kurfürstentum Mainz*, 293.

[107] Blécourt, "The Making of the Female Witch."

[108] Rowlands, "Witchcraft and Gender in Early Modern Europe"; Opitz-Belakhal, *Böse Weiber*, 9–41; Schwerhoff, "Hexerei, Geslecht und Regionalgeschichte."

[109] Bever, *The Realities of Witchcraft*, 47–59.

[110] Dillinger, *Böse Leute*, 300–305, 341–49; Midelfort, *Witch Hunting in Southwestern Germany*, 164–78.

a career out of it. But whether greed was the crucial underlying aim remains questionable even in those cases. Robin Briggs thinks that "even greedy witch-hunters probably convinced themselves that they were doing holy work, producing a typical amalgam of idealism and self-interest."[111] Moreover, such people could not have achieved much without more comprehensive support in society, making them facilitators rather than initiators.[112] Goodare concludes about the money-motive that "in origin it is probably a modern fantasy."[113]

What about the theories that presupposed social and psychological functions? Some aspects of the functionalist approach still inspire historians today. Thomas's and Macfarlane's observation that accusations could arise in contexts of refused charity has also been found in continental cases. At a general level, witch trials indeed tended to occur at times of psychological and social stress.[114] However, a fundamental problem with the functionalist model is that witch-hunts did not so much release those tensions as aggravate them. In his pioneering study on witch trials in southwestern Germany, Erik Midelfort concluded that "major panics served no valid social function. Society was not made stronger and more cohesive by such trials, but weaker and more torn by suspicion and resentment."[115] Midelfort accepts that small trials may have served a social function of restricting eccentric behaviour. Still, the shock and fear of the more extensive hunts wrecked regular bonds of social trust, broke down friendships, made respected members of communities vulnerable to attack, and left behind families that were consumed with grief and self-accusation.[116] Other historians confirmed Midelfort's findings, and even the functional effects of small trials are highly questionable.[117] Witchcraft accusations and persecutions tended to draw people apart rather than bring them together.

As an overall observation, it has to be noted that the aforementioned explanatory models presupposed one or just a few systematic causes, needs, intentions, or functions behind the witch-hunts. But *if* this had

[111] Briggs, *Witches & Neighbours*, 342.

[112] Briggs, 189.

[113] Goodare, *The European Witch-Hunt*, 385.

[114] Goodare, 95, 241; Dillinger, *Hexen und Magie*, 129–30.

[115] Midelfort, *Witch Hunting in Southwestern Germany*, 4.

[116] Midelfort, 93–190.

[117] Sharpe, *Witchcraft in Early Modern England*, 46; Goodare, *The European Witch-Hunt*, 95, 166.

been the case, we should expect the persecutions to display regular patterns, targeting specific classes of individuals rather than others. However, there was quite a lot of variation in who took the initiative and became the victim.[118] The persecutions could erupt quite suddenly, take many victims, disappear for a long while, and erupt elsewhere later. Again, it looks more like genuine panics rather than an intelligent and systematic campaign stemming from underlying motives. Historical experts still find general factors such as religious warfare, misogyny, or social stress significant for the understanding of witch-hunts but as preconditions and no longer as essential causes. After all, in the overwhelming majority of cases, such factors did not result in any witch trials at all.[119] Large waves of persecution were also hardly a product of central coordination. Around the year 1590, there was an intense outburst of witch trials in parts of Germany, and Wolfgang Behringer analyses:

> In all territories, the witch persecutions began unreflectively. Nowhere was there a plan that envisaged in advance how single trials would be extended elsewhere. Nor was there any arrangement between territories in which people agreed on how to cooperate, and not a single time were there special lawsuits or preliminary considerations (...). The 'outbreak' of persecution around 1590 took the region by surprise. Reflections on the course of events only occurred after the extraordinary scale of the trials had become apparent.[120]

Behringer also makes a striking observation about officials who executed the trials. They were doing it, as he writes:

> although the work was very time-consuming and not very enjoyable. The boundless misery of the witch trials burdened most of the commissioners psychologically, which makes the assumption that the processes were carried out from 'lower' motives (enrichment, etc.) seem absurd."[121]

What Behringer argues here is that not only the victims of trials suffered but even many of the persecutors!

[118] Thurston, *The Witch Hunts*, 16.
[119] Briggs, *Witches & Neighbours*, 188.
[120] Behringer, *Hexenverfolgung in Bayern*, 122.
[121] Behringer, 320.

Despite the problems with their hypotheses, it still needs to be mentioned that the works of people like Trevor-Roper, Thomas, Muchembled, and the feminist authors have helped the field of witchcraft scholarship advance. These authors put the topic of witch-hunting higher on historians' agenda. They highlighted relevant aspects, such as the Reformation context, social dimensions of village accusations, interactions between popular and elite cultures, and pervasive misogyny. Notably, the view that witch trials were instrumentalized for hidden purposes or fulfilled certain underlying functions is still in vogue among various German historians, who have developed a more refined version of the argument—we shall have an extensive look at their ideas after we have delved more deeply into several German persecutions. But for now, suffice it to say that the international trend over the past decades has been to abandon theories that presuppose underlying functions or purposes. The failure of the grand explanatory frameworks has shaped a "state of the art" in witchcraft historiography that will now be further examined.

6.4 The "State of the Art"

Many modern scholars have searched for the rationality behind the apparently irrational. A hunt for a sect of flying, fornicating, magic-using, and devil-worshipping witches appears so illogical to an audience today that it seems obvious to look for an underlying rationale that makes more sense to us. But it turns out that this muddies the waters even further. Suppose we assume that witch-hunting served the equilibrium of society, the eradication of peasant culture, or the oppression of women. In that case, many aspects of witchcraft texts and trials become even stranger, less understandable, and increasingly illogical. A key lesson of witchcraft scholarship over the past decades has been to take at face value what its proponents actually wrote and believed. If we assume that the initiators and executors of trials genuinely meant what they said, their actions do not look illogical after all. Witch-hunting was not a means towards a hidden end; it was an end in itself.[122]

Historical scholars have found various ways of formulating this viewpoint. "That the ostensible motives for trying witches were real ones is a virtually inescapable conclusion from a scrutiny of the more detailed trial

[122] Gibson, "Thinking Witchcraft"; Gaskill, "The Pursuit of Reality"; Machielsen, *Martin Delrio*, 212.

records," Geoffrey Scarre writes.[123] "Witch beliefs were not a trick; the fear of witches was very real," states Johannes Dillinger.[124] Martine Ostorero highlights that "people really believed it and without a doubt in good faith."[125] Or, in Julian Goodare's words: "a witchcraft accusation was not 'really' about something else; it was really about witchcraft."[126] Christina Larner summarizes this current viewpoint most wonderfully: "Witch-hunting was actually witch-hunting."[127]

This is not to say that persecution was unrelated to other conflicts or aims that people had. These historians are willing to accept that animosities between families or between individuals stimulated accusations and trials in many instances, as could other factors like greed or jealousy. James Sharpe writes, "It is possible to find an enormous range of antagonisms, personal feuds, and areas of competitiveness underlying a witchcraft accusation."[128] But it does not imply that the beliefs were not real or that witch-hunting was used as a vehicle for a different purpose. Johannes Dillinger developed a subtle model for how witch trials and other conflicts were interrelated, which he calls the "evil people paradigm."[129] Dillinger observes that allegations of witchcraft often focused on people the accusers already disliked for other reasons, such as social upstarts, loudmouths, or strangers. The terms "witches" and "evil people" were often used interchangeably. This might seem to suggest that allegations served the purpose of harming ostensibly antisocial people. Still, Dillinger proposes a different causal connection: people's allegedly asocial behaviour indicated their identity as a witch. After all, Satan and his demons were trying to seduce as many people as they could, and obviously, it was evil individuals who most easily succumbed to their will. When looking for witches, people tended to think of disagreeable people from the local community first.

Today, ironically, we see the people who initiated and executed these murderous trials as the evil ones. But we should not forget that from the witch-hunters' own perspective, they were doing honourable work. This observation does not imply moral relativism—even by the standards of the time, persecutions were seen as excessive, cruel, and judicially

[123] Scarre, *Witchcraft and Magic*, 47.
[124] Dillinger, *Hexen und Magie*, 127.
[125] Ostorero, *Le diable au sabbat*, 9.
[126] Thurston, *The Witch Hunts*, 204; Goodare, *The European Witch-Hunt*, 115, 384.
[127] Gibson, "Thinking Witchcraft," 169–70.
[128] Sharpe, *Witchcraft in Early Modern England*, 49.
[129] Dillinger, *Böse Leute*; Dillinger, *Hexen und Magie*, 133.

irresponsible, and witch-hunters were labelled as zealots by their opponents.[130] Yet, it does not alter the fact that the trying of witches could be motivated by genuine intentions. Some infamous trials took place in the German Prince-bishopric of Würzburg under the rule of prince-bishop Julius Echter von Mespelbrunn, and they have tainted his reputation ever since. But this same prince-bishop was also the man who turned Würzburg into one of the most beautiful baroque cities of Germany, who gave his name to a new hospital that was renowned for its care for orphans, pilgrims, and the mentally ill, and who regularly went there himself to distribute food and to wash the pensioners' feet.[131] From Mespelbrunn's own perspective, it consistently served the well-being and salvation of Christian communities.

Why has there been this continuing tendency to look for other purposes? Several historians have reflected on that.[132] It has been highlighted that the beliefs about witches look so strange to a modern audience, while at the same time, it is so apparent that these beliefs were illusory, that we almost inevitably tend to think that they must have been about something else. Hence, we look for more universal and less time-bound explanations that make sense to our enlightened frameworks, such as greed or power play. Thus, we impose our modern assumptions on the early modern world. The rebuttal of the cynical interpretations of witch-hunting does not make the horrors of the trials any less awful, however. Geoffrey Scarre comments: "It was something even more depressing – a frightful example of how morally motivated action can lead to massive suffering, and good intentions produce the direst consequences."[133]

The failure of the earlier explanatory models in witchcraft historiography has also sparked some idiographic reflections on the methodology of history. It is argued that the past, once again, is too messy and too complex, while there are always too many exceptions for a general rule to work. Jan Machielsen: "Any scholar of early modern witchcraft should be sceptical of grand narratives – and remind themselves that the devil is in the details."[134] It is emphasized that every witch trial had its own particular dynamics, resulting from its own complex interplay of innumerable factors

[130] Behringer, *Witches and Witch-Hunts*, 175; Briggs, *Witches & Neighbours*, 407.

[131] Roper, *Witch Craze*, 23–24.

[132] Gibson, "Thinking Witchcraft," 169; Goodare, *The European Witch-Hunt*, 385; Briggs, *Witches & Neighbours*, 383–85.

[133] Scarre, *Witchcraft and Magic*, 63.

[134] Machielsen, "Thinking with Montaigne," 452.

related to their unique historical and regional context.[135] On closer scrutiny, witch persecutions were not even a single phenomenon, making the search for any single set of laws pointless from the outset.[136] "Sweeping generalizations about it are either false or so banal as to lack any analytical power," comments Robin Briggs.[137] Johannes Dillinger renders it a big step forward that mono-causal interpretations are no longer worthy of serious consideration.[138] In an essay with the illustrative title "Many Reasons Why," Robin Briggs proposes to look instead for "a more finely grained picture" and a "greater diversity at local level."[139] This also bears upon the relationship between history and the natural sciences:

> I suspect that there is an explanatory law which dictates that complex problems of this type never have simple or precisely identifiable causes. (…) Once we start aggregating, the variables multiply so fast that chaos theory, with its patterns of unpredictability, is the scientific model which best fits the case.[140]

Historians have repeatedly restated this trope and with good reason. Yes, we should appreciate the diversity of witch beliefs and trials and place them within their highly complex historical context. The interpretation of the meaningful symbolic world the involved actors lived in must also be addressed carefully. Deterministic, mono-causal models look ill-suited indeed. But simultaneously, there is something unsatisfying about this trope. The idiographic tendency to announce the end to all large explanatory frameworks is too easy, as it leaves so much unexplained. The empirical findings of historians about witch beliefs and persecutions have unearthed at least two intriguing theoretical puzzles, neither of which has been appropriately recognized yet. What historians have found out about witch-hunts should not lead us towards a theoretical dead-end but, rather, to a new beginning.

The first theoretical puzzle is the striking level of adaptedness that the cumulative concept of witchcraft contained. Even if we accept that witch-hunters did not deliberately create witch-hunts for an underlying purpose

[135] Levack, *The Witch-Hunt in Early Modern Europe*, 3; Blauert, *Frühe Hexenverfolgungen*, 23–24, 114.

[136] Kieckhefer, "Mythologies of Witchcraft in the Fifteenth Century," 105–7.

[137] Briggs, *Witches & Neighbours*, 7.

[138] Dillinger, *Hexen und Magie*, 73.

[139] Briggs, "Many Reasons Why," 51.

[140] Briggs, 53.

and that they genuinely believed in it, one earlier observation about the witch-hunt system still stands: Marvin Harris was right to call it "too well-designed." Just think of how some of the deepest anxieties of the period, such as fear of the devil, anti-Christian communities, harmful magic, and uncontrollable female sexuality, were ingeniously arranged into one frightening concept. Or think of additional elements that stimulated the continuation of trials with an ever-increasing circle of suspects, like the sabbath, nightly flight, torture, or the focus on children. Such notions endowed the concept of witchcraft with remarkable robustness, resilience, and a striking capacity to trigger witch-hunts that spiralled out of control. It is not by accident that earlier scholars assumed that it contained "built-in structures" that were "fabricated," "braided," "designed," and "perfected" to serve particular goals. So, how did this apparent "design" of the concept of witchcraft come about?

Even some of the idiographic historians cannot help but be impressed by how ably the concept of witchcraft spread within its historical contexts. Johannes Dillinger analyses how this elastic concept attached itself to a broad spectrum of social conflicts: "What made the notion of the witch so very explosive was its ability to translate almost every form of negative behaviour into an indication of witchcraft. The range of behaviours that potentially triggered suspicions was entirely open-ended."[141] Robin Briggs maintains that "witchcraft was peculiarly malleable, available to fit any kind of discord, because the link between ill-will and physical effects did not need to be demonstrated."[142] Drawing inspiration from psychoanalytic theory, Briggs additionally argues that the notion of the witch played on people's unconsciousness by appealing to some of their profound anxieties, as well as feelings of guilt and repressed libido. At the psychological level, it "allowed fantasy to run riot," making it a "brilliantly successful fiction."[143] Yet, the question that stares us in the face is: how did the concept develop these insidious capacities?

There is a second intriguing question that has remained unanswered. We saw that modern scholars have been searching for the beneficiaries of witch trials in every nook and cranny but have not found them in considerable numbers. Meticulous empirical research has indicated that, in many cases, there were no interested groups of people who substantially

[141] Dillinger, *Böse Leute*, 350, 442.

[142] Briggs, *Witches & Neighbours*, 265.

[143] Briggs, 164-7. 250, 265; Briggs, *The Witches of Lorraine*, 384.

benefitted from the witch-hunts and that communities as a whole were often ravaged by them. The persecution of witches was often destructive for nearly all and everyone. But if this is so, how could the witch-hunting phenomenon survive for such a long time? Why was it so difficult to control an outbreak of witch panic once it was set in motion? It seems strange that communities would tolerate such harmful practices if no tangible benefit could offset the cost of human life. In other words, we observe forms of socio-cultural "design," apparently without intentional designers, that did not significantly benefit anyone. How can Darwinian theory help us out?

6.5 Darwinian Attempts

Darwin had the good fortune of developing his theory with data from colleagues who were, as Daniel Dennett calls it, "theoretically untainted."[144] Researchers always run a risk of seeing only those aspects of reality that match preconceived conceptions, and current evolutionary biology would be an unlikely exception to this rule.[145] Yet, while the works of naturalists before 1859 were undoubtedly characterized by all sorts of biases, it was not a pro-Darwinian bias. According to Dennett, this posed an essential check on Darwin's enthusiasm: the facts he worked with had not been collected to prove his theory right. Still, many of those naturalists' findings, like the analyses of divine "design" or the morphological similarities between internal structures, provided his model with much-needed support. Subsequently, the principle of evolution by natural selection did not make those earlier insights redundant. On the contrary; within their new theoretical context, they took on a whole new meaning.

The relationship between Darwinian cultural evolution and the scholarship of European witch-hunting may work similarly. When witchcraft historians collected their data and developed their interpretations, none of them thought of substantiating a model of Darwinian cultural evolution. Regarding Darwinian theory, it was all worked out in theoretical innocence. Only unintentionally do so many of their findings offer clues for developing a Darwinian scenario. Some historical scholars highlighted how the witch-hunt system appears to have been designed, while others retorted that sources provide no substantial evidence for shrewd designers

[144] Dennett, *From Bacteria to Bach and Back*, 238.
[145] Dennett, *Breaking the Spell*, 30–32.

operating behind the scenes. This looks exactly like the "design without designer" phenomenon that cultural evolutionists envisaged. Moreover, the harmful consequences of witch trials on people as well as on societies suggest a "selfish" cultural phenomenon.

We should not immediately adopt a Darwinian explanation, however. Robin Briggs thought chaos theory to be the best model, and even scholars acquainted with cultural evolution stress that socio-cultural phenomena are not "adaptive" by default—they can be a by-product of something else or result from processes of cultural random drift.[146] Still, the reason for adopting an adaptive research strategy here is that the clustered cumulative notions of diabolical witchcraft appear too fine-tuned and well-disposed to have the kind of effects that they so often had within their historical context: the dynamics of a witch-hunt. It looks like something conspicuously functional, and chaos theory or randomness cannot explain that. Since intentional design seems unable to do the explanatory job either, Darwinism may provide an outcome.

The study of witchcraft is not an entirely pristine area for Darwinian theory. The pervasiveness of witchcraft beliefs worldwide indicates that evolved psychological propensities may lie at the root of it.[147] This does not automatically imply that humans are born with a witch-detection mechanism, and Briggs and Goodare suggest a scenario in which broader innate social skills, like detecting antagonisms or deceit, helped trigger witch persecution under certain conditions. Some neurologists, for instance, claim that fear is located in the amygdala region of the brain, so according to Goodare it is conceivable that this brain area was activated during outbreaks of witch-hunting. These historical reflections also correspond to what scientists from other disciplines say: specific mental modules, such as threat perception, coalitional psychology, or the punishment of defectors, were selected independently and not for witch persecution. But when persecution occurred, such modules were activated.[148]

The witchcraft historian who has addressed humans' psychobiological traits most extensively is Edward Bever, for instance, when explaining the

[146] Salganik, Dodds, and Watts, "Experimental Study of Inequality and Unpredictability in an Artificial Cultural Market."

[147] Briggs, *Witches & Neighbours*, 373–74, 392–94; Goodare, *The European Witch-Hunt*, 2, 374.

[148] Parren, "The (Possible) Cognitive Naturalness of Witchcraft Beliefs"; Koning, "Witchcraft Beliefs and Witch Hunts."

overrepresentation of women amongst victims.[149] In his account, men are more likely to express aggression in interpersonal conflicts through physical means, while women more often resort to indirect means like gossiping, poisoning, and other forms of concealed attack. These differences would stem from Palaeolithic times but could still make women more apparent targets for witchcraft accusations. Bever does not deny the impact of cultural expectations but thinks that such expectations may be enhancements of older biological disparities.[150] The gender dimension of witchcraft might also trigger further interdisciplinary considerations. One striking aspect of early modern witch beliefs is that women were thought to have a stronger sexual appetite than men, which flies in the face of current evolutionary psychology. This either means that people at the time were dead wrong about that or that evolutionary psychology has mistaken a current stereotype—women are pickier than men—for a biological fact. Lyndal Roper's finding, however, that male witches were thought to be provided with multiple partners instead of one diabolical lover is more in line with the current evolutionary outlook.[151]

Yet, let us also reemphasize the limitations of evolutionary psychology. What a biological perspective *cannot* tell us is why people in this particular period in European history had this detailed and culturally complex fear of diabolical witchcraft. We have no "witches fly to the sabbath on a broomstick module" or a "the devil sometimes appears as a goat module" located in our brain. True, people in Bamberg in 1626 were possibly walking around with activated amygdala areas, but a hundred years earlier or later, they also had amygdalae, but no witch-hunt occurred. Preconditions do not equal explanations.

The cumulative concept of witchcraft was something thoroughly *cultural*, so it is also at the semi-autonomous level of cultural transmission and selection that the origin of its apparent design must be sought. Some tentative Darwinian efforts in this direction have been undertaken. One is to look at the cultural evolution of witch beliefs from a cognitive viewpoint.[152] The current evolutionary field assumes gene-culture coevolution has endowed our brains with learning biases that make some cultural vari-

[149] Bever, "Witchcraft, Female Aggression, and Power in the Early Modern Community."

[150] Bever, *The Realities of Witchcraft*, 49–55.

[151] Roper, *Witch Craze*, 91.

[152] Keitt, "Rethinking with Demons"; Singh, "Magic, Explanations, and Evil"; Parren, "The (Possible) Cognitive Naturalness of Witchcraft Beliefs."

ants more easily adopted and transmitted. We might, for instance, think of the idea of female initiation into witchcraft through diabolical intercourse. Cognitive scientists highlighted that our brains eagerly adopt information about threats, sexual contacts, and rumours, so this variant's reproductive success looks unsurprising.[153] However, explanatory limitations still need to be emphasized. As a lecturer, I often notice that the story of diabolical intercourse still attracts students' immediate attention. This cultural variant's ability to play on our minds has not abated yet. But unlike so many early moderns, students today tend to find the story amusing, and none of them is prepared to burn others for it. Hence, to understand how this particular story spread so well within its historical context, we also need to look beyond universal cognition. It requires questions like: what did female sexuality at this specific period in time *mean* to people? What did diabolical evil *mean* to them? One of the merits of the cognitive view on cultural evolution is that it takes an essential step towards cultural complexity. However, it still suffers from evolutionary psychology's problem of universality. Stuart Clark complains that cognitive approaches bleach out historical specificity, transforming "the rich colour and texture of lived experience" into "the drab monotony of brain functions."[154] Here, he has a point.

Another option is to apply the idea of cultural group selection to witch-hunting. Older functionalist accounts lacked an explanatory model for how alleged social functions of witch trials came about, and this is precisely what cultural group selection may provide: beliefs in witches and practices of trying them had positive effects on the functioning of cultural groups, also resulting in the survival of these variants themselves. Several group-selectionists recently proposed that witch-hunts, in some societies at least, functioned as an incentive for prosocial behaviour.[155] It made people cautious not to be labelled as witches, facilitated conflict resolution, and softened social and economic inequalities. If applied to the European case, we might presume that cultural groups that persecuted witches out-competed other groups *because of that*.

[153] Parren, "The (Possible) Cognitive Naturalness of Witchcraft Beliefs"; Blackmore, *The Meme Machine*, 121.

[154] Clark, "One-Tier History," 89.

[155] Campbell, "On the Conflicts between Biological and Social Evolution," 196; Joseph Bulbulia et al., "The Cultural Evolution of Religion."

However, the failure of older functionalist models should make us wary of this approach. Village witchcraft may have been subject to such mechanisms, perhaps. Still, the elaborate notions of diabolical witchcraft and the large trials resulting from them look ill-suited to any such approach. Interestingly, anthropologists reached similar conclusions about the damaging societal effects of witch-hunting regarding non-western case studies.[156] "The widespread fear that witchcraft is running wild has shocking effects in everyday life – tearing families apart and setting people against each other," Peter Geschiere wrote about witch-hunting in post-colonial Cameroon.[157] Another problem is that the alleged underlying mechanism requires repeated selection rounds in which group-functional variants are accumulated. Regarding the European case study, this conflicts with another key finding of historians: witchcraft panics did not often occur at the same place more than once or twice.[158] This would have provided any blind selective process with too little time to steer witch beliefs and persecutions into a group-functional direction. The group-selectionist paper indeed argues that the extent to which witch beliefs and persecutions had positive social effects depends on "the relative strength of competition between religious groups versus the competition of subgroups or individuals within religious groups." When the latter forces are more substantial, witch-hunting may become socially destabilizing. They maintain that "to evaluate such questions, a new collaborative evolutionary history is needed."[159] This makes the European case study all the more relevant.

The remark about "competition of subgroups" hints at a power-oriented perspective, in which persecutions survived because they served the interests of one group within society at the expense of others. However, considering the stark limitations of older power perspectives in witchcraft historiography (see Sect. 6.3), this line of research does not look very promising either. What about the option of "competition of individuals"? The authors do not elaborate, but this remark suggests a tendency to explain witch-hunting as a tool that serves individuals' interests. To a certain degree, this suggestion is more in line with the current witchcraft scholarship, as one of its key findings has been that proponents of trials

[156] Gershman, "Witchcraft Beliefs and the Erosion of Social Capital"; Knauft, *Good Company and Violence*.

[157] Geschiere, *The Modernity of Witchcraft*, 21.

[158] Roper, *Witch Craze*, 11, 133; Behringer, *Witches and Witch-Hunts*, 149.

[159] Joseph Bulbulia et al., "The Cultural Evolution of Religion," 395–96.

thought to have good reasons for what they were doing: combating harmful magic and enhancing chances of salvation. What this observation does not tell us yet, however, is how people came to believe in diabolical witchcraft in the first place, especially considering its largely illusory character. What is more, the trials often *subverted* people's interests. Think of the inhabitants of Zeil who called for a witch-hunt but, in many cases, ended up at the stake or in the witch oven themselves. Historical research provides many more examples of the witch-hunting phenomenon killing its initiators or their families.[160] This makes it hard to see how a blind evolutionary process, operating at the level of competition between individuals, would have occurred.

Here, the explanatory power of the "selfish culture hypothesis" comes to the fore. Instead of looking at the competition between cultural groups, subgroups, or individuals, we could highlight the competition between cultural variants and try to look at the whole witch-hunting phenomenon from *their* perspective. Perhaps beliefs and trials did not so much evolve to serve the interests of society, the powerful, or individuals but rather to serve their *own* self-reproduction. Nobody has explored this possibility in detail, let alone integrated the detailed insights of witchcraft historians.[161] It may have worked as follows: innumerable beliefs and practices regarding witchcraft appeared, but most did not reproduce well. Yet, within this ocean of variants that were too poorly adapted for survival, sometimes variants appeared that *did* have significant positive effects on their own reproduction, like the beliefs in the witches' sabbath, nightly flight, or child witches. In places where such notions were more prominent in people's minds, the variants often triggered larger persecutions. Subsequently, the unusual scale of those trials may have attracted much attention, leading to the adoption of those variants by other people and similar events elsewhere. Through repeated selection rounds, notions conducive to witch-hunting thus accumulated into an ever more explosive cluster of beliefs. This scenario does not require involved actors who comprehended these causes and effects.

[160] Dillinger, *Kinder im Hexenprozess*, 135; Midelfort, *Witch Hunting in Southwestern Germany*, 93–94, 137; Decker, "Die Hexenverfolgungen im Herzogtum Westfalen," 367; Rummel and Voltmer, *Hexen und Hexenverfolgung*, 103, 112.

[161] For very short references, see: Bever, *The Realities of Witchcraft*, 182; Aunger, *The Electric Meme*, 1–2.

Several historians have already come close to a scenario quite like this. Older models often treated witch-hunting as a top-down product of demonology, but more recent studies stress that it frequently happened the other way around.[162] Contemporary publications on witchcraft were, in many cases, a response to, or a report of, significant outbreaks of witch-hunting. But as soon as the new demonological ideas were put in print, they helped trigger similar trials elsewhere. Robin Briggs describes how persecutions were sporadic and localized for a long time, and did not get going in earnest. However, "where it became intense, of course, there was a serious danger that it would reinforce popular fears and thus raise anxiety levels further."[163] Trevor-Roper stated that "every spectacular episode increased the power of the myth," while Julian Goodare thinks trials and demonologies were "mutually reinforcing."[164] Especially interrogations of suspects have been identified as possible breeding grounds for new demonological ideas. Lyndal Roper argues that the confessions of alleged witches continually fed back into works of demonology, which then enlarged the interrogators' idea of what crimes witches committed.[165] Brian Levack: proposes a cumulative scenario in which an inquisitor uses information from manuals to formulate questions during trials, while, at the same time, they "might use some of the specific charges against the accused, or his own imagination to give a new twist to the standard charges." Subsequently, such new twists slip into new manuals and treatises, impacting investigations elsewhere.[166]

Such remarks are intriguing but remain fragmentary and disconnected from a broader theoretical framework. But looking at them from an evolutionary point of view, they can take on a whole new meaning. The remarks become essential ingredients for a Darwinian scenario in which new cultural variants provide the raw material for a selective process that builds its own adaptations from it. This is not mere tautology. If we take the example of the witches' sabbath, the proposed model requires several assumptions that can be examined empirically and might be false to varying degrees. (1) Many cultural variants regarding witchcraft appeared, most of which did not reproduce well. (2) The belief in the witches'

[162] Richard Kieckhefer, "The First Wave of Trials," 166; Dillinger, *Hexen und Magie*, 53.

[163] Briggs, *Witches & Neighbours*, 291.

[164] Trevor-Roper, *The European Witch-Craze*, 115; Goodare, *The European Witch-Hunt*, 85.

[165] Roper, *Witch Craze*, 106–7.

[166] Levack, *The Witch-Hunt in Early Modern Europe*, 49.

sabbath triggered larger and more spectacular persecutions. (3) This effect contributed to the survival of the belief in the witches' sabbath itself. (4) This enhancing effect was neither intended nor recognized as such by the actors involved; it was a chance hit. Unlike the case of the Hopi rain dance, we have many sources at our disposal that allow for such investigations.

Empirical research into these questions is only fragmentary. Hypothesis (1) has been addressed in the sense that historians have devoted attention to how concepts of witchcraft continually varied. Still, there is hardly any research into what kind of variants lacked enhancing effects on trials and may have disappeared because of that. Hypothesis (2) is probably best examined, as a broad consensus has taken shape that the notion of the sabbath triggered chains of accusations.[167] Especially trial records show how sabbath beliefs made interrogators ask suspects for accomplices. Hypothesis (3) has been studied less explicitly, although there are many hints in this direction.[168] Briggs, for instance, writes that "ideas about the sabbath were also perpetuated through the trials themselves. It was routine practice to read out the confessions of the accused at their executions, itself a well-attended public event."[169] Lastly, hypothesis (4) has not been examined at all. Modern historians explicitly identify the sabbath as a critical trigger for large trials, but did contemporaries also understand that? And did supporters of witch trials introduce the idea for that reason? We could ask similar questions about the enhancing effects of notions like nightly flight or child witches.

We shall now proceed as follows. In the next chapter, we look at escalating interactions between witch beliefs and trials in Germany, especially from 1560 to 1630. This period saw the most significant rise in trials, providing us with opportunities to see whether, and if so, how the anticipated Darwinian mechanisms operated. We particularly look at outbreaks of witch-hunting in Wiesensteig in 1562, the Trier region around 1590, and the town of Ellwangen in 1611. The last chapter offers critical reflections on the proposed Darwinian model. It was already mentioned that various German historians have not given up on the idea of witch-hunts being instrumentalized for hidden goals. In contrast to earlier models,

[167] Rummel and Voltmer, *Hexen und Hexenverfolgung*, 43; Behringer, *Witches and Witch-Hunts*, 59; Goodare, *The European Witch-Hunt*, 13, 16.

[168] Trevor-Roper, *The European Witch-Craze*, 115; Behringer, *Hexenverfolgung in Bayern*, 224; Schormann, *Hexenprozesse in Deutschland*, 32–33.

[169] Briggs, *Witches & Neighbours*, 43.

these historians do not look for just one or a few underlying aims but maintain that witch-hunts were used as a *multifunctional* instrument. This is the most challenging counter-model, or complementary model, to the proposal made here. It will thus receive careful consideration *after* we have gained more knowledge of how German witch-hunts occurred. In this light, the empirical question of what contemporaries did or did not understand about the enhancing effects of cumulative notions of witchcraft becomes even more urgent. In subsequent sections, I thus investigate apparent levels of comprehension that both proponents and opponents of trials seem to have had. Then, I address the prospects and pitfalls of the "viral" analogy, and I finish with an inescapable question: if concepts of witchcraft were so well-built for self-reproduction, then why did witch-hunting cease? But first, we move to Wiesensteig in the year 1562. In this southern German town, ominous events were about to initiate witch-hunting's most lethal phase.

CHAPTER 7

Cases

7.1 Wiesensteig 1562

We do not know precisely what happened during the witch-hunt of Wiesensteig in 1562. A fire in 1648 destroyed two-thirds of the town, probably taking the trial records with it. We do know, however, that the events in this small Lutheran principality were sensational, particularly by the standards of their time.[1] The occurrence of a witch trial in that place and historical instance was not very surprising as such. After a terrifying hailstorm ravaged the region in August 1562, many calls for witch-hunting were heard, soon followed by trials. But while persecutors in towns nearby put at most a few individuals on trial, Wiesensteig took things a lot further. The persecutors here thought that witches had met at a crowded assembly, where they adored the Devil and planned their crimes. A *Flugblatt* (news-sheet) from 1563 reported the witches' alleged misconducts in gruesome detail: they had blinded and lamed people, killed both young and old, harmed cattle, dug up buried children to make poisons, and caused the recent hailstorm.[2] What the *Flugblatt* further broadcasted was a staggering number of witches killed: sixty-three.

[1] Midelfort, *Witch Hunting in Southwestern Germany*, 88–90; Waite, *Eradicating the Devil's Minions*.

[2] Behringer, *Hexen und Hexenprozesse in Deutschland*, 138–40.

S. Hofhuis, *The Evolutionary History of Witch-hunting*, https://doi.org/10.1007/978-3-031-66936-1_7

Witch-hunts had already been going on in Germany for more than a century.[3] After the diabolical witchcraft mythology had appeared in the Western Alps around 1430, cumulative notions penetrated Germany, triggering trials in various places.[4] Yet, the spread of elaborate witchcraft concepts remained patchy, and trials remained sporadic and quite small. This holds not only for Germany but for much of Europe—only a small percentage of people killed as diabolical witches were executed before 1562.[5] From the 1520s onwards, trials even declined. By 1532, Martin Luther remembered how there had been quite some witches in his youth, but added: "Now that the gospel is here, you do not hear so much about it because the gospel drives the devil away."[6]

Why did witch-hunting fail to make much of an impact before 1562? This is not because many cumulative notions that would make later persecutions so deadly were missing.[7] When diabolical witch trials began in Alpine districts like Valais, Vaud, Aosta, and Dauphiné, the first reports already contained allegations about diabolical pacts, nightly gatherings, devil worship, trampling on the cross, banquets, dancing, illicit sex, the cannibalism of children, as well as magical crimes. From its inception, the concept also interacted with actual trials based on confessions extracted through torture. Stories about large numbers of witches likewise circulated.[8] The first description of a witches' gathering stems from a chronicler of the Swiss city of Luzern, Hans Fründ. Around 1430, he wrote that the witches numbered seven hundred people "of whom more than two hundred were burned within one and a half years, while more are convicted and burned every day."[9] Medieval statistics were often vague and inflated, but fragmentary archival sources indicate that Fründ's reports on the trials' size had at least some grounding in reality.[10] It appears that hundreds of people were burned in the early Alpine period alone.[11]

[3] Behringer, 72–129.

[4] Dillinger, *Böse Leute*, 110; Behringer, *Hexenverfolgung in Bayern*, 78–79; Schmidt, *Glaube und Skepsis*, 23–32.

[5] Levack, *The Witch-Hunt in Early Modern Europe*, 184–91.

[6] Rampton, *European Magic and Witchcraft*, 335.

[7] Broedel, "Fifteenth-Century Witch Beliefs," 38–41; Ostorero, Bagliani, and Tremp, *L'imaginaire du sabbat*.

[8] Ostorero, *Le diable au sabbat*, 27.

[9] Fründ, "Rapport sur la chasse aux sorciers et aux sorcières," 43.

[10] Ostorero, *Le diable au sabbat*, 27; Goodare, *The European Witch-Hunt*, 42.

[11] Behringer, *Witches and Witch-Hunts*, 68.

Past interpretations, like Joseph Hansen's or Hugh Trevor-Roper's, presumed that inquisitors had fabricated those first cumulative notions from above.[12] Clerics were already experienced in persecuting Jews and Christian sects, and allegedly then transferred their attention to a new target. More recent research, however, offers a more differentiated picture.[13] True, the origin of witch trials was crucially related to the persecution of heretics, as these Alpine valleys had previously witnessed murderous hunts against Waldensians. But instead of being a deliberate "design," witchcraft concepts and trials only gradually emerged from these persecutions—older stereotypes and practices regarding heretics merged with popular and learned notions about practitioners of harmful magic. Secular rather than clerical authorities conducted many of the first trials, and extensive descriptions of the new sect only appeared afterward.[14] In a critique of the old top-down model, her Swiss colleague Andreas Blauert argues that:

> Witch trials were not simply 'invented' and then conducted alongside the persecution of heretics and sorcerers. Instead, it can be demonstrated that the underlying concept of witchcraft, as it was developed from the regional heresy and sorcery trials, tended to absorb and transform its predecessors.[15]

A further characteristic of the cumulative concept apparent in the early days was its propensity to be disseminated. To some extent, its spread was fortuitous. Around the time of its appearance, principal clergymen from all over Europe gathered in the nearby city of Basel for the most important Roman Catholic event of the time: the ecumenical council of Basel 1431–1437.[16] Hundreds of abbots, priors, theologians, and other clerics travelled to this significant city along the Rhine to discuss important issues, like papal supremacy and the Hussite heresy in Bohemia. No formal sessions were devoted to the new witchcraft sect, but the topic was discussed in detail outside of the official programme. Among the many people who then took the news abroad was Johannes Nider, a Dominican theology

[12] Hansen, *Zauberwahn, Inquisition und Hexenproceß im Mittelalter*; Trevor-Roper, *The European Witch-Craze.*

[13] Richard Kieckhefer, "The First Wave of Trials."

[14] Ostorero, Bagliani, and Tremp, *L'imaginaire du sabbat*, 509–24; Ostorero, *Le diable au sabbat*, 49.

[15] Blauert, *Frühe Hexenverfolgungen*, 48–9.

[16] Bailey and Peters, "A Sabbat of Demonologists."

professor from Vienna. In 1437, he published his *Formicarius* (The Ant Hill), warning people against spiritual hazards of any kind, including diabolical witchcraft.[17] Nider's description included sabbaths and magical crimes and it would be the most influential text on witchcraft for many decades.

Over the next hundred years, the mythology revealed its ability to generate trials.[18] A notorious case occurred in the northern French city of Arras in 1459–1460.[19] One of Arras's inquisitors attended a witch trial in the eastern French town of Langres, where he learned that two people from his region were mentioned as accomplices. These accusations sparked a hunt that killed twelve people and many more arrests. The proceedings attracted widescale attention, producing a wave of paranoia, dramatically affecting the town. Three current experts write: "The city's business had been disrupted, its commercial networks damaged, and its reputation deeply tarnished."[20] The trials only ended when persecutors overstepped their mark by attacking wealthy and powerful individuals. Arras was only one example. The new witchcraft mythology, for instance, also instigated numerous trials in northern Italy.[21] Most persecutions there were limited, but in the city of Como, a runaway panic seems to have resulted in at least thirty people being killed.

Still, the witch-hunting phenomenon did not succeed in spreading very effectively, and historians have already highlighted some plausible explanations. One is that central cumulative notions often failed to operate *in combination*.[22] A typical example is Nider's *Formicarius*. The idea of witches flying to sabbaths was probably one of the concept's most incendiary traits, but Nider did not connect flight to sabbath attendance and seems to have thought of witches' gatherings as nearby, local affairs.[23] He even dismissed flight as a diabolical delusion. This was no accident, as flight remained a somewhat ancillary element in witchcraft theory for

[17] Nider, "Formicarius"; Bailey, *Battling Demons*.

[18] Behringer, *Witches and Witch-Hunts*, 68–72.

[19] Gow, Desjardins, and Pageau, "Introduction: The Arras Witch Treatises in Context."

[20] Gow, Desjardins, and Pageau, 2–3.

[21] Herzig, "Witchcraft Prosecutions in Italy," 250–52; Behringer, *Witches and Witch-Hunts*, 78–79.

[22] Broedel, "Fifteenth-Century Witch Beliefs," 42–49; Richard Kieckhefer, "The First Wave of Trials."

[23] Nider, "Formicarius," 134–35, 139; Bailey, *Battling Demons*, 46–49.

some time.[24] Before 1562, the connection between harmful magic and the idea of a heretical sect was often missing, as many trials focused on either collective heresy or magic. Misogynous overtones were not consistently well developed either, and the early Western Alpine trials did not markedly target women more than men.[25]

The witchcraft mythology became particularly unstable as soon as it spread beyond the Alpine heartlands. The cumulative concept tended to pare down as various elements went missing. The Alpine version also became immersed in local paradigms about witchcraft and magic, thus losing much coherence.[26] The hunt of Arras provides an example of a relatively successful transmission, but aspects of the Alpine paradigm, like child killing or cannibalism, were absent, and suspicions of harmful magic occurred in only one case.[27] When analysing the spread of witch-hunting beyond the Alps, Richard Kieckhefer notes that it concerned "a highly fluid mythology, in which no two trials resemble each other very closely."[28] On top of that, cumulative notions of witchcraft and practices of trials faced resistance wherever they appeared. Medieval political and religious authorities tended to reject beliefs in witchcraft or flying women as superstitious and pagan.[29] So, when proponents of witch-hunting warned the world about the danger, they often found their beliefs dismissed as diabolical trickery.

These obstacles to witchcraft beliefs and trials were also present in German-speaking parts of the Alps and beyond.[30] Here, witchcraft concepts were integrated with older, local witchcraft lore. As Laura Stokes put it, the traditional German Alpine witch looked more like a solitary woman who was "driven by anger to raise hail and smite her enemies."[31] Witches tended to go out into the forest, perform weather magic, and ally themselves with wolves. Accusations and confessions in the area mainly concerned women who operated alone, or in twos or threes, and met with personal devils. This also manifested itself in Heinrich Kramer's *Malleus*

[24] Bailey, "The Medieval Concept of the Witches' Sabbath," 125–26.
[25] Goodare, *The European Witch-Hunt*, 46; Bailey, *Battling Demons*, 49–51.
[26] Richard Kieckhefer, "The First Wave of Trials," 169–75.
[27] Goodare, *The European Witch-Hunt*, 47; Mercier, *La vauderie d'Arras*, 61–85.
[28] Kieckhefer, "Mythologies of Witchcraft in the Fifteenth Century," 102.
[29] Goodare, *The European Witch-Hunt*, 71–73.
[30] Richard Kieckhefer, "The First Wave of Trials," 172–75; Stokes, *Demons of Urban Reform*.
[31] Stokes, *Demons of Urban Reform*, 29.

Maleficarum.[32] During a relatively intense wave of trials around 1480, this Dominican friar operated as a witch-hunter in the Alps and southern Germany. Kramer was familiar with the learned discourse on diabolical witchcraft. Still, his German background manifested itself in the way he connected witchcraft to misogynous stereotypes, as well as in his many references to female village witches as likely culprits. Kramer hardly stressed witchcraft's collective dimension.

Other German adoptions were also patchy. An influential preacher in Strasburg, Johann Geiler von Kaysersberg, warned his flock about the diabolical danger in the early sixteenth century, but he did not treat it as a collective super crime, with witches meeting at large sabbaths, and he was ambiguous about the reality of nightly flight.[33] While acknowledging the diabolical pact and harmful magic, Martin Luther was sceptical or dismissive about witches' gatherings and flight.[34] Two newssheets from this period describing witchcraft cases—from Schiltach (1533) and Wittenberg (1540)—also presented a relatively simple picture of the crime and did not mention a sabbath.[35] This patchiness impacted trials. In some cases, the idea of witches forming networks helped activate a snowballing of accusations, but persecution mostly remained restricted to only a few women.[36] Furthermore, resistance to cumulative witchcraft beliefs and trials was paramount. Not long after the *Malleus* was published, a lawyer named Ulrich Molitor wrote a critique of witch-hunting that gained a far bigger audience.[37]

The subsequent decline of witch-hunting varied.[38] In some places like the Low Countries and northern Italy, trials did not diminish much, but in most of Europe, including Germany, the trend was unmistakable. Illustratively, Kramer's book was reprinted at least fourteen times before 1521 but then waned in popularity.[39] Historians have pointed out some

[32] Broedel, "Fifteenth-Century Witch Beliefs," 44–49; Institoris, *The Hammer of Witches*, 1–58.

[33] Voltmer, "Preaching on Witchcraft?"

[34] Haustein, "Martin Luther als Gegner des Hexenwahns."

[35] Zika, *The Appearance of Witchcraft*, 182–85.

[36] Schmidt, *Glaube und Skepsis*, 23–32; Behringer, *Hexenverfolgung in Bayern*, 78–79; Waite, *Eradicating the Devil's Minions*, 66.

[37] Behringer, "Witchcraft and the Media," 219–20.

[38] Behringer, *Witches and Witch-Hunts*, 78–82.

[39] Levack, *The Witch-Hunt in Early Modern Europe*, 188.

plausible causes.[40] The economic and environmental crises were mild compared to the second half of the sixteenth century. Renaissance intellectuals like Erasmus or Andrea Alciato criticized witchcraft theory, and there was the distraction of astonishing developments, like the Reformation, the peasant revolts, and the rise of Anabaptism. The cumulative concepts of witchcraft also warrant consideration. Past scholarship has assumed that the concept was fully developed when the *Malleus* was published.[41] Yet, significant developments were still about to happen. Apart from being patchy, some of the concept's elements that would help trigger the "superhunts" of the seventeenth century were still largely missing, like standardized questioning, child witches, and highly elaborate sabbaths. Admittedly, some early commentators already described sabbaths as huge events.[42] But these were outliers. The idea of witches' gatherings remained relatively infrequent, and sabbaths tended to be considered quite small.[43] The social profile of alleged witches also often remained restricted, like a scholarly Italian Renaissance dialogue on witches, which spoke of "women who are mostly rough and boorish."[44] Hence, by the 1550s, the concept of witchcraft had not yet evolved into a virulent form, and little indicated that the worst was still to come.

Things began to change, however, around 1550. One factor was the Little Ice Age.[45] The first half of the sixteenth century had been characterized by relatively mild winters, and populations in Germany were growing. But by the 1550s, the weather had started to do strange things. A climatic contraction of the number of fertile months stimulated fiercer competition over scarce resources, putting immediate pressure on people's livelihoods. In the Lutheran city of Esslingen, located close to Wiesensteig, a furrier named Dionysius Dreytwein kept a chronicle of his town and, from around 1550, began to report frightening events. Terrifying hailstorms damaged crops, a flood swept away many houses, and winters produced more snow than people could remember. Dreytwein also told of increasing

[40] Goodare, *The European Witch-Hunt*, 158–59; Levack, *The Witch-Hunt in Early Modern Europe*, 185–89.

[41] Hansen, *Zauberwahn, Inquisition und Hexenproceß im Mittelalter*; Trevor-Roper, *The European Witch-Craze*, 41.

[42] Franc, "Le Champion des Dames," 455; Ostorero, Bagliani, and Tremp, *L'imaginaire du sabbat*.

[43] Zika, *The Appearance of Witchcraft*, 11–35, 70–98.

[44] Mirandola, *La sorcière*, 148.

[45] Behringer, "Climatic Change and Witch-Hunting."

cases of witchcraft.[46] In 1559, two distinguished women in Horb were found responsible for weather-making. One year later, two women in Maulbron were "burned to ashes" for making storms, floods, and killing children.[47]

So-called *teufelsbücher* (devil books) flooded the German book market, expressing concerns about sin being rampant and the devil preparing for the final battle.[48] The religious drama of the Reformation was utterly confusing to many people. Western Europe splintered into Roman Catholic, Lutheran, and Calvinist blocks, but the situation was less than evident in the mid-sixteenth century. Many religious movements competed for domination, and Anabaptist Christian sects attracted many followers.[49] Their passionate believers rejected the religious authorities of other denominations and uniquely rejected child baptism. Some Anabaptists were relatively peaceful, but others tilted towards violence and apocalyptic hysteria. At some of their clandestine meetings, consecrated hosts and other sacramentals were dishonoured. There were promises of miracles, like the healing of the sick and the resurrection of the dead. In 1534, Anabaptists took over the city of Münster, where they authorized polygamy. While waiting for the end of times, they declared their leader the new king.[50]

The overall response was one of disgust and repression.[51] After the reconquest of Münster, Van Leiden and his allies were brutally tortured and burned in public. Anabaptists became the most heavily persecuted group in Europe. The bizarre stories about them helped fuel a poisonous climate of polemics against subversive groups allegedly aided by the devil. Many princes in Germany secularized heresy trials to make persecution more effective, and the argument of a "crimen exceptum" was brought forward to ease the use of torture.[52] This repression forced Anabaptists to meet in even more secretive places, like barns, attics, forests, or fields, which in turn stimulated further paranoia. In the near vicinity of the city of Esslingen, authorities discovered that dozens of Anabaptists were meeting in a wooded ravine near a fortress.[53] When the group was captured in

[46] Dreytwein, *Esslingische Chronik*, 216–17, 234–35.
[47] Dreytwein, 234.
[48] Midelfort, *Witch Hunting in Southwestern Germany*, 69.
[49] Goertz, *Deutschland 1500–1648*, 149–56.
[50] Waite, *Eradicating the Devil's Minions*, 17.
[51] Goertz, *Deutschland 1500–1648*, 149–56.
[52] Waite, *Eradicating the Devil's Minions*, 199.
[53] Waite, 144–46.

July 1562, people were upset to learn that some Anabaptists had travelled up to sixty kilometres to attend the meeting.

Then, late morning, 3 August 1562, shortly after the discovery of the Anabaptist meeting, black clouds suddenly appeared in the sky. It almost seemed like midnight. Thunder broke out throughout the region of Esslingen and Wiesensteig, as well as an unprecedented hailstorm.[54] The destruction was horrific. There were severe floods, many windows were broken, and vineyards and harvests were destroyed. The populace was exasperated. Such misfortune was traditionally explained as divine retribution for ubiquitous human sin, and many pastors again sent this message from their pulpits. But considering the scale of the disaster, numerous people in the region no longer believed sin to be a sufficient explanation. They resorted to something else: witchcraft!

Stories about witches soon passed from mouth to mouth. In nearby Stuttgart, it was reported that several witches had been seen flying on pieces of wood towards a nearby moor a few days before the storm, to return one hour later.[55] Some were identified by name, which resulted in their arrest and execution. In the city of Esslingen, a preacher named Thomas Naogeorgus incited the population against witchcraft to the detriment of a woman named Berchta Schreinerin.[56] A few years earlier, Schreinerin had already been accused of witchcraft by a neighbour, but authorities refused a trial. The neighbour's tongue was even cut off for defamation. This time, however, the officers were more undecided. Guests at a wedding feast claimed to have seen Schreinerin sleepwalking on a roof, and when a few of them went to check her out, no trace of the woman was found. Instead, a black cat suddenly jumped away. Officials warned Naogeorgus not to incite the people, but Schreinerin was arrested, along with a few other women. The suspects were tortured, but since they refused to confess, the authorities remained unsure about the case.

Meanwhile, things took a different course in the nearby county of Wiesensteig.[57] In this small territory of around five thousand inhabitants, the widespread calls for witch-hunting were answered with strident

[54] Dreytwein, *Esslingische Chronik*, 327; Behringer, "Climatic Change and Witch-Hunting," 335–36.

[55] Dreytwein, *Esslingische Chronik*, 238, 244; Waite, *Eradicating the Devil's Minions*, 149–50.

[56] Jerouscheck, *Die Hexen und ihr Prozess*, 65–80.

[57] Midelfort, *Witch Hunting in Southwestern Germany*, 88–90; Waite, *Eradicating the Devil's Minions*, 144–53.

persecution. From August to December, witch burnings were held regularly, adding up to sixty-three killings. An exchange of letters between officials in Wiesensteig and Esslingen shows that persecutors in the former town were unusually obsessed with the case.[58] Unlike Esslingen, Helfenstein's persecutors believed to be eliminating a network of witches who regularly held meetings at a large sabbath. The investigators in Wiesensteig also provided Esslingen with some essential information: one suspect confessed to having seen Schreinerin at the Wiesensteig sabbath, and she was alleged to have flown to it on a black devil's back. Esslingen was about to free Schreinerin and the others, but the news encouraged them to continue the case.

Still, renewed torture in Esslingen failed to elicit confessions, and Barbara Schreinerin proclaimed to have rock-solid confidence in God.[59] When new evidence did not appear, all were freed. Wiesensteig was disgruntled, reminding Esslingen of its duty to obey "the strict and serious order of God."[60] That Esslingen still decided in the suspects' favour may have been related to counter-voices from the nearby Lutheran University town of Tübingen. Two distinguished preachers, Matthaeus Alber, and Wilhelm Bidenbach, passionately opposed the new surge in witch-hunting.[61] In their view, people were mistaken to attribute the recent misfortune to witchcraft, as the real cause lay in divine punishment for our own sins. Later on, authorities in Esslingen fired its preacher, Thomas Naogeorgus, and his earlier incitements against witches seem to have been among their considerations.

Yet, despite the local resistance, Wiesensteig was still a persecution of importance. The events in the county were considered newsworthy, reaching audiences far and wide—the *Flugblatt* that reported the sixty-three killings is a case in point.[62] During the early days of the Reformation, such newssheets had already demonstrated their polarizing potential by spreading sensationalist news.[63] An essential source of information to the broader population, they were relatively cheap, broadly consumed, and often read

[58] Jerouscheck, *Die Hexen und ihr Prozess*, 80–86, 268–69.

[59] Jerouscheck, 81–86.

[60] Jerouscheck, 85–6.

[61] Alberum and Bidenbach, *Ein Summa etlicher Predigten von Hagel und Unholden*; Midelfort, *Witch Hunting in Southwestern Germany*, 39.

[62] Behringer, "Hexenverfolgungen im Spiegel zeitgenössischer Publizistik," 356.

[63] Zika, *The Appearance of Witchcraft*, 179–209; Voltmer, "Hört an neu schrecklich abentheuer," 91.

aloud in communal settings. Topics usually included dramatic events like celestial apparitions, gruesome murders, monstrous births, military battles, and religious scandals. But the news about the Wiesensteig persecution also spread through other means. Dreytwein's chronicle of Esslingen displays a vivid interest in the trials.[64] Archival sources from the nearby city of Reutlingen reveal that the persecution was discussed in a local inn, and even as late as 1589, a *Flugblatt* from Ulm referred to Wiesensteig as the place where the eradication of the witches' sect had started.[65]

Notably, the events attracted the attention of one of Germany's most charismatic Catholic preachers, Petrus Canisius.[66] This priest from the Low Countries was a member of the recently founded "Society of Jesus," better known as the Order of Jesuits. After the Council of Trent 1545–1563, a new strident Catholicism aimed to restore the old faith after the onslaught of Protestantism, and Jesuits were the embodiment of this Counter-Reformation zeal. The order was especially involved in public preaching and education. As the head of its German section, Canisius attracted the crowds with dramatic exorcisms and sermons about constant fights with the devil. Well acquainted with Wiesensteig, the witch trials of the town made a lasting impression on him. In 1563, he wrote to a colleague about how the number of witches was "strangely multiplying."[67]

The episode in Wiesensteig was, as the Canadian historian Gary Waite puts it, "a critical moment in the history of European witch-hunting."[68] Thomas Fritz said it fulfilled a "role model function," and Behringer calls it "a visible sign of a *paradigm shift*."[69] The hunt helped stimulate some critical changes to how people imagined witchcraft.[70] The transformation is nicely illustrated in Dreytwein's chronicle. Before 1562, he only reported witchcraft as an endeavour of women who operated individually or in small numbers. After 1562, Dreytwein considered it a conspiracy of people

[64] Dreytwein, *Esslingische Chronik*, 240–46.

[65] Fritz, "Hexenverfolgungen in der Reichsstadt Reutlingen," 190–92; Behringer, *Hexen und Hexenprozesse in Deutschland*, 219.

[66] Voltmer, "Jesuiten und Kinderhexen," 214; Behringer, *Hexenverfolgung in Bayern*, 112–13.

[67] Midelfort, *Witch Hunting in Southwestern Germany*, 88; Behringer, *Hexen und Hexenprozesse in Deutschland*, 140.

[68] Waite, *Eradicating the Devil's Minions*, 157.

[69] Behringer, *Witches and Witch-Hunts*, 83; Fritz, "Hexenverfolgungen in der Reichsstadt Reutlingen," 190.

[70] Midelfort, *Witch Hunting in Southwestern Germany*, 70–71; Behringer, "Witchcraft and the Media," 223.

assembled by the dozen. Importantly, the exceptional events in Wiesensteig hardly look preplanned. The hunt depended on a mix of accidental events, like the hailstorm, the heightened paranoia, and the disclosure of the Anabaptist meeting nearby. Gary Waite argues that this gathering had made people in Weisensteig more inclined to think of witchcraft as a substantial group that secretly met each other and travelled long distances.[71] It makes the Wiesensteig hunt look like an example of a cultural "chance hit."

While the immediate causes of the Wiesensteig persecution were quite random, it was not at random that this particular hunt received so much attention. Dreytwein's chronicle wasted far fewer words on the trials at Maulbron or Horb, plausibly because they were less sensational. The relatively modest persecutions of Stuttgart or Esslingen also largely faded into oblivion, as did other trials in the region, such as those in Rottweil (1561), Spalt (1562), and Illereichen (1563).[72] In the latter case, the persecution started after the populace had pressured the local count to kill a witch. However, since the local ideas mainly concerned harmful magic and not collective meetings or flight, a chain of accusations failed to occur.[73] When looking across Germany we can find many more examples, like Paderborn (1555 and 1556), Velturns, Villanders and Amt Neustadt (1558), Bürgel (1559), Osnabrück (1561), Rockenhausen (1561–1562), Winterberg (1562), Bayreuth and Bingen (1563), and Horn (1564).[74] These trials differed in scale—in Paderborn, the suspects were released quickly, while in Osnabrück, sixteen were burned—but none was as newsworthy as Wiesensteig. Here, we see the proposed Darwinian mechanism: some fortuitous cultural variants helped make the Wiesensteig hunt large, but because of that effect, those particular variants achieved wide dissemination. Wiesensteig created the potential for more sizeable trials.

[71] Waite, *Eradicating the Devil's Minions*, 145.

[72] Zeck, *Im Rauch gehn Himmel geschüggt*, 43,123; Behringer, *Hexenverfolgung in Bayern*, 99.

[73] Behringer, *Witches and Witch-Hunts*, 85–86.

[74] Decker, "Die Hexenverfolgung im Hochstift Paderborn," 323; Waite, *Eradicating the Devil's Minions*, 188; Moeller, *Dass Willkür über Recht ginge*, 49; Pohl, *Zauberglaube und Hexenangst im Kurfürstentum Mainz*, 51–52, 143; Rügge, *Die Hexenverfolgung in der Stadt Osnabrück*, 26; Schmidt, *Glaube und Skepsis*, 82; Grawlich, "Der Hexenkommisar Heinrich von Schultheiß," 306; Bradford Smith, *Reformation and the German Territorial State*, 167–68; Meier, "Die Anfänge der Hexenprozesse in Lemgo," 90.

For the Wiesensteig variants to become successful, more was needed than just making their appearance. We saw that notions about large sabbaths were already present earlier but had failed to stabilize. Reproductive success in a Darwinian process never solely depends on intrinsic qualities; it is the environment that counts. The environment before 1562 was suboptimal, but the following decades became more auspicious. According to Edward Bever, weather disasters and economic deprivation worsened, producing an "intensification of interpersonal conflicts and their manifestation as suspicions and acts of *maleficium* among the common people."[75] In addition, religious and political changes were taking shape. The spiritual landscape fractured into Lutheran, Catholic, and Calvinist regional blocks, and within their domains, religious elites strived to bring the populace more strictly under religious control. Cultural life became more pious, disciplined, and polarized. Wolfgang Behringer speaks of a general "darkening" and "hardening" of mentalities and conditions.[76]

Witch trials tended to come and go in waves.[77] After the surge of the early 1560s had abated, many new trials in Germany started around 1570.[78] The new Wiesensteig model of witch-hunting proved influential: ideas about witchcraft progressively focused on sabbath meetings, nightly flights, and growing numbers of witches. A small treatise on witchcraft printed in Saxony in 1573 that called for more persecution described how "the number of witches, as we have learned sufficiently from the experience of the trials, grows bigger every year."[79] Around this time visual depictions of witches increasingly focused on somewhat larger gatherings. A German map from 1570 depicts a witches' dance near the town of Laub that was attended by fifteen witches.[80] In the Swiss city of Zurich, a Protestant pastor named Johann Jakob Wick amassed a magnificent news collection between 1560 and 1588, and his stories and pictures also reveal the typical trends. The many reports on witchcraft included in the collection show that small trials received only modest attention, while a sensational persecution in Geneva from 1571 was covered widely.[81] One report from 1571 about Geneva said that twenty-one witches were executed,

[75] Bever, *The Realities of Witchcraft*, 386.
[76] Behringer, *Hexenverfolgung in Bayern*, 112–21.
[77] For the early waves, see: Blauert, *Frühe Hexenverfolgungen*.
[78] Behringer, *Hexen und Hexenprozesse in Deutschland*, 149–59.
[79] Behringer, 158.
[80] Haas, *Hexen und Herrschaftspolitik*, 62.
[81] Wick, *Die Wickiana*, 78–79, 161, 179; Zika, *The Appearance of Witchcraft*, 194–209.

while thirty-six remained imprisoned. A later newssheet on the Geneva hunt spoke about "an incredible number of witches" there, and yet another presented an elaborate drawing of a sabbath that was attended by many demons and seven witches.[82]

The revival of trials not only occurred in the German-speaking world but also in countries like France, Scotland, and England, generating a resurgence of witchcraft literature.[83] The *Malleus* was reprinted again, and many new texts appeared. Not all arguments were supportive, as the new wave probably helped motivate Johann Weyer to write his sceptical book of 1563.[84] His book encountered passionate support but equally passionate rebuttals. Strange as it may sound, those rebuttals were, in specific ways, more forward-looking. Weyer's argument was innovative, but in many ways, it represented the older orthodoxies that had interpreted witch beliefs as diabolical trickery and delusions. The new defences of witch-hunting challenged those orthodoxies.[85]

The revival of witchcraft writing especially manifested itself in France. In 1574, the Calvinist jurist and theologian Lambert Daneau published his influential *Les sorciers*, which was superseded in 1580 by Jean Bodin's *De la démonomanie des sorciers*.[86] This "great Renaissance polymath, jurist, political theorist, historiographer, religious thinker, natural philosopher, and demonologist," as Virginia Krause calls him, had become especially concerned about diabolical witchcraft after he learned of a young woman who had voluntarily confessed to being a witch.[87] Bodin incorporated new information from the trials in his book and made the sabbath a more critical element than Kramer had previously done.[88] The book became the new go-to manual.[89] Thus, by the 1580s, ideas about witchcraft had become more conducive to the creation and spread of large trials.

However, concepts of witchcraft had not yet evolved into a form that would enable the super-hunts of the seventeenth century. The Bamberg hunt crucially resulted from indictments coming from children, endless

[82] Wick, *Die Wickiana*, 184, 188–89, 196–98.

[83] Levack, *The Witch-Hunt in Early Modern Europe*, 189–90; Behringer, *Hexen und Hexenprozesse in Deutschland*, 130–37.

[84] Valente, "'Against the Devil, the Sublte and Cunning Enemy.'"

[85] Schmidt, *Glaube und Skepsis*, 81.

[86] Daneau, *Les Sorciers*.

[87] Krause, *Witchcraft, Demonology, and Confession in Early Modern France*, 8.

[88] See, for instance: Bodin, *De la démonomanie des sorciers*, 60–62, 72–73, 223–34.

[89] Krause, "The Will to Know and the Unknowable," 120.

chains of accusations, a striking bureaucratic efficiency, and the targeting of people from almost all walks of life. In 1580, child witches were not a prominent topic, and the focus on large sabbaths was not yet as well developed. After the first Alpine period, the witches' social profile also became quite restricted. Admittedly, Daneau reported that amongst witches, one could even find people "of quality, like gentlemen, ladies, scholars," and he also stressed that it could be "men and women, young and old."[90] Bodin did not exclude the idea of highly placed witches either, but overall, witchcraft theory and trials remained more limited.[91] A southern German lawyer in 1580 typically remarked about witches that "such women are normally poor."[92] Only during a long wave of trials beginning in the 1580s did some of these elements shift significantly. We now move to the German city of Trier in 1582 to see how that happened.

7.2 Trier 1582

In 1582, some worrisome rumours circulated in Trier.[93] In the duchies of Lorraine and Luxemburg, not far away to the west, witch persecutions had already been going on for some time, and more recently, trials had occurred in the nearby town of Saarburg. Trier was not yet affected by this new wave of persecution, although it had witnessed trials before. What set off the recent rumours was that some of the female suspects from Saarburg had fled to Trier and were hiding in the city walls. Trier had a reputation for being relatively lenient on witchcraft, which may be why these suspects sought shelter there. The rumours alerted the authorities, though, and they arrested the suspects, sending them back to their hometown. This did not end Trier's involvement, as the proceedings in Saarburg led to two alleged visitors of the witches' sabbath from Trier being mentioned. It concerned Margarethe Braun, who already suffered from a bad reputation in the community, and her daughter. The accusations came at a conspicuous moment for the city, plagued by epidemic disease, bad weather, failed harvests, and rising food prices.[94] The indictments fuelled an atmosphere

[90] Daneau, *Les Sorciers*, 5.

[91] Opitz-Belakhal, "Der Magistrat als Hexenjäger," 48.

[92] Behringer, *Hexen und Hexenprozesse in Deutschland*, 156; Dillinger, *Böse Leute*, 225; Dillinger, "Richter als Angeklagte," 123.

[93] Voltmer, "Zwischen Herrschaftskrise, Wirtschaftsdepression und Jesuitenpropaganda," 53–67; Voltmer, "Germany's First 'Superhunt'?," 240–41.

[94] Voltmer, "Krieg, uffrohr und teufelgespenst," 23.

where these "evil women" seemed plausible culprits for misfortunes.[95] The authorities took the news from Saarburg in earnest, and Margarethe Braun was arrested and tried under the supervision of the town bailiff.

The bailiff's name was Dietrich Flade. His family had settled in the town recently, but Flade had quickly made an incredible fortune, becoming the wealthiest person in the city. He possessed much land and three magnificent houses and served in a series of prominent posts, like electoral councillor, vice governor, professor, and president of Trier's University.[96] Trier was also a religious and intellectual centre and the seat of the Archbishop of Trier—a post taken by Johann VII von Schönenberg—who ruled as prince-archbishop over a significant electoral state. Yet, Dietrich Flade was not particularly liked.[97] It was widely thought that his incredible wealth had been assembled too quickly. He was known for being exploitative and corrupt, and many owed him debts. Moreover, in a previous conflict between the city and the prince-archbishop, Flade had taken sides with the prince.[98] Trier had tried to become a free imperial city within the Holy Roman Empire, but Von Schönenberg's predecessor had used military means to keep it confined within the electoral state. Flade's choice made him a traitor in the eyes of many.

A harsh edge to Flade's character became apparent during his trial of Margarethe Braun.[99] He made sure that many alleged witnesses were heard to testify against Braun. Flade also had her imprisoned and ensured she was subjected to torture. When the torments did not produce the confessions that he hoped for, Flade had her relocated to a worse prison cell. Eventually, Braun was freed, and her trial failed to trigger additional persecution. That further escalation was prevented was not so much due to any mildness on Flade's part as it was to Braun's unwillingness to confess and her impressive ability to endure torture.

While witch trials in Trier abated, subsequent years did not bring the city much fortune.[100] The weather remained freezing and wet, so fertile harvests became an exception. Shortages began to be felt in nearly all

[95] Voltmer, "Zwischen Herrschaftskrise, Wirtschaftsdepression und Jesuitenpropaganda," 63.

[96] Dillinger, *Böse Leute*, 215–19; Zenz, *Ein Opfer des Hexenwahns*.

[97] Dillinger, *Böse Leute*, 209–12, 215–27; Voltmer, "Die hexenverfolgungen im Raum Trier," 721–22.

[98] Voltmer, "Zwischen Herrschaftskrise, Wirtschaftsdepression und Jesuitenpropaganda," 43–53.

[99] Voltmer, 60–66, 78.

[100] Voltmer, "Krieg, uffrohr und teufelgespenst," 23; Dillinger, *Böse Leute*, 90, 239–40.

domains of socio-economic life. The religious atmosphere was characterized by an ardent Catholicism, with expectations of the end of times. Jesuits increasingly began impacting the city's spiritual life through their sermons, visitations, and especially their control over the educational system.[101] Pupils in the gymnasium and students at the university were educated in an atmosphere of passionate religiosity. The Jesuits performed plays with them, carrying titles like "The Anti-Christ" and "The Last Judgement." Typically, they also incited the population against diabolical witchcraft; one Jesuit warned that "through sorcery, female evil is exacerbating public scarcities."[102]

Accusations of witchcraft kept circulating in the city. The usual religious means of avoiding misfortune no longer seemed to suffice, so the angry populace increasingly called for persecution. In early 1586, an old woman suspected of bewitching milk was thrown off a bridge into the Moselle River by an angry mob. The atmosphere made the authorities revive investigations, and under Flade's supervision, it did not take long before the first woman was burned.[103] A remarkable development was that hatred of witches merged increasingly with the stereotypes about an allegedly corrupt elite.[104] Trials in the nearby countryside brought about accusations against rich people, like a woman from Trier who was alleged to have appeared at the witches' dance on a black horse, wearing expensive clothing. Later testimonies stated that *many* wealthy citizens from Trier were present at the sabbath and that a dignitary from the city had spoken there. Then, surprisingly, accusations from nearby villages found an unexpected target: Dietrich Flade![105]

These extraordinary charges against a high official might have been dismissed had they not coincided with another atypical development: accusations coming from boys.[106] Already in 1585, a boy from the Jesuit college had told stories about visits to a witches' dance, where he played

[101] Franz, "Hexenprozesse in Der Stadt Trier," 334–38, 348; Voltmer, "Jesuiten und Kinderhexen," 215–16.

[102] Zenz, *Ein Opfer des Hexenwahns*, 12.

[103] Voltmer, "Zwischen Herrschaftskrise, Wirtschaftsdepression und Jesuitenpropaganda," 67–73.

[104] Dillinger, *Böse Leute*, 123–30, 223–29, 320–9.; Biesel, "Les descriptions du sabbat," 189–95.

[105] Zenz, *Ein Opfer des Hexenwahns*, 17–20.

[106] Voltmer, "Jesuiten und Kinderhexen," 215–17; Voltmer, "Zwischen Herrschaftskrise, Wirtschaftsdepression und Jesuitenpropaganda," 69–69, 76–77.

the timpani. Around the time of the accusations against Flade, new voluntary confessions appeared, now coming from two schoolboys. The possibility of children being witches was not wholly new.[107] During some earlier trials, children had already played roles both as convicts and witnesses. The topic garnered some attention in demonological literature, like Kramer and Bodin, who incidentally discussed how adult witches seduced their children, especially daughters, into witchcraft.[108] Yet, when Trier was confronted with confessions from boys, there was not much of an established tradition to rely on.

Still, the voluntary confessions were met with serious consideration.[109] The boys' alleged impartiality would make their testimonies trustworthy, and at some point, the boys were also presented to the prince-archbishop. In his presence, they confirmed that rich people from Trier were involved in the diabolical sect. Additionally, they professed that one high official had even tried to poison the archbishop himself. The prelate had indeed been ill, so the stories impressed him. Soon after that, one of the boys came across Flade and exclaimed that this was the man he had seen at the witches' dance. Meanwhile, accusations against Flade from the countryside continued to pile up, and high officials around the prince-archbishop began to wonder in earnest whether the allegations might be trustworthy.[110] The news about Flade spread like wildfire. While Jesuits continued to incite the public, pressure from many layers of Trier society made the authorities favour more persecution.[111]

The torture sessions in the city, as well as in the nearby countryside, presented an increasingly detailed picture of what this network was supposedly capable of.[112] Witches were meeting on moors nearby, where they held festive dances and devoured child corpses, while dissonant music was performed on instruments made from horse skulls and cow horns. Under Satan's supervision, the witches had been responsible for weather disasters, illnesses, and the killing of people and livestock. The confessions presented loathsome images of rich witches, who dictated the weather magic

[107] Behringer and Opitz-Belakhal, "Hexenkinder - Kinderbanden - Straßenkinder," 21–24; Voltmer, "Jesuiten und Kinderhexen," 214–16.

[108] Walinski-Kiehl, "The Devil's Children," 172–73; Behringer, "Kinderhexenprozesse," 33.

[109] Voltmer, "Jesuiten und Kinderhexen," 215–17; Dillinger, *Böse Leute*, 253–55.

[110] Zenz, *Ein Opfer des Hexenwahns*, 15; Voltmer, "Zwischen Herrschaftskrise, Wirtschaftsdepression und Jesuitenpropaganda," 81–84.

[111] Dillinger, *Böse Leute*, 242–45; Voltmer, "Germany's First 'Superhunt'?," 232.

[112] Dillinger, *Böse Leute*, 119–33, 223–29; Biesel, "Les descriptions du sabbat," 189–95.

to ensure their future wealth through rising prices. Worryingly, the network members were planning new magical attacks, like damaging the nearby woods so that no wood would be left for witch burnings. Some suspects confessed that there were even plans to eradicate the city.[113]

The sense of anger and panic amongst the population heightened, and church bells were rung throughout the night to protect the city against further magical attacks. Flade could no longer move safely through the streets as he was harassed by boys and students from the Jesuit colleges, some of whom also assembled outside his house.[114] Throughout his career, Flade had been a protégé of the prince-archbishop, but this time support was not forthcoming. Von Schönenberg had misgivings about witchcraft accusations, but he was a weak and physically feeble ruler with little energy to resist popular pressure. Moreover, the boys' confessions made him concerned about his safety. Hence, on 4 July 1588, he ordered an investigation into the Flade case.[115] The new suspect did everything in his power to turn the tide rising against him. Flade wrote Von Schönenberg a letter in which he pleaded his innocence and mobilized friends who argued that Flade could not possibly be a witch as he had initiated and supervised witch trials. Yet, the number of accusations continued to grow, and at some point, Flade tried to flee, not just once but twice. His attempts were thwarted, and two mayors of Trier ordered him to stay in the city. Flade again beseeched the prince-archbishop to let him go, promising to spend the rest of his life in a convent and to donate all his money to the electoral state. Von Schönenberg, however, interpreted his offer as an implicit confession of guilt. On 23 March 1589, he had his former ally imprisoned in the town hall, and Flade was subjected to torture.

Soon, Flade admitted that avarice, greediness, scientific curiosity, and "lust of the flesh" had made him the devil's prey.[116] He had served the devil, forsaken God, attended the dances, and helped destroy harvests. The death verdict was inescapable. On 18 September 1589, Flade was first strangled and then burned in public in view of almost the entire city population. It would not be the last sensational burning in Trier, as the attack

[113] Voltmer, "Zwischen Herrschaftskrise, Wirtschaftsdepression und Jesuitenpropaganda," 100.

[114] Dillinger, *Böse Leute*, 157, 247–48, 260–64; Voltmer, "Zwischen Herrschaftskrise, Wirtschaftsdepression und Jesuitenpropaganda," 80.

[115] Zenz, *Ein Opfer des Hexenwahns*, 20–35; Voltmer, "Zwischen Herrschaftskrise, Wirtschaftsdepression und Jesuitenpropaganda," 77–97.

[116] Voltmer, "Germany's First 'Superhunt'?," 255.; Zenz, *Ein Opfer des Hexenwahns*, 37–55.

on the city elite did not stop with Flade. In his confessions, he had accused other highly placed individuals, including the two mayors who had supervised his arrest.[117] Their burnings followed soon. Within a few years, a significant number of the Trier elite had been executed. The carnage did not follow the lines of the older conflict between the city and the electoral state, as the killing now happened quite randomly.[118] The number of people executed in the town was not excessive, a few dozen, but what was unparalleled was their social profile. While a majority of the victims were female, rich men had become a new category to suffer disproportionally.

That so many accusations had come from the surrounding countryside was not accidental, as the whole region teemed with witch trials.[119] Witch-hunting reached unprecedented levels, especially in the electoral state territories and neighbouring principalities like Luxemburg and Sponheim. The region was characterized by traditions of self-government by the populace, and initiatives from those social classes mostly drove forward persecution. When the anger against witches came to a head, communal village meetings decided to form ad hoc witch-hunt committees, so-called *Hexenauschüsse*, which then instigated persecutions on a semi-democratic basis.[120] Participants showed their determination against the present evil through vivid symbolic ritualism. Oaths were taken under the linden tree in the village square, or knives were stabbed into a table to be grabbed by all members as proof of unconditional allegiance. Official authorities were pressured not to intervene, and many contemporary observers noticed that the atmosphere resembled a popular revolt.[121]

Through the condonement, or partial cooperation, of the government, the *Hexenauschüsse* turned into merciless killing machines. Whenever they received names of alleged witches, investigations often resulted in execution. Resistance could be interpreted as an indication of witchcraft,

[117] Franz, "Hexenprozesse in Der Stadt Trier," 345; Voltmer, "Zwischen Herrschaftskrise, Wirtschaftsdepression und Jesuitenpropaganda," 84–86; Behringer, *Witches and Witch-Hunts*, 97.

[118] Voltmer, "Die hexenverfolgungen im Raum Trier," 721–22; Dillinger, *Böse Leute*, 97–102.

[119] Voltmer, "Gott ist tot, und der Teufel jetzt Meister!"; Rummel, "Phasen Und Träger Kurtrierischer Und Sponheimischer Hexenvervolgungen," 283–84; Dillinger, *Böse Leute*, 305–57.

[120] Rummel, "Das 'ungestüme Umherlaufen der Untertanen'," 151; Rummel, *Bauern, Herren und Hexen*, 38–39.

[121] Rummel, "Phasen Und Träger Kurtrierischer Und Sponheimischer Hexenvervolgungen," 283.

offering grounds for additional arrests. Confessions of the executed, as well as the names of accomplices they had mentioned, were read aloud in plain language during well-attended witch burnings.[122] This created new streams of rumours, accusations, and arrests. A contemporary observed that the mere mention of a possible suspect was the beginning of "agitation, conflict, quarrel, wrangle, scolding and smearing" within their community.[123] Gerhard Schormann has argued that the public appeal of these burnings "was ideal for the further spread of the witchcraft doctrine."[124] Around 1590, hundreds of people were burned in the electoral state alone, and their social profile was similar to those persecuted in the city. Around thirty percent of the victims were now male, and elites were not spared.[125]

In the nearby territory of the imperial abbey of St. Maximin, a hunt unfolded that was worse than anything Europe had seen so far.[126] Typically, involvement in persecution started after community members had been mentioned as sabbath accomplices elsewhere. An accusation in 1587 against a female suspect set a chain of accusations in motion. Like the surrounding region in that year, St. Maximin's population had become highly sensitized to witchcraft, so a *Hexenauschuss* was installed that proceeded zealously. A preserved notebook from a judge shows names of the interrogated listed carefully alongside the accomplices they had mentioned, often numbering around twenty, but in some cases leading up to about sixty accusations. A boy operated as a "star witness" and accused about one hundred and fifty people during the four years of his imprisonment.[127] The hunt lasted from 1587 to 1596, leading to four to five hundred deaths—one-fifth of the imperial abbey's population—and the social profile was comparable to that of the electoral state. Complete families were wiped out, and some villages of the territory were largely eradicated.[128]

The persecutions of the Trier region reached their zenith around 1590, but from then on, things de-escalated. Complaints about miscarriages of

[122] Schormann, *Hexenprozesse in Deutschland*, 32–33; Rummel, *Bauern, Herren und Hexen*, 110.

[123] Quoted in: Voltmer, "Gott ist tot, und der Teufel jetzt Meister!," 195.

[124] Schormann, *Hexenprozesse in Deutschland*, 33.

[125] Voltmer, "Die hexenverfolgungen im Raum Trier," 739–40.

[126] Voltmer and Weisenstein, *Das Hexenregister des Claudius Musiel*; Voltmer, "Die hexenverfolgungen im Raum Trier," 731–32.

[127] Voltmer, "Jesuiten und Kinderhexen," 218; Dillinger, *Böse Leute*, 254.

[128] Voltmer, "Gott ist tot, und der Teufel jetzt Meister!," 197–210, 223.

justice were more vocal, and questions regarding trial payment became a source of conflict.[129] Von Schönenberg implemented a new law condemning the proceedings as disorderly and prejudiced. The new law shows an awareness that things had gotten out of control and that the authorities' position was in peril. Yet, while the law's effects were soon noticeable, trials in the electoral state still lingered for some time, only coming to a halt around 1596.[130] The new law should not be interpreted as a rejection of witch persecution per se, however.[131] Trier's suffragan Bishop Peter Binsfeld was one of the key people involved in the trials, and in 1589, he published his witchcraft treatise, the *Tractatus de confessionibus maleficorum et sagarum* (Treatise on Confessions of Sorcerers and Witches). This was an ardent plea for more witch-hunting, mirroring many of Binsfeld's recent experiences. However, one of Binsfeld's colleagues in Trier, the Dutch-German cleric Cornelius Loos, soon prepared a rebuttal. This member of the Trier theological faculty claimed that witch-hunts were killing the innocent. Binsfeld responded vehemently, and with government support, he ensured Loos was imprisoned. Loos was forced to recant his arguments, and his book remained unpublished.

In retrospect, the city of Trier, the electoral state, and the abbey of St. Maximin do not seem to have done themselves a great favour with these trials. While some people may have benefitted financially, horrendous fears gripped these communities; their social bonds were unravelled, economic activity was destroyed, numerous lives were annihilated, and taxpayers were wiped out. One of the reasons for the trials' slowdown after 1591 was that people had become physically and emotionally exhausted.[132] By giving in to popular demands for persecution, the authorities had also imperilled their positions and endangered precarious balances between jurisdictions—a disagreement about a trial between Trier and Cologne at some point nearly triggered a military confrontation.[133]

[129] Dillinger, *Böse Leute*, 84–86, 248, 332, 438; Rummel, *Bauern, Herren und Hexen*, 35–42.

[130] Voltmer, "Die hexenverfolgungen im Raum Trier," 726–32; Rummel, "Phasen Und Träger Kurtrierischer Und Sponheimischer Hexenvervolgungen," 267–3.

[131] Dillinger, *Böse Leute*, 225, 260, 367; Voltmer, "Demonology and Anti-Demonology"; Voltmer, "Zwischen Herrschaftskrise, Wirtschaftsdepression und Jesuitenpropaganda," 90–97.

[132] Voltmer, "Die hexenverfolgungen im Raum Trier," 731–32; Rummel, "Phasen Und Träger Kurtrierischer Und Sponheimischer Hexenvervolgungen," 273–74.

[133] Dillinger, *Böse Leute*, 377.

What is more, the city's reputation was severely damaged.[134] Trier became known as a contemporary Sodom, and especially Protestants seized the opportunity to discredit this overly Catholic city as being unusually susceptible to Satan's schemes.[135] In the early seventeenth century, Trier's city chronicler Johann Linden looked back in anguish at what had happened: "A plague or a cruel enemy could hardly have ravaged the region more excessively than this boundless inquisition and persecution." Linden observed that not enough people were left to cultivate the land in some areas.[136] The city chronicler appears not to have been the only one in the region to have realized that things had gone astray. While the Trier region still experienced some smaller trials, it would never let things get out of hand this terribly again.[137]

While the extraordinary course of the trials seems to have hindered further persecution in the region, consequences elsewhere were different. First, there were immediate spillover effects in other principalities.[138] Accusations from other places had sparked Trier's trials, but it soon became an inflammatory hotbed for new indictments itself. Names provided by Trier's persecutors helped initiate trials in places like Sponheim, the Eifel region, and around the Moselle and the Rhine. Second, information about what happened spread throughout Germany and beyond.[139] Trier´s hunts had been sensational. A teacher of the Duke of Bavaria's son happened to be in Trier around the time and reported, "What we hear about these pernicious witches in that region borders on the unbelievable. Everywhere in the area, one can see almost more stakes of burned witches than green trees, but like the Hydra, only more witches seem to be growing back."[140] An official from Hessen who resided in the electoral state informed his landgrave about "wondrous" events there. The confessions of witches had uncovered dances with "many hundreds of sorcerers." The witches had

[134] Dillinger, 360–63.

[135] Rummel, "Phasen Und Träger Kurtrierischer Und Sponheimischer Hexenvervolgungen," 264; Meder, *Acht Hexenpredigten*, 38; Schmidt, *Glaube und Skepsis*, 73, 264–65.

[136] Zenz, *Die Taten der Trierer*, VII:14.

[137] Rummel, "Phasen Und Träger Kurtrierischer Und Sponheimischer Hexenvervolgungen," 273–82.

[138] Dillinger, *Böse Leute*, 240, 369–70; Rummel, "Phasen Und Träger Kurtrierischer Und Sponheimischer Hexenvervolgungen," 261–71.

[139] Dillinger, *Böse Leute*, 359–60; Behringer, "Das 'Reichskhündig Exempel' von Trier."

[140] Dillinger, *Böse Leute*, 360.

even penetrated the prince-archbishop's sleeping room to poison him.[141] In 1590, a newssheet printed in Ulm provided an overview of Germany's current trials and devoted much attention to the vast number of executions in the electoral state. In one village, "so many women had been burned that only two very pious ones remained."[142] Probably most important of all was Binsfeld's treatise. Soon translated into German, it became a standard reference.

In combination with other factors like the unusual climatic and economic hardship, the Trier hunts helped trigger an unprecedented wave of witch-hunts sweeping Germany in the years around 1590.[143] In addition to sabbath accusations spreading from place to place, another mechanism made trials spread: when communities learned about witch-hunts nearby, this often activated older suspicions against individuals in their midst, followed by calls for their persecution.[144] To many contemporaries, the flood of trials proved that alarmists like Bodin and Binsfeld were right.[145] Cumulative notions of witchcraft also seem to have penetrated more layers of the population: A witness in Augsburg testified before the court that she had "never believed that such people existed, but only saw it now after it had become so loud and clear."[146] Historians today mostly rely on written documents, but Edward Bever reminds us of all the stories about witchcraft that must have been "transmitted across the countryside during work breaks and business socializing, over dinner tables in towns and taverns, and during the evening discussion and storytelling that spread news and ideas from household to household in the countryside."[147]

The importance of the Trier trials lay not only in the way they enhanced the fear of witchcraft; they also stimulated some shifts in what witchcraft was supposed to look like. As we saw, one of the hunt's striking features was its size. Trier came to stand for a new and dramatic type of persecution that did not kill dozens but hundreds.[148] One of today's key experts, the

[141] Voltmer, "Germany's First 'Superhunt'?," 234–5.

[142] Behringer, *Hexen und Hexenprozesse in Deutschland*, 221.

[143] Behringer, 180–84.

[144] Rummel, *Bauern, Herren und Hexen*, 26–27, 317.

[145] Behringer, *Hexenverfolgung in Bayern*, 145, 224; Labouvie, *Zauberei und Hexenwerk*, 115.

[146] Behringer, *Hexenverfolgung in Bayern*, 224.

[147] Bever, *The Realities of Witchcraft*, 70.

[148] Dillinger, *Böse Leute*, 359–60; Behringer, "Das 'Reichskhündig Exempel' von Trier," 437–39.

German historian Rita Voltmer, cautions us that this notion of a "super-hunt" in Trier was, to a certain extent, a contemporary construct.[149] Reports at the time often lost sight of distinctions between the city of Trier, the electoral state, and St. Maximin, merging and inflating them into one massive event. Yet, as Voltmer admits, the actual number of executions was still unprecedented and boosted the idea of witches forming large and growing networks. The trend is distinctly noticeable in visual representations of the sabbath.[150] As mentioned above, sabbath crowds were already getting bigger, but in 1594, a newssheet printed in Erfurt portrayed the Trier sabbath as an event with dozens of visitors; the most elaborate depiction so far.

The growing focus on elaborate sabbaths was not solely linked to Trier. Trial records from Rostock 1584 already spoke of six hundred attendants.[151] Persecution starting in Lutheran Osnabrück in 1583 led to an exceptional number of killings, as within half a year, 121 women had been burned there, and by 1592, the death count reached 180.[152] A newssheet printed in Jena in 1588 reported the striking events, adding that witches were meeting on the Blocksberg in Saxony, where no less than 8000 witches had assembled. The new focus on the sabbath was noticeable in other countries.[153] Before 1587, the prestigious Parliament of Paris, questioned over half of the people about the sabbath before they were executed as witches. After 1587, such questions occurred in nearly ninety percent of cases. The growing attention is also perceptible in the many demonological works that appeared around the time. Nicolas Rémy gave the sabbath extensive consideration in his *Daemonolatreiae* from 1595, and Martin Delrio, in 1600, was the first to present the sabbath as a highly elaborate event.[154] It was about to set a new standard.

Another shift Trier helped engender was an expansion of the witch stereotype, bringing to the attention witches who were male, rich, and powerful.[155] It was mainly this cultural "mutation" that struck contemporaries

[149] Voltmer, "Germany's First 'Superhunt'?"

[150] Goodare, *The European Witch-Hunt*, 74; Voltmer, "Germany's First 'Superhunt'?," 248–49, 253.

[151] Moeller, *Dass Willkür über Recht ginge*, 320.

[152] Rügge, *Die Hexenverfolgung in der Stadt Osnabrück*, 21–38, 103–5.

[153] Monter, "Witch Trials in Continental Europe 1560–1660," 7–10; Goodare, *The European Witch-Hunt*, 73–76.

[154] Rémy, *Demonolatry*, 47–65; Rio and Maxwell-Stuart, *Investigations into Magic*, 83–98.

[155] Behringer, "Das 'Reichskhündig Exempel' von Trier," 436.

as noteworthy. When the jurist and philologist Franciscus Modius consulted St Maximin's book collection in the late 1580s, he wrote in his diary about the extraordinary number of witch burnings there, also mentioning the city of Trier, "where an extremely learned Doctor Flade was arrested, bearer of an excellent name, who had presided over the city in the archbishop's name."[156] Likewise, a report on Trier in a Dutch newssheet printed in Antwerp mentioned the burning of a "powerful and rich man named Flade."[157] In 1590, a *Fuggerzeitung*—an influential German news report—stressed that the Trier hunt "spared no one, neither the rich nor the poor," and the Erfurt newssheet highlighted the presence of rich witches in word and image.[158] Demonologists did not fail to notice. Binsfeld remarked that people of "high standing" could be witches too, and Martin Delrio, in 1600, referred to the Flade case.[159]

Through Trier, the concept of witchcraft attached itself to antielitist sentiments, thereby enlarging the potential scope of persecution. The effects already became apparent during the wave of 1590.[160] Within the heated popular mood of the time, alleged governmental protection of rich witches was considered one of the worst forms of corruption, and wealthy individuals were burned in more places.[161] Of similar importance was the role played by children. The events in the Trier region helped turn the child witch into a significant topic in witchcraft theory, as well as in actual trials.[162] In his book, Peter Binsfeld addressed the subject on various occasions, highlighting that children were invaluable informants about the people attending sabbaths. In neighbouring Dietz, a child was brought to the court whose self-accusations might have been ignored just a few years

[156] Rummel, "Phasen Und Träger Kurtrierischer Und Sponheimischer Hexenvervolgungen," 266.

[157] *Warachtighe ende verschrickelycke beschryuinge van vele tooueanaers ende toouenerssen.*

[158] Behringer, "Das 'Reichskhündig Exempel' von Trier," 443; Voltmer, "Hört an neu schrecklich abentheuer," 124.

[159] Binsfeld, *Tractat Von Bekantnuß der Zauberer unnd Hexen*, 84; Dillinger, *Böse Leute*, 225–26.

[160] Behringer, *Hexenverfolgung in Bayern*, 202–4, 410; Haas, *Hexen und Herrschaftspolitik*, 87, 113–14.

[161] Pohl, *Zauberglaube und Hexenangst im Kurfürstentum Mainz*, 229–30, 263; Briggs, *The Witches of Lorraine*, 139; Dillinger, *Böse Leute*, 328–29, 350.

[162] Dillinger, *Kinder im Hexenprozess*, 104–6; Behringer and Opitz-Belakhal, "Hexenkinder - Kinderbanden - Straßenkinder," 22–29; Voltmer, "Jesuiten und Kinderhexen," 203–4, 218–32.

earlier.[163] The impact was lasting, as many demonologists included the Trier events in their works.[164] As late as 1625, Bamberg's preacher Friedrich Förner still referred to Trier's child witches—an important detail considering the role that children would play in the subsequent hunt in his town.

The increased focus on children has attracted historians' attention. Dillinger describes children as "door openers," often providing the first essential confessions.[165] What exactly made children so very vulnerable? Undoubtedly, they were easier to influence and had fewer inhibitions about providing imaginative testimonies. In contrast to adults, they probably did not fully grasp the possible consequences of their words, let alone understand how these might be turned against themselves. Perhaps children also enjoyed the sudden power and attention they got. Either way, their impact on further witch-hunts is indisputable.

In sum, Trier was of pivotal relevance. Rita Voltmer calls it "an unprecedented but exemplary witch-hunt."[166] Johannes Dillinger writes, "Electoral Trier became the superregional paradigm of a new type of persecution," while Wolfgang Behringer says that "The Trier persecutions were of decisive importance because the previous pattern of witch-hunting broke down."[167] With its new focus on large sabbaths and rich, male and child witches, the events of the Trier region gave the concept of witchcraft a new "virulence." Importantly, the events giving rise to these shifts were hardly preconceived. After the first accusations arose in 1582, it took a long time before the trials started in earnest. That they finally took off at all was crucially related to a rich mixture of coincidental occurrences, like exceedingly bad weather, heightened widespread aggression, the weakness of Von Schönenberg's rulership, the unusual boys' confessions, and specific local beliefs about witches being numerous, and often rich and corrupt.

It is tempting to assume that such spectacular events must have been designed and created by some primal movers or architects. Yet, many experts stress that this is not what happened. Binsfeld's treatise, for

[163] Binsfeld, *Tractat Von Bekantnuß der Zauberer unnd Hexen*, 60, 110–11; Dillinger, *Böse Leute*, 254.

[164] Walinski-Kiehl, "The Devil's Children," 173–74; Rio and Maxwell-Stuart, *Investigations into Magic*, 269; Förner, *Dämonenglaube und Zauberei im Jahre 1625*, 623–3.

[165] Dillinger, *Kinder im Hexenprozess*, 255.

[166] Voltmer, "Gott ist tot, und der Teufel jetzt Meister!," 186.

[167] Dillinger, *Böse Leute*, 359; Behringer, *Witches and Witch-Hunts*, 97.

instance, was only a report of and a response to events already unfolding. Dillinger argues that these persecutions developed their own momentum, in which "trials themselves aroused more fear of witches."[168] Günther Franz states that "the wave of trials was neither a conspiracy nor a manufacture."[169] The same applies to the wave of persecution sweeping Germany around 1590; in a passage quoted earlier (Sect. 5.3), Wolfgang Behringer argued that its many trials took people by surprise.

Yet again, while the *causes* of the extraordinary scale of the Trier hunts were rather accidental, things look less fortuitous when considering their *effects*. Like the Wiesensteig trials, it was hardly by accident that this set of persecutions gained so much traction. The Trier persecutions contained variants that made them huge, which is precisely what caused the sensation. Due to this widescale attention, some insidious cultural variants underlying the trials managed to lodge themselves into many more brains, inciting similar events elsewhere. The proposed Darwinian mechanism is visible again: quite coincidental, new cultural variants are preserved because of non-accidental effects on their continuation. What also bears upon the proposed Darwinian scenario is that the Trier persecutions were a nightmare for the communities involved. Any approach searching for the "functions" or "adaptations" of the trials may thus seem misguided. But when we look at the matter from the reproductive interests of the witch-hunting phenomenon *itself*, the Trier hunt suddenly appears like a prodigious "mutation."

Still, concepts of witchcraft had not yet evolved into their most inflammatory form. Beliefs and practices regarding witch-hunting were still not yet fully adapted to create the kind of persecution that occurred in Bamberg 1626–1632 and other places around that time. What was lacking was their efficient judicial machinery, their more professionalized set of persecutors, and the even higher number of victims. Some crucial transformations to witch-hunting were still to come. To understand what changed, we now move to the small southern German town of Ellwangen, where an unfortunate event during a church service unleashed one of Germany's most terrifying witch-hunts.

[168] Dillinger, *Böse Leute*, 240.

[169] Franz, "Hexenprozesse in Der Stadt Trier," 349.

7.3 Ellwangen 1611

In Spring 1611, churchgoers in the town of Ellwangen celebrated the Easter Service. Its elaborate religious ritual usually provided a sense of unity and belonging. But this time, an eerie incident provoked a vicious stream of rumours, triggering a chain of events that eventually brought the town's social fabric to breaking point. A seventy-year-old woman from the nearby village of Rindelbach, Barbara Rüfin, was observed to take the consecrated host out of her mouth just after receiving it from the priest.[170] The pastor noticed the uncanny act and instantly washed his hands to cleanse them of his physical contact with her. Some churchgoers saw the incident and were appalled. Did Rüfin's action not suggest a disdain for the body of Christ? After the service, the story went from mouth to mouth, and added to it was something else known about Rüfin: in her village, she had a reputation for witchcraft. It was said that she used salves to kill livestock. Was this Easter Service incident not an indication that Ellwangen had a devil's ally in its midst?

Ellwangen's authorities learned about the rumours and found them sufficient grounds for an investigation. Several people from Rüfin's local community were heard, and the inquiry laid bare the fact that Rüfin's reputation for witchcraft went back a few decades. Not only villagers scorned the seventy-year-old woman for witchcraft, but even her own family. Her husband was alleged to have called her a witch who should give herself up to the authorities.[171] On top of that, a search of Rüfin's house revealed mixtures that resembled witch potions.

This all happened at an unfortunate time for Ellwangen. The small town was the capital of a minor *Fürstpropstei* (prince-provostry), which had been under a lot of stress for some decades. The many crises of the late sixteenth and early seventeenth centuries had made themselves felt in the form of strange weather, ruined harvests, agricultural shortages, economic decline, epidemics, and price inflation.[172] Moreover, in religious and political matters, Ellwangen was a spearhead of Tridentine reform. Protestant peasants had temporarily overtaken the city in 1525, but after Catholicism was restored, the prince-provosts implemented strident policies of

[170] Mährle, "O wehe der armen seelen," 377–80; Midelfort, *Witch Hunting in Southwestern Germany*, 101.

[171] Midelfort, *Witch Hunting in Southwestern Germany*, 101.

[172] Midelfort, 98–100; Mährle, "O wehe der armen seelen," 342–48; Pfeifer, *Ellwangen*, 39–50.

confessionalization. Jesuits gained influence, and Petrus Canisius regularly attended the town to rouse the flock. By 1603, a new zealous prince-provost was installed, Johann Christoph von Westerstetten, who had been educated by the Jesuits.[173] His activities often accompanied paranoia about laypeople or clerics who sabotaged the holy policies.

Typically, Ellwangen had not escaped the wave of trials around 1590.[174] Cumulative notions about flight, sabbath meetings, and the *crimen exceptum* had helped trigger the execution of around twenty people within four months. The hunt had been quite traditional in its almost exclusive targeting of women, but it also showed some typical traits of the 1590 wave: it was organized in quite an ad hoc manner, starting with arrests of people at the lower level of society, but then moving up the social ladder—both rich and poor were killed.

The persecution starting in 1611 was different, although this was not apparent from the beginning. The inquiry into the Rüfin case lingered for some time as the investigators were undecided.[175] What initially dampened persecutory zeal was that not all witness statements from Rindelbach reinforced the charges. Some of the people heard stated that there was nothing serious going on with Rüfin. Her husband said that he had only called his wife a witch when angry and had not genuinely meant a word of it. Bursting into tears, he begged the investigators not to harm his wife. Barbara Rüfin vehemently denied all charges, professing that she did not know how to bewitch cattle. She also explained the incident during the Easter service: due to her age, she had trouble swallowing, so she had taken the host out of her mouth to put it in more deeply with her fingers. Rüfin said she knew this was inappropriate and planned to express her regret to the pastor.

Despite such uncertainties, the investigators subjected Rüfin to torture. From 20 April onwards, she was stretched twice a day for fifteen minutes. It did not immediately bring about a confession, but after a few days, she could no longer bear the torments and broke down emotionally. Rüfin provided a complete, elaborate confession: she had been seduced by a demon, made the pact, performed harmful magic, and attended the sabbath. As soon as the torture stopped, she recanted her confessions, but

[173] Durrant, *Witchcraft, Gender, and Society in Early Modern Germany*, 13; Mährle, "O wehe der armen seelen," 369–75.

[174] Mährle, "O wehe der armen seelen," 354–68.

[175] Mährle, 379–81; Midelfort, *Witch Hunting in Southwestern Germany*, 102–3.

once the painful questioning resumed, she again confessed. After many more sessions, the investigators observed that Rüfin had become "completely confused."[176] This did not prevent them from taking her testimonies in dead earnest. After all, how could she confess such terrifying details about witchcraft if she had not committed the crimes? On 16 May, Barbara Rüfin was executed with a sword and then burned at the stake.

Due to her alleged sabbath attendance, Rüfin was also forced to name companions. Thus, new arrests, new torture sessions, and new names of alleged sabbath accomplices soon followed.[177] In the case of the Trier trials, it had taken a long time before things became earnest, and proceedings had remained quite ad hoc in their implementation. The Ellwangen persecution, too, was initially hesitant but then proceeded with more thoroughness and discipline. By the end of May, the prince-provost Von Westerstetten installed a professionalized *Hexen Deputation*, a witches' deputation, mainly consisting of people from his entourage. With full government support, the committee was allowed to advance under its own terms, making some pivotal innovations to the practice of witch-hunting.[178] Soon after Rüfin's trial, they dispensed with witness reports and a search for circumstantial evidence. Instead, they went straight to enforcing confessions through torture, using standardized questionnaires to inquire more efficiently about fundamental cumulative notions. Suspects were not allowed to mention the names of people who had already been arrested or executed, and three new denunciations against individuals were considered sufficient grounds for arrest.

This unusual judicial machinery was already preceded in some respects by a witch-hunt from 1603 to 1606 in the prince-abbey of Fulda in Central Germany. The prince-abbot Balthasar von Dernbach had done something special there, by giving full authority over the persecutions to an ally of his, Judge Balthasar Nuss. He pursued the trials under his own conditions in cooperation with just a few professional colleagues. Within three years, an estimated 276 people were burned.[179]

[176] Midelfort, *Witch Hunting in Southwestern Germany*, 103.

[177] Mährle, "O wehe der armen seelen," 381–82; Midelfort, *Witch Hunting in Southwestern Germany*, 103–5.

[178] Mährle, "O wehe der armen seelen," 381–90, 399–404; Midelfort, *Witch Hunting in Southwestern Germany*, 103–7; Behringer, *Witches and Witch-Hunts*, 113.

[179] Robisheaux, "The German Witch Trials," 186; Behringer, *Witches and Witch-Hunts*, 110–13.

The example gained traction in Ellwangen, where the local *Hexen Deputation* was about to take things even further. Anyone who fell into their hands was almost certainly doomed—there was not a single release.[180] Withstanding torture was pointless, as there was no end to the agonies. A few people managed to escape, but most others confessed and provided names of accomplices, soon followed by their burning. The time between arrest and execution only lasted around two weeks. In the summer of 1611, there was a burning of about ten people every two or three weeks, slowing down to about five to eight people each time later that year. As Erik Midelfort put it, this was a "runaway hunt."[181] In that year alone, 126 people were killed, to be followed by at least 143 burnings in 1612. Like so many trials of the period, the arrests soon fanned out over multiple segments of society. Women were still overrepresented, but about thirty percent were men, and nearly all social categories were hit. There were also burnings of clerics; three priests confessed to having baptized children in the devil's name for many years, which explained why Ellwangen had so many witches.[182]

The *Hexen Deputation* presented its efforts as a holy pursuit against an acute and present evil, but judicial abuses were rampant.[183] Some suspects were raped in prison, and one guard offered advice on how to escape torture to a woman suspected of witchcraft, asking for her jewellery and other goods in return. Possessions of witches were confiscated, and in some cases, the suspects' properties were already inventoried before they had made a first confession. Such blatant miscarriages of justice, in combination with the swelling number of victims, stimulated resistance. Rumours spread about many of the incarcerated being innocent. However, one of those most insidious traits of the witch-hunting phenomenon also made itself felt, as criticisms were interpreted as indications of guilt. At some point, one of Ellwangen's judges was horrified to see his wife convicted, but his protest soon ended in his arrest and execution.

On 11 July 1613, a sensational event seemed to prove the persecutors right.[184] A sixteen-year-old girl named Maria Ostertegin voluntarily

[180] Mährle, "O wehe der armen seelen," 382, 413–18; Midelfort, *Witch Hunting in Southwestern Germany*, 104–12.

[181] Midelfort, *Witch Hunting in Southwestern Germany*, 100.

[182] Midelfort, 105.

[183] Mährle, "O wehe der armen seelen," 392–99, 413–27; Midelfort, *Witch Hunting in Southwestern Germany*, 107–10.

[184] Dillinger, *Kinder im Hexenprozess*, 123–34; Mährle, "O wehe der armen seelen," 430–31.

handed herself in at the city gate, declaring herself a witch. As her story went, she had been lured into witchcraft as a child by a friend, who by then was already burned. Since that point, Maria claimed, she had subjected herself to diabolical intercourse, harmful magic, and sabbath attendance. By now, she could no longer endure the snares of her demon and demanded her incarceration. Ostertegin voluntarily mentioned thirty accomplices by name, and on 21 August, she was killed by the sword. There are no indications that the girl was mentally impaired, raising the question of why she provoked her death. Dillinger speculates that it might have been an early modern variant of today's "suicide by cop," where people seek death through violent encounters with the police.[185]

Ellwangen trials spilled over into other territories. The proceedings against Barbara Rüfin had already produced a name from nearby Zöbingen, bringing about a small chain of accusations there that led to a few deaths.[186] Later, the *Hexen Deputation* aided nearby territories by assembling a book listing sabbath accomplices from elsewhere alphabetically by the town where they lived. Once handed over, these names led to trials in places like Ellingen, Dinkelsbühl, and especially Schwäbisch Gmünd, where it helped trigger a hunt that probably killed around fifty people.[187]

Meanwhile, the persecution in Ellwangen slowed down a bit in 1613, partly due to the departure of prince-provost Von Westerstetten and a more lenient successor. Still, persecution remained intense, and since Ellwangen enjoyed much judicial and political independence from external control, it could proceed much as it liked. When trials finally came to an end in 1618, the overall death toll had reached an estimated 430 people.[188] Like other massive hunts, the events attracted much attention elsewhere. Some commentators expressed admiration, like the renowned Jesuit philologist Jacob Gretser, who dedicated a book to Von Westerstetten.[189] A newssheet printed in Nurnberg in 1615 praised Ellwangen's "venerable prince" who executed without consideration of person both "rich and poor, simple and dignified, men and women."[190] The pamphlet further revelled in what it saw as the laudable punishment

[185] Dillinger, *Kinder im Hexenprozess*, 131–34.

[186] Haas, *Hexen und Herrschaftspolitik*, 125–26.

[187] Midelfort, *Witch Hunting in Southwestern Germany*, 112–20; Mährle, "O wehe der armen seelen," 431–34.

[188] Mährle, "O wehe der armen seelen," 382, 427–31.

[189] Durrant, *Witchcraft, Gender, and Society in Early Modern Germany*, 42.

[190] *Zwo Hexenzeitung, Die Erste, Von Dreyen Hexen Pfaffen*.

of the three witch-priests: first, incisions were made in their right hands, tonsured scalps, and foreheads. Then, salt and vinegar were rubbed into the wounds, their fingernails pulled out, and they were stripped of religious clothing. By the time of the burning, all traces of priestly dignity had been removed.

But there were also misgivings. Many towns in its vicinity did not follow up on the accusations that were provided to them against their subjects.[191] In 1618, the neighbouring territory of Oettingen went so far as to accuse Ellwangen of judicial misconduct officially.[192] One of Oettingen's charges was that the *Fürstpropstei* had used confiscations of the executed to enrich itself. Ellwangen responded derisively, dismissing all accusations. After all, its persecutions had solely been directed at protecting its subjects against the dangers of witchcraft and reconverting the seduced to save their souls. It was acknowledged that the properties of the executed were confiscated, but this happened to compensate for the harm they had done while the money was spent on pious causes.

By the time of this exchange, the hunt was already drawing to a close. The intensity of Ellwangen's response may well have been, as Erik Midelfort nicely put it, "the kind of passionate self-defence that springs from glimmers of self-doubt."[193] What seems to have brought the Ellwangen trials down was the incredible damage they were causing. In 1613, the Jesuit Johann Finck already fearfully wondered: "Where this matter will lead us or what end it might have, I do not know." He predicted that "this hideous evil has become so rampant, and has infected so many like a plague, that if the magistrate continues to exercise its power along this path, the whole city will languish in misery."[194] By 1618, the city was languishing in misery indeed.[195] The hunt had destroyed much of the city's social fabric, and the grief was overwhelming. Significant parts of the population were annihilated. Properties and houses were abandoned, contracts had been broken, civic debts had soured, and guild and civil offices were left vacant. The care of the many orphans left behind was a pressing problem. Like the Trier region in the 1590s, the damage seems

[191] Mährle, "O wehe der armen seelen," 434.

[192] Haas, *Hexen und Herrschaftspolitik*, 121–26; Mährle, "O wehe der armen seelen," 425–27.

[193] Midelfort, *Witch Hunting in Southwestern Germany*, 112.

[194] Duhr, *Geschichte der Jesuiten in den Ländern Deutscher Zunge*, 2:1:489.

[195] Mährle, "O wehe der armen seelen," 375–76, 434–38, 448; Midelfort, *Witch Hunting in Southwestern Germany*, 111–12.

to have engendered an awareness that things had gone too far. If the city was to survive, this hunt had to stop. And so it did. Later on, in the persecutory wave of the late 1620s, the *Fürstpropstei* still experienced some trials but avoided snowballing.

Yet, the effects of the Ellwangen hunt on other places were different, as its model found fertile grounds elsewhere.[196] Wolfgang Mährle maintains that it created a "new persecutory paradigm" characterized by a more assertive top-down initiative, a more professionalized investigation, and a more lethally efficient bureaucracy.[197] Behringer argues that it helped bring about "a new type of persecution" that "gained momentum within weeks and led to the burning of hundreds of witches within a few months."[198] Its potential for transplantation soon became apparent in the southern German prince-bishopric of Eichstätt.[199] Von Westerstetten left Ellwangen to take a position as prince-bishop there, bringing the new witch-hunting model with him. When accusations against witches occurred in the near vicinity of the town in 1617, it initiated a hunt that was aggressively implemented from above by professional persecutors who enjoyed the support of the highest government. Its focus on women was quite traditional—they constituted an estimated eighty-eight percent of victims—but like in Ellwangen, the heart of society was attacked.[200] The households of brewers, butchers, innkeepers, cobblers, smiths, tailors, and almost all local elite families were hit. At least around two hundred, but perhaps even up to seven hundred people, were executed.[201]

The Ellwangen hunt had started at a moment of relative peace in witch-hunting history, but the Eichstätt trials were part of a more significant flood. In the electoral state of Mainz, things re-escalated between 1616 and 1618, and an estimated 361 people were killed;[202] 1616 was also the year when the first extensive Bamberg hunt began, killing around two hundred people (see Sect. 5.4). Around the same time, trials in the neighbouring Franconian prince-bishopric of Würzburg produced an even higher number of three hundred estimated deaths. Wolfgang Mährle has

[196] Schwerhoff, "Hexenverfolgungen in der frühneizeitlichen Germania Sacra," 18; Voltmer, "Hexenbrenner Und Hexenbischöfe," 202, 217.

[197] Mährle, "O wehe der armen seelen," 438.

[198] Behringer, *Witches and Witch-Hunts*, 112–13.

[199] Durrant, *Witchcraft, Gender, and Society in Early Modern Germany*.

[200] Durrant, xxii.

[201] Durrant, 20; Behringer, "Westerstetten, Johann Christoph von (1563–1637)."

[202] Robisheaux, "The German Witch Trials," 187.

argued that the persecutory authorities of these Franconian trials "drew their inspiration from the Ellwangen example from the beginning, both in their organization, as well as in the legal implementation of proceedings."[203] It has been retorted that Mährle's claims lack conclusive evidence, but the Franconian hunts have eerie similarities to the Ellwangen model. It is telling that Friedrich Förner, the charismatic suffragan bishop of Bamberg, dedicated his witchcraft sermons to Von Westerstetten.[204]

The trials of the late 1610s had largely waned by the early 1620s, only to be superseded by the most lethal wave beginning around 1626.[205] By that time, the Ellwangen model was widely emulated. Popular pressure for witch-hunting remained paramount as an initial trigger for persecution. But as soon as authorities became involved, professional awe-inspiring *Hexenkommissare* (witchcraft investigators) took control. Armed with extraordinary legal powers, they pushed persecutions towards an unprecedented scale. Questionnaires were standardized, torture was relentless, and escape was nearly impossible. Also typical of this wave, as noted in contemporary reports, was its even higher involvement of children and an enhanced attack on men and elites.[206] The new model blossomed, especially in ecclesiastical territories where relative judicial independence coincided with vigorous Counter-Reformation Catholicism.[207] The Bamberg hunt of 1626–1632 offers a prime example. It not only outdid Ellwangen in the way it gained momentum but also in the eventual death toll. The simultaneous persecution in Würzburg killed around a thousand people.[208] Notable was Würzburg's unusually high number of child witches executed, and its bloodbath amongst clerics—forty-three were burned.

Electoral Mainz also experienced renewed trials, leading to an estimated 768 people being burned.[209] Runaway hunts swept parts of southern Germany in places like Mergentheim, Ortenau, Baden-Baden,

[203] Mährle, "O wehe der armen seelen," 439.

[204] Förner, *Dämonenglaube und Zauberei im Jahre 1625*, 1–2.

[205] Robisheaux, "The German Witch Trials," 187–88; Behringer, *Witches and Witch-Hunts*, 108–20.

[206] Midelfort, *Witch Hunting in Southwestern Germany*, 179; *Zwo Warhafftige, und doch Männiglich zuvor bekante Newe Zeitung*; *Ein Warhafftige und gründtliche Beschreibung*.

[207] Dillinger, "The Political Aspects of the German Witch Hunts," 68.

[208] Walinski-Kiehl, "The Devil's Children," 175–77; Rowlands, "Würzburg, Prince-Bishopric Of."

[209] Robisheaux, "The German Witch Trials," 187.

Oberkirch, and Offenbach.[210] In some cases, even the families of the people who had instigated or perpetrated the trials in the first place were hit. Yet, the electoral state of Cologne, to the northwest, would experience Germany's worst witch-hunt.[211] Throughout its scattered territories, probably over two thousand people were burned within a few years.[212] A past interpretation presumed that this hunt was a top-down "war against the witches," but recent research indicates that it was more of a complex mixture of bottom-up and top-down initiatives.[213] Our focus on Catholic areas should not let us forget the Protestants. Principalities like Mecklenburg, Pomerania, and Sachsen-Coburg experienced persecutions that altogether killed thousands of people. They also show similar patterns, with broader segments of society targeted.[214]

The Ellwangen hunt constitutes a critical moment in a broader trend towards the rationalization of witch-hunting. The tendency is discernible in the electoral state of Mainz. Herbert Pohl observed that in the late sixteenth century, arrests depended upon mixed evidence: accusations, house warrants, witness statements, and investigations into people's reputations. Questioning tended to be somewhat improvised and open-ended. But by the late 1620s, the mere mention of somebody as a sabbath accomplice was sufficient grounds for arrest, and examinations focused more efficiently on enforcing cumulative confessions.[215] Pohl writes:

> The range of what was considered orthodox in witchcraft theory was still wide in the sixteenth century and only experienced a certain fixation after the emergence of the new interrogatory practices of the seventeenth century, which, in certain ways, also expanded popular beliefs in witches.[216]

The confiscation of witches' property was also rationalized, and especially the execution of rich people created possibilities for further persecution. In the town of Dieburg half the city council was burned.

[210] Midelfort, *Witch Hunting in Southwestern Germany*, 121–54.
[211] Behringer, *Witches and Witch-Hunts*, 115.
[212] Behringer, 113–17; Decker, "Die Hexenverfolgungen im Herzogtum Westfalen."
[213] Schormann, *Der Krieg gegen die Hexen*; Becker, "Hexenverfolgung in Kurköln."
[214] Rummel and Voltmer, *Hexen und Hexenverfolgung*, 117–19; Moeller, *Dass Willkür über Recht ginge*.
[215] Pohl, *Zauberglaube und Hexenangst im Kurfürstentum Mainz*, 170–76, 227–30.
[216] Pohl, 295.

The continuation of another trend marked the early seventeenth century: both in textual and visual representations; sabbaths became more crowded. In 1612 the French judge Pierre de Lancre published his *Tableau de l'inconstance des mauvais anges et démons* (Description of the Inconstancy of Evil Angels and Demons), which surpassed the extravagance of earlier sabbath descriptions.[217] A subsequent version also included a baroque picture, showing more than fifty witches assembled.[218] By 1620, a widely reproduced German picture named "Zauberey" (Sorcery) went as far as to present the sabbath as a mass event with hundreds of visitors.[219] Let us remember the Bamberg interrogator Ernst Vasoldt, who extracted 126 names of sabbath companions from one suspect. In the early 1560s, that would have been a highly implausible thing for an investigator to do. In the 1590s, it was already somewhat more conceivable, but by the 1620s, it had become quite obvious.[220]

Figures like Vasoldt can also help us understand how those German trials of the 1610s and 1620s became so big. The professionalization of witch persecution increasingly required professional persecutors. Some moved through Germany to offer services in various *Hexen Kommissionen*, transmitting practical knowledge about how to proceed.[221] It is also worth further reflecting on the significance of children. In Ellwangen, their role happened to be quite limited, but by the 1620s, there were hardly any runaway panics that were not initiated or enhanced by child confessions.[222] Adding to the gruesomeness, alleged child witches were increasingly executed. By now, children also took centre stage in witchcraft theory, and a German pamphlet from 1626 was entirely devoted to the subject.[223]

In his 1972 masterpiece on witch-hunting in southwestern Germany, Erik Midelfort contemplated the extraordinary scale of the late 1620s trials. "The reason witch hunting developed to such extreme dimensions during this period has always been unclear," he remarked.[224] The many disasters, like epidemics, warfare, failed harvests, and economic ruin, played their part, but other periods faced similar challenges. After the

[217] Lancre, *On the Inconstancy of Witches*.
[218] Krause, *Witchcraft, Demonology, and Confession in Early Modern France*, 98–99.
[219] Voltmer, "Herren und Hexen," 15.
[220] Goodare, *The European Witch-Hunt*, 236–38.
[221] Rummel and Voltmer, *Hexen und Hexenverfolgung*, 110–12.
[222] Behringer, *Witches and Witch-Hunts*, 142; Walinski-Kiehl, "The Devil's Children," 173.
[223] *Newer Tractat von der Verführten Kinder Zauberey*.
[224] Midelfort, *Witch Hunting in Southwestern Germany*, 124.

1620s, these things even aggravated, but witch-hunting in the 1630s and 1640s began to decline. Hence, environmental disasters alone cannot explain the events, so Midelfort proposed looking elsewhere, too: "At least one element is missing if we consider merely the superstitious witchcraft beliefs of a region and the catalytic disasters that may have triggered panic. That element is mood – the mood of villagers, townspeople, lawyers, governors, and judges."[225] To understand the "mood" at that time of the Thirty Years' War, Midelfort thought of the despair amongst Protestants and the religious euphoria amongst Catholics.

This argument appears valid, but one crucial element still seems missing: the cumulative concept of witchcraft's adaptedness at that historical juncture. By the late 1620s, the doctrine was more predisposed than ever to trigger runaway panics that could kill hundreds of people. Through the enhanced focus on huge sabbaths, child witnesses, professionalized investigations, and standardized questionnaires, concepts of witchcraft had evolved in ways that boosted the reproduction of witch-hunting. The general "mood" and environmental hardship remain essential, but we must add the cumulative concept's ability to exploit such conditions.

The origin of this "design" is not to be found in grand intentional strategies or central preparation. The hunts of Wiesensteig, Trier, and Ellwangen became unusually big due to a blend of incidents, contingent conditions, and specific ideas about witchcraft coming together at a particular place and moment. Yet, the new local cultural variants then became a transregional standard precisely because of their ability to create considerable persecution. The big hunts had the most cultural "offspring." Beliefs and practices mutually enforced each other. These persecutions did not grow large because the communities or actors involved really benefitted. Ellwangen further illustrates how witch-hunting could ruin communities. But again, when considering the reproductive interests of the witch-hunting phenomenon *itself*, Ellwangen looks like yet another lucky "mutation."

Some essential elements of the proposed Darwinian scenario are still missing, though. Until now, our focus has been on new ideas and practices that *enhanced* witch-hunting. But a Darwinian model requires another side to that coin. The cumulative preservation of accidentally adaptive variants can only produce adaptations if those variants appeared within oceans of competitors that failed to survive as successfully. We could think

[225] Midelfort, 124.

of ideas about witchcraft that hindered the continuation of persecution or were too insignificant to reproduce. Moreover, attention so far has primarily focused on elements of the early modern context that boosted trials, like weather disasters, anxieties over Anabaptists, or Counter-Reformation zealotry. However, a Darwinian model also presumes forces that hinder it. It was said above that there is no such thing as a cultural free lunch. If a cultural phenomenon is to survive, it must compete with other forms of culture for human attention. Was the early modern environment also like a hostile jungle for the witch-hunting phenomenon? Did enemies lurk in the dark, waiting to pounce? This proposed cultural "struggle for life" will now be explored.

7.4 Culture's Struggle for Life

An illustrated *Flugblatt* in 1555 reported an alarming case of witchcraft in Derenburg, a town in central Germany.[226] "A terrifying story" ran the headline of an account of what had happened to three female witches and two men in October that year. The newssheet began with some conventional warnings about the devil's hatred of humankind, women as his "weak tool," and the imminent last judgment that was making Satan furious.[227] The report then moved to the case, and an exciting storyline unfolds. Two witches named Gröbische and Gißlersche were burned on the first of October. However, when the fire of Gröbische's stake was ignited, Satan came down from the air and took her away before everyone's eyes. Two days later, the witches returned to Gißlersche's place, where they violently ejected her husband from the house. A neighbour, who was alerted by all the noise, saw the witches dancing around a fire in the house, with Gißlersche's husband lying dead on the ground. A visual illustration of the *Flugblatt* depicted the events in gruesome detail. The newssheet also reported that on the fifteenth of October, a third witch was put to the stake for harmful magic that had lamed a man and killed his cattle.

This newssheet about the case of Derenburg was typical for its time: it highlights witchcraft as a feminine crime, and its overall description is rather unelaborate. The sexual diabolical pact and harmful magic are present, but the sabbath and nightly flight are absent. What makes this

[226] Zika, *The Appearance of Witchcraft*, 180–82.
[227] Zika, 180–1.

newssheet remarkable, though, are some strange additions, like the devil saving a witch from the stake, the witches' return, the killing of the husband afterward, and the dance around the fire. We see four unusual cultural variants about witchcraft, or "mutations," in public discourse at an auspicious moment in witch-hunting history. The details made the story sensational and contained much that the mid-sixteenth-century German public craved: the supernatural, dangerous, and diabolical. No wonder a publisher was happy to print it. Hence, these four variants had ample opportunity to become standard notions of witchcraft mythology.

However, none of them was a reproductive success. In contrast to notions like nightly flight, large sabbaths, or the child witch, these "mutations" hardly survived. If we contemplate their features, this should not surprise us. Take the example of the devil saving a witch from the fire while many people were watching. One problem of this variant was that it was probably too vulnerable to refutation. The diabolical witchcraft mythology faced continuous scepticism, and people who had attended a burning where such supernatural rescues allegedly occurred might well retort that they had not seen it happen. It is probably not by accident that supernatural events that *did* become familiar tropes, like flights, sabbaths, or magic, were usually thought to occur in secret. This circumvented any refutations. Another problem of the diabolical rescue was that it made witch-hunting futile. What is the point of a witch trial if Satan saves his witches just before they get burned? The idea of witches returning afterward and killing a person was even more hazardous. For the reproduction of witch-hunting, people needed to believe that it reduced and not enhanced the danger. The variant about the witches dancing around a fire in a house appears too trivial to have survived the cultural "struggle for life" amongst witch beliefs.

If the Darwinian model proposed here is warranted, such "failed" cultural variants must have been ubiquitous. The cultural "lucky mistakes" that became common tropes must have stood out against a continuing flood of variants that fell into obscurity. Witchcraft experts now and then refer to strange notions about witchcraft, like "an odd tale," "an unusual picture," "a highly uncommon element," an "odd statement," or a "quite original" notion.[228] However, as mentioned, systematic empirical research

[228] Richard Kieckhefer, "The First Wave of Trials," 174–75; Briggs, *The Witches of Lorraine*, 141; Dillinger, *Kinder im Hexenprozess*, 129; Briggs, *The Witches of Lorraine*, 144; Schmidt, *Glaube und Skepsis*, 44.

has not been undertaken so far into what such "unfit" variants looked like. How did they appear? Why were they unsuccessful? How many of them were there? Something else that has not been addressed is their potential theoretical significance: such variants may have been essential ingredients of an evolutionary process that helped create the socio-cultural "design" of the cumulative concept of witchcraft.

If just a few sentences in one newssheet already provide four failed variants, we can safely assume that further empirical research into trial records, scholarly works, pamphlets, chronicles, sermons, or visual representations would deliver innumerable other examples. This empirical question lays bare an almost endless research agenda, requiring more work than any scholar could perform in a lifetime. Notably, the failed variants that historians can still retrieve only represent a small percentage of all the variants that once appeared. Many witchcraft notions that were discussed in local taverns, during sermons, in sessions of torture, or at the village square will not have made it into any document. The variants written down or printed may have disappeared because the documents were lost—when the Wiesensteig trial records vanished, they probably took many unique cultural variants with them.

Examples of visible supernatural events during witch burnings form one interesting category of failed variants. As it turns out, the Derenberg *Flugblatt* is not the only source to mention a case. An extensive newssheet about a burning in Dillingen, October 1579, recounts how a witch was brought to the stake and threatened to set the executioner on fire. The executioner was not impressed, but two ravens appeared when the witch was about to burst into flames, "and in front of many people (a great number of them were watching) they took her away in the air."[229] Subsequently, the executioner's clothes caught fire, and he died cruelly. A newssheet from 1588 on the persecution in Osnabruck contains another account.[230] When many witches were about to be burned, the devil appeared to take four of them with him. When the executioner ignited the fire, an awful wind suddenly arose, threatening to douse the fire. Many onlookers were terrified, and the executioner feared for his safety. Nevertheless, he maintained the fire, so the other witches received well-deserved punishment. Dreytwein's chronicle contains another case.[231] During one of the witch

[229] Kuntzen, *Newe Zeitung von einer Erschrecklichen That.*

[230] Rügge, *Die Hexenverfolgung in der Stadt Osnabrück*, 105.

[231] Dreytwein, *Esslingische Chronik*, 245–46; Waite, *Eradicating the Devil's Minions*, 150.

burnings in Wiesensteig 1562, twenty witches were burned while more than three thousand people watched. Then, sensationally, an angel of God descended against a red sky, urging people to stay faithful to God to avoid punishment.

In Jacob Wick's news collection, we can find other examples of unusual supernatural events. During a trial in 1568, seven witches were executed in the Swiss town of Erlach, of whom one was a young woman who confessed and repented profoundly. When she was convicted in court, "a smell spread through the courtroom that was so very penetrating that all judges and all others had to pinch their noses. People thought that Satan had left the smell behind because he was unsure about his possession of the girl's soul."[232] Another report tells of a case in Luzern in 1570, where an imprisoned witch confessed to several crimes. When investigators re-entered her cell to extract more confessions, they "found nothing but her skin, that was bloated like a bladder. They saw neither bones nor flesh", so the devil had taken her interior and soul away. [233]

As mentioned above, nightly flights and sabbaths were usually thought to happen secretly, but there were occasional deviations from this pattern. Investigations in the Saar region produced a story about a twenty-year-old farmhand who fell asleep in his master's barn but noticed the barn was full of light when he woke up.[234] The witches were having their dance there. The *Malleus Maleficarum* maintained that the common people often sighted witches flying like shepherds who saw a sorceress from Waldshut fly on a demon just after the witch and the demon had stirred up a hailstorm.[235]

While observable supernatural events seem to have been somewhat of a trope, none of these specific variants became standard elements of witchcraft narratives. Again, these variants were probably too vulnerable to refutation. The sudden wind in Osnabruck or the sudden smell in Erlach possibly happened through natural causes, but these variants still seem "unfit" for robust survival because their actual occurrence at that specific stage of a witch trial must have been infrequent at best. The Dillingen and Osnabruck variants about the witches being saved, or the Dillingen addition about the executioner's burning, tend to discourage further hunting.

[232] Wick, *Die Wickiana*, 161.
[233] Wick, 179–80.
[234] Labouvie, *Zauberei und Hexenwerk*, 228.
[235] Institoris, *The Hammer of Witches*, 292.

The sudden smell at Erlach or the Wiesensteig angel appearing against a red sky may also have been too insignificant in their effects on trials to reproduce further. As argued above, the secrecy of many of the witches' acts was one of the cumulative concept's most ingenious traits. Yet, what these stories indicate is that the supporters of witch-hunting were not necessarily the ones who deliberately designed the concept that way. *They* continually introduced stories about visible supernatural events. More plausibly, a hidden selective process weeded such unhelpful variants, thus creating a more unfalsifiable concept of witchcraft.

Even the relatively well-organized and successful persecution of Ellwangen produced a variant with much potential for undermining further hunting. During the middle years of the persecution, the trial records began to report unusual witches' confessions: witches were not really killed during their burning, but continued to appear at sabbaths and lived in a diabolical palace, a sort of witches' heaven, where they enjoyed a pleasant afterlife and continued their diabolical activities.[236] Mährle and Dillinger notice that these confessions threatened to make persecution a farce: witch-hunting would neither save people from the witches' crimes nor ensure the witches' divine salvation. No wonder that this belief was an evolutionary dead-end.

Continuing variations likewise characterized successful notions about witchcraft. There was a lot of variation regarding the objects that witches were alleged to fly on: amongst the ordinary people, there were alleged sightings of flying cats and dragons.[237] Regarding witches' meetings, we saw that small sabbaths lost ground to large sabbaths due to the latter's effect of creating more extensive hunts. But there were also stranger deviations. Take a claim from a fifteenth-century Arras treatise about witches' assemblies being held "on the same night in various forests and places."[238] Kieckhefer maintains that in an age without artificial lighting, it conveyed "a powerful sense of the supernatural and, in this case, demonic." But he is probably also correct to render it an "apparently inconsequential detail."[239] This notion of sabbaths moving from place to place within one night did not take hold, which might be related to it adding nothing to

[236] Dillinger, *Kinder im Hexenprozess*, 129–31; Mährle, "O wehe der armen seelen," 435.

[237] Blécourt, "The Flying Witch"; Schmidt, *Glaube und Skepsis*, 388; Groß, *Hexerei in Minden*, 280–81.

[238] Anonymous of Arras, "A History of the State, Case, and Condition of the Waldensian Heretics," 47.

[239] Kieckhefer, "Mythologies of Witchcraft in the Fifteenth Century," 91.

the sabbath's essential implication of witches knowing each other—the variant was superfluous. Interrogations in Lorraine produced some variants that were even hazardous for witch-hunting. A female suspect said she could see another sabbath nearby while at a sabbath.[240] This idea may have hindered the number of identifiable accomplices, as it reduced the number of witches present at one sabbath. Another confession said that witches did not appear in their regular bodies but as cats.[241] This "mutation" could undermine the sabbath's reproductive mechanism, as witches could no longer identify each other.

There were also many variants regarding what happened at witches' meetings. Some sources describe the devil in a surprisingly subservient role.[242] A newssheet printed in Cologne in 1596 tells of a conflict between witches and devils at an assembly on the Lüneburg Heath.[243] The witches wanted some money back from Lucifer, and they beat demons so hard that the spirits fled. Many witches then flew to hell, where they scolded the devil for not paying them. Lucifer gave in, and a dragon appeared who brought the witches their treasures. Other variations describe how certain women chaired the sabbaths, in some cases being the devil in disguise.[244] Several early Italian accounts maintained that the witches' assembly was presided over by Oriente, a woman who instructed witches on healing and other matters. A confession in southern Tirol in 1506 produced a story about an ugly woman named Jostin, elected "queen of the land of angels," wearing a golden robe and a multicoloured crown.[245]

Unfamiliar motifs also appeared regarding what was being venerated. An early Alpine account discussed witches adoring a statue of Jupiter, and the newssheet from Erfurt 1594 shows witches dancing around a column with a toad sitting at the top.[246] Toads were also the topic of some odd accounts in the Basque Country. There was nothing strange about toads attending sabbaths as such, as they were considered filthy, diabolical

[240] Briggs, *The Witches of Lorraine*, 141.

[241] Briggs, 144.

[242] Moeller, *Dass Willkür über Recht ginge*, 321–22.

[243] *Zwo erschreckliche und unerhörte Geschicht.*

[244] Kieckhefer, "Mythologies of Witchcraft in the Fifteenth Century," 97; Moeller, *Dass Willkür über Recht ginge*, 320.

[245] Behringer, *Hexen und Hexenprozesse in Deutschland*, 111–2.

[246] Ostorero, *Le diable au sabbat*, 7–8; Voltmer, "Hört an neu schrecklich abentheuer," 132, 151.

animals.[247] However, some reports stressed how witches were maltreating them to extract poisons. "They take the skin of the toads, which they bite into with their teeth, and the toads protect themselves as best they can with their feet and scratch them on the face hideously," De Lancre recounts.[248] Another Bask description mentions how toads were whipped with switches and pressed on the ground with a foot to make them spew.[249] In today's world, such a report might infuriate animal rights activists, but it was not what would motivate the early moderns to act. The maltreatment of toads did not become a central motif.

Initiation into witchcraft was also a topic of continuing variation. There was much diversity in what the devil wore when he appeared to people, which names he used to introduce himself, and which inhuman traits betrayed his real identity.[250] Some variants more strongly diverged from the familiar narratives. A news report about the persecution of Geneva in 1571 tells the story of a servant who rode a carriage through a forest and came across a witches' synagogue there. A demon named Moreth then brought him to the devil, who sat on a chair, and before Satan, the servant was initiated into the sect.[251] The *Malleus Maleficarum* provides a confession of a young woman in the German town of Breisach who was seduced into witchcraft by her aunt in the following manner:

> One day she had to go upstairs with her aunt and enter a room at her command. There, she saw fifteen young men in green-colored garments after the fashion in which knights go about, and the aunt said to her, "Well, then! From among these young men, I will hand over to you the one that you want, and he will take you as his bride."[252]

In comparison with the idea of initiation into witchcraft through diabolical intercourse, these variants look meagre. The Geneva variant seems dull, and the Breisach variant contains trivial and arbitrary details—why upstairs, why fifteen, and why green clothes? The idea of female diabolical intercourse was more shocking and perverse and better adapted to crucial

[247] Wilson, *The Magical Universe*, 417–18; Labouvie, *Zauberei und Hexenwerk*, 216.
[248] Lancre, *On the Inconstancy of Witches*, 403.
[249] Henningsen, *The Salazar Documents*, 116.
[250] Pohl, *Zauberglaube und Hexenangst im Kurfürstentum Mainz*, 352–59, 286; Roper, *Witch Craze*, 82–103.
[251] Wick, *Die Wickiana*, 196–98.
[252] Institoris, *The Hammer of Witches*, 283–84.

obsessions of the age, like diabolical evil and uncontrollable female sexuality. However, a more common aspect of the Breisach story was that witches introduced other witches to witchcraft. That has remained a significant competitor to initiation through diabolical contact.[253] This may be due to two important implications that enhance witch-hunting's spread. If one has found one witch, there might be others, and witches should be stopped from introducing even more people.

Retaining some diversity and malleability was in the cumulative concept's best reproductive interests, especially concerning harmful magic. The concept was not restricted to just a few magical acts, like weather-making or bewitching of livestock. This open-endedness ensured that almost any unusual misfortune could provoke new trials.[254] This is not to say that *all* magical acts were helpful, as a charge from Württemberg against the town of Bretten may illustrate. As we saw earlier, the sceptic Johann Weyer ironically wondered why witches were not being used in military conflicts, but that was precisely what the Württembergers in 1504 accused their opponents of.[255] The Bretteners were alleged to have used witches to cause a storm that damaged the Württemberg army. The German historian Hans Jürgen Schmidt provides a plausible explanation for why it did not reproduce well: "The idea of using witches for a community was completely nonsensical because witchcraft was characterized precisely by the fact that it was directed against the community. Why should witches help a city where they were burned?"[256]

As explained above, the cultural Darwinian "struggle for life" should not only be interpreted as competition between various ideas and practices regarding witchcraft. There was also competition coming from other forms of culture. Take alternative explanations for unusual misfortune. A crucial one was divine punishment.[257] In our discussion of Wiesensteig, we already met the Lutheran preachers Alber and Bidenbach, who claimed that people had only themselves to blame for the hailstorm; their sins had provoked God's fatherly reprimand. After all, had people not been swearing, disregarding the poor, and sinning with wine? No wonder God

[253] See, for instance: Kieckhefer, *Hazards of the Dark Arts*, 38–41; Voltmer, "Preaching on Witcchraft?," 206; Wick, *Die Wickiana*, 161.

[254] Voltmer, "Herren und Hexen," 14.

[255] Schmidt, *Glaube und Skepsis*, 44.

[256] Schmidt, 44.

[257] Midelfort, *Witch Hunting in Southwestern Germany*, 34–66; Behringer, *Witches and Witch-Hunts*, 7, 88–89.

destroyed the vineyards. By rejecting the possibility of weather magic and attributing the hailstorm exclusively to God, these Württemberg preachers stood in an old and well-established intellectual tradition, which nicely dovetailed with a central theological tenet of Christianity: God's omnipotence. The belief that God and the devil were equal opponents was regarded as a dualistic heresy, and Alber and Bidenbach framed witch beliefs as such.[258]

However, for the witchcraft doctrine, this theological obstacle was not insurmountable. A viable counter-strategy was not to contest the omnipotence of God—that would have been a heresy indeed—but to frame witchcraft as a tool of God's punishment.[259] Human sin was indeed ubiquitous, but that was precisely why God had given the devil permission to use witches as a harmful instrument. This explanatory model was rather complex and, arguably, counter-intuitive. But after 1560, it gained ground at the expense of explaining misfortune as solely and directly resulting from divine punishment. Wolfgang Behringer argues that the "Iron Century" disasters put the latter model under pressure. Ecological and socio-economic pressures were becoming so dramatic and inexplicable that solely referring to divine punishment no longer seemed sufficient. Nevertheless, the providential explanation of direct divine punishment remained a formidable challenge for the witch-hunting phenomenon. Even during its peak around 1600, many theologians in Württemberg continued to uphold the older tradition, which seems to have played a part in the relatively low number of burnings there.[260]

Then, there was a second dangerous competitor to witchcraft as an explanatory model. Like people today, early modern Europeans could interpret misfortunes like hailstorms, or illnesses as merely incidental or natural events. At a theological level, the idea of natural causes did not necessarily compete with divine intervention. In a sermon from 1602, the Lutheran German preacher Johann Georg Sigwart, for instance, accepted natural causes but hastened to add they were only secondary to the primal cause of God's judgment.[261] Sigwart presented natural causes as God's way of achieving his ends. Nevertheless, during the entire period of witch-hunting, natural causes often operated as a somewhat separate explanatory

[258] Alberum and Bidenbach, *Ein Summa etlicher Predigten von Hagel und Unholden.*

[259] Meder, *Acht Hexenpredigten*, 91; Bodin, *De la démonomanie des sorciers*, 270.

[260] Midelfort, *Witch Hunting in Southwestern Germany*, 45.

[261] Sigwart, *Ein Predigt Vom Reiffen und Gefröst*, 3–4.

model, posing a threat to the witchcraft doctrine. Take Johannes Kepler, who defended his mother against witchcraft accusations in a trial in Leonberg in 1620. This central figure of the scientific revolution stressed that his mother's superstitious accuser had mistaken the natural causes of her misfortune for an act of witchcraft.[262] A striking example of an uneducated person making a similar point stems from a late sixteenth-century record from the Dutch Peel region. A daughter of a woman who was convicted of witchcraft remembered a past event involving the death of a horse, whose stomach later turned out to be full of worms: "At the time, nobody spoke of sorcery. Now, people would say that it results from witchcraft, but I don't believe a thing of it. Not that I have much understanding of such matters, but if I may be so bold to say something about it: the illness and death of cattle, that's just nature."[263]

Kepler was not the only one to render certain witchcraft beliefs superstitious. However, this does not imply that the domain of superstition, as certain contemporaries already called it, was a safe haven for the witchcraft doctrine. Here, other competitors reared their heads. One alternative explanation for misfortune was provided by astrology.[264] The age-old belief that the position of the planets, the Sun, and the Moon affected events on Earth was common in this period. Proponents often saw astrology as an explanation for natural causation, and it competed with witchcraft beliefs for precisely that reason. After the Franconian hunts of the late 1620s, the physician Cornelius Pleier wrote that "if the witches are alleged to have caused the thunderstorm and the frost, then why were astrologists already able long before to predict their occurrence out of merely natural causes?"[265] Further competition was to be found at the level of "superstitious" beliefs of a more popular and folkloric kind. The fact that Celtic areas of the British Isles experienced relatively few trials has been linked to pervasive beliefs in fairies as agents of misfortune.[266]

As soon as competing explanations for misfortune had been shaken off, the witch-hunting phenomenon found other challenges in its path. Even if people saw witchcraft as the cause of their troubles, this did not automatically imply that persecution was the answer. People might, for instance,

[262] Rublack, *The Astronomer & the Witch*, 9, 248.

[263] Otten, *Duivelskwartier*, 187.

[264] Midelfort, *Witch Hunting in Southwestern Germany*, 34–35; Thomas, *Religion and the Decline of Magic*, 756–57.

[265] Pleier, *Malleus Judicum*, 9.

[266] Hutton, *The Witch*, 215–42.

resort to beneficent magic. In her study on Rothenburg ob der Tauber, a German town that witnessed only a few trials, Alison Rowlands observed that "as long as the system of beneficent magic retained its resilience, there was less reason for people to look to the law as a response to suspected witches."[267] Then, there was the possibility of deciding not to punish witches but to bring them to remorse and repentance to regain them for Christianity.[268] An influential theologian, Jesuit Adam Tanner, argued that the fight against Satan was not to be conducted by the burning of witches.[269] Tanner accepted that witches existed but thought that persecution was a hazardous affair that too quickly led to the deaths of the innocent. Witches should be won back for God, and eventually, it was up to God to judge them. At the village level, alternative ways of dealing with alleged witches existed. Accusations could threaten a community's social stability, and all sorts of rituals existed to avoid social escalation.[270] Only when such routine procedures had failed would the appeal to the forces of law become a more conceivable next step.

Even if people were convinced of a specific witch's guilt and favoured legal punishment, they could still decide not to undertake any action because of the personal risks. Initiating a trial was a way of making dangerous enemies among the accused and their allies.[271] Authorities could also respond harshly to what they saw as cases of defamation—think of the accuser in Esslingen who had his tongue cut off.[272] In addition, there seems to have been a certain awareness that witch persecution could create unpredictable dynamics that could spiral out of control to the detriment of friends, family, or even oneself. Legal persecutors potentially also suffered the consequences. They, too, could end up at the stake, as happened to Flade, or otherwise be punished for their miscarriages of justice. In 1601, the parliament of Paris sentenced an executioner from Rocroi to lifelong labour on the galleys because of gross abuses during trials.[273]

[267] Rowlands, *Witchcraft Narratives in Germany*, 75; Thomas, *Religion and the Decline of Magic*.

[268] Behringer, *Witches and Witch-Hunts*, 4.

[269] Dillinger, "Friedrich Spee und Adam Tanner," 48–52.

[270] Goodare, *The European Witch-Hunt*, 92–114; Rowlands, *Witchcraft Narratives in Germany*, 38.

[271] Briggs, *The Witches of Lorraine*, 383.

[272] Walz, *Hexenglaube und magische Kommunikation im Dorf der frühen Neuzeit*, 515; Briggs, *The Witches of Lorraine*, 176–78.

[273] Behringer, *Witches and Witch-Hunts*, 182.

If the support for witch persecution within the populace was explicit and substantial, accusations might still lead nowhere if authorities refused to proceed. Research into witch persecution focused on places where trials became big, but over the past decades, more attention has been paid to places where they remained small or absent.[274] Alison Rowlands's study on Rothenburg ob der Tauber is a case in point. Many accusations from below could have triggered panics there, like a woman who mentioned twenty-two names of sabbath accomplices.[275] Nevertheless, doubts about the trustworthiness of accusations, risks of harming the innocent, and fears of social havoc and damage made councillors end such dynamics quickly.

If authorities did decide to try witchcraft suspects, the further reproduction of trials met new hindrances. An acute threat was decent judicial procedures.[276] Take the illustrious city of Venice. The fact that the Venetian Republic failed to become a hotspot of witch burnings was not due to the absence of legal proceedings there; it was just that judges weighed the evidence carefully and usually rendered allegations unproven.[277] Another major problem was financial costs.[278] A hunt's initial religious fervour and communal solidarity could begin to wane as soon as the bills piled up, with nobody interested in paying them. There was often insufficient money or prison capacity to warrant continued persecution. As soon as the killing of alleged witches began, victims themselves also resisted.[279] This does not necessarily imply that these people were opposed to witch trials as such, but obviously, they were unlikely to favour their own execution. This resulted in tenacious refusals to confess, the mobilization of allies, escapes from prisons, or appeals to higher courts. Opposition could even extend to the executions themselves. During a burning in German Werdenfels, some women at the stake yelled to the audience: "Fellow pious women, flee over the mountains, for whoever falls into the hands of the tormentors faces severe torture and must die."[280]

[274] Schmidt, *Glaube und Skepsis*.

[275] Rowlands, *Witchcraft Narratives in Germany*, 56–57.

[276] Haas, *Hexen und Herrschaftspolitik*, 73–77.

[277] Seitz, *Witchcraft and Inquisition in Early Modern Venice*.

[278] Rummel, *Bauern, Herren und Hexen*, 193–202, 218; Labouvie, *Zauberei und Hexenwerk*, 127.

[279] Behringer, *Hexenverfolgung in Bayern*, 206–9; Briggs, *The Witches of Lorraine*, 162–63.

[280] Behringer, *Hexenverfolgung in Bayern*, 207–8.

Of course, there were also all sorts of other things people could do with their time. Julian Goodare observes, "People in early modern Europe had lives to get on with, and they did not usually regard dealing with the witchcraft threat as the most important thing in their lives."[281] One such alternative activity was other forms of persecution. Witch trials were a product and a beneficiary of the broader persecutory machinery of Christian Europe, but there are indications that forms of persecution also competed.[282] Cornelius Loos warned that witch trials were a distraction from the more urgent threat of Protestantism.[283] Some historians argue that the decline of trials against Waldensians in the fifteenth century or against Anabaptists in the second half of the sixteenth century created more room for witch-hunting. Moreover, warfare and popular revolt were generally much higher on the authorities' list of concerns.[284] We also need to think of all the other things people could do with their lives, like working for their daily bread, visiting taverns, or regularly attending church. That Eichstätt did not experience significant witch-hunting in the years before Von Westerstetten's appointment was probably related to his predecessor being more interested in things like gardening and humanism.[285]

Lastly, we must not underestimate the corroding effect of scepticism. People today often find it extraordinary what Europeans at the time believed about witchcraft, but contemporaries could express similar amazement.[286] The extravagant notions about sabbaths, flights, and metamorphoses were often considered contrary to nature and ridiculous. In a critique of witch-hunting named *Christliche Erinnerung* (Christian Reminder), the Lutheran theologian Johann Meyfart stated that even a child or a court jester would be ashamed of telling such things. "Where is reason? Where is people's sense?"[287] Advocates of witch-hunting were well aware of such scepticism and anticipated the critiques.[288] The Lutheran preacher David Meder wrote that witchcraft was "terrifying to speak about

[281] Goodare, *The European Witch-Hunt*, 259.

[282] Waite, *Eradicating the Devil's Minions*, 63–65, 113, 187; Goodare, *The European Witch-Hunt*, 158, 170.

[283] Eerden, "Cornelius Loos und die 'magia falsa,'" 156–60.

[284] Briggs, *Witches & Neighbours*, 342, 405–6.

[285] Durrant, *Witchcraft, Gender, and Society in Early Modern Germany*, 10–13.

[286] Duni, "Doubting Witchcraft," 224–25; Schmidt, *Glaube und Skepsis*, 308.

[287] Meyfart, *Christliche Erinnerung*, 160–1.

[288] See, for instance: Mirandola, *La sorcière*, 130–31; Rémy, *Demonolatry*, 74.

and to hear of and that it may appear unbelievable to many."[289] The defence of witchcraft theory could also damage one's reputation. Jean Bodin's intellectual stature was already recognized at the time, but interestingly, his career never developed beyond a middle-ranking position at a provincial subalternate court. Amongst other factors, it may have been related to misgivings about his uncompromising views on witchcraft.[290]

Now that we have learned about the many hurdles to reproduction that witch beliefs and trials encountered, it is obvious why witch-hunting did not, in Julian Goodare's words, "hurtle onwards until it had engulfed everyone in its path."[291] Early modern Europe was a competitive cultural jungle, with reproductive threats and hazards looming all around. But that may leave us even more in awe of how ingeniously the cumulative concept spread. By inflating people's fears and by efficiently evading its counter-forces, witchcraft concepts, and trials still killed tens of thousands of people. The key argument here is that these ingenious traits did not necessarily result from the shrewdness of the actors involved, but, to a significant extent, from a blind selective process amongst cultural variants. Moreover, witch-hunting was often highly damaging to the actors involved, being primarily adapted for the benefit of its own "selfish" reproduction.

Or did it? The preceding sections mainly focused on findings confirming this interpretation. But has the possibility of shrewd human agents operating behind the scenes not been dismissed too quickly? Earlier on, it was said that some German historians have not given up on the idea of witch persecutions fulfilling hidden goals or purposes. To them, the conclusions drawn here would indeed look short-sighted. Their arguments were not developed in response to this Darwinian model, as this model is new. But as we shall see, their viewpoints are still directly relevant. The interpretations of these historians are more well-thought-out than earlier arguments about hidden uses, and we shall now explore how they pose a challenge to the model.

[289] Meder, *Acht Hexenpredigten*, 50.

[290] Behringer, *Witches and Witch-Hunts*, 181–82; Krause, Martin, and MacPhail, "Introduction," 14–16, 44.

[291] Goodare, *The European Witch-Hunt*, 317.

CHAPTER 8

Reflections

8.1 Multifunctional Instrumentalization?

In his critique of witch-hunting, Cornelius Loos professed that trials had brought about "a new form of alchemy, in which gold and silver are made out of blood."[1] An essential motivation behind it all was the pursuit of money. Loos was not the only contemporary to express such suspicions. Trier's city chronicler Johan Linden lamented that the Trier persecutions had been supported by "many officials, who hoped to gain money and wealth from the burnings." Aldermen, hangmen, judges, notaries, scribes, and innkeepers had enriched themselves, and the executioner even "rode on a magnificent horse through the streets, dressed in gold and silver like a courtier, while his wife competed with noblewomen in the luxury of her clothing."[2] Friedrich Spee recounted how it was "discussed everywhere by many that the quickest and most convenient way of getting rich was through the burning of witches."[3]

In earlier sections, we saw how many historians have played down alleged ulterior motives behind witch-hunts. The pursuit of money is amongst the motivations they dismiss most eagerly, like Julian Goodare, who renders it "probably a modern fantasy." He explains: "The idea of

[1] Quoted in: Eerden, "Cornelius Loos und die 'magia falsa,'" 139.
[2] Zenz, *Die Taten der Trierer*, VII: 13–4.
[3] Spee, *Cautio criminalis oder rechtliches Bedenken wegen der Hexenprozesse*, 296, 257.

S. Hofhuis, *The Evolutionary History of Witch-hunting*,
https://doi.org/10.1007/978-3-031-66936-1_8

witch-hunting for money is attractive because it attributes to the witch-hunters a motive that is readily understood in the modern world."[4] The line of reasoning has become pervasive: witch-hunting seems so very bizarre to an audience today that we presume that it must have been about something else. Hence, we impose our "enlightened" frameworks upon the early modern period. However, if this is so, what should be made of the quotes of Loos, Linden, and Spee? Apparently, it made perfect sense within their early modern frameworks to assume that witch-hunts were used for material goals. So why should historians today refrain from considering this motive?

Some German historians have misgivings about the trope of witch-hunting as an end in itself. They believe that the idea of trials fulfilling hidden functions or being used for hidden purposes has been dismissed too radically, and they use contemporary comments to support that.[5] Take, for instance, a merchant from Cologne, Hermann von Weinsberg, who in 1589 expressed misgivings about the existence of witches and the validity of trials: "One could not get rid of old women and despised people more quickly than in this way."[6] Or take a study on trials in Mecklenburg by Katrin Moeller, who learned that many charges were dismissed because the accusers were already suspected of hatred, envy, and vengefulness.[7]

These German historians acknowledge the value of their colleagues' careful interpretation of subjective intentions.[8] We can better understand what happened by taking people's magical fears and beliefs seriously. But, they argue, it is precisely by interpreting the sources carefully that we can learn that not everything at the time was perceived as magical.[9] Early modern people knew that money could buy them things. They, too, wanted to get rid of enemies, were eager to advance socially, and rulers liked power. One does not need an enlightenment to recognize that. Both proponents and opponents of witch-hunting already understood that accusations and trials were instrumentalized for such ends. Supporters did not profess their ulterior motives in print because that would have subverted their strategy.

[4] Goodare, *The European Witch-Hunt*, 385.

[5] Rummel and Voltmer, *Hexen und Hexenverfolgung*, 99; Voltmer, "Die hexenverfolgungen im Raum Trier," 741–45; Moeller, *Dass Willkür über Recht ginge*, 251–52; Rummel, "Das 'ungestüme Umherlaufen der Untertanen'," 154, 161.

[6] Rummel, "Das 'ungestüme Umherlaufen der Untertanen'," 154.

[7] Moeller, *Dass Willkür über Recht ginge*, 296.

[8] Groß, *Hexerei in Minden*, 22–23; Schmidt, "Ein Politisches Ausrottungsprogramm?," 147.

[9] Voltmer, "Die politischen Funktionen der frühneuzeitlichen Hexenverfolgungen," 96–98.

But if we look at circumstantial evidence and the comments of contemporary critics, it becomes apparent that official motivations were not the whole story. Rita Voltmer states that it was often less about the honour of God or the protection of the community than about "power, greed, bribery, envy, and revenge."[10]

This is not to say that these historians return to the kind of grand, sweeping explanations that finally aim to crack the nut.[11] Neither do these historians presume that witch-hunts were usually conducted in a consciously cynical manner, although such cases can indeed be found.[12] What they also admit is that accusations sometimes escalated into panics that became largely dysfunctional for all and everyone.[13] However, such dysfunctionality is seen as the exception since belief in witches usually amalgamated with concrete and opportunistic purposes. Mundane uses can be found at all levels, by all participants, and at all stages of trials. Instead of a monocausal or monofunctional explanation, these experts see witch-hunting as a multifunctional phenomenon.

Some early versions of this German approach were still close to old-school functionalism, presuming that witch trials fulfilled "latent functions" for society.[14] More recent supporters of the idea of multifunctional instrumentalization, however, noted that this has become unfashionable. Authors like Voltmer, Rummel, and Barbara Groß thus emphasize competent actors as the driving force behind beliefs and persecutions.[15] They argue that in local circumstances, witch-hunting could be an attractive course of action due to religious, political, social, and economic factors. Mundane motivations played their part at both conscious and subconscious levels, and Groß links the approach to the sociology of Pierre Bourdieu. Action is "pre-structured by largely unconscious, incorporated perception, thought and action schemes, that serve the actor's orientation

[10] Voltmer, "Von der besonderen Alchimie, aus Menschenblut Gold zu machen," 131.

[11] Schmidt, "Ein Politisches Ausrottungsprogramm?," 147.

[12] Voltmer, *Hexen. Wissen was stimmt*, 83–84; Rummel, "Vom Umgang mit Hexen und Hexerei," 100.

[13] Walz, *Hexenglaube und magische Kommunikation im Dorf der frühen Neuzeit*, 52, 515; Voltmer, "Von der besonderen Alchimie, aus Menschenblut Gold zu machen," 132.

[14] Walz, *Hexenglaube und magische Kommunikation im Dorf der frühen Neuzeit*; Labouvie, *Zauberei und Hexenwerk*.

[15] Voltmer, "Die politischen Funktionen der frühneuzeitlichen Hexenverfolgungen," 94–96; Groß, *Hexerei in Minden*, 24–34.

within their social world."[16] In contrast to the "witch panic" narrative of the "genuine belief historians," these authors stress that trials were usually created and fostered by individuals who weighed their options and (sub) consciously knew what they were doing.[17] While accepting that witch trials often damaged the social fabric, the other two angles on the *cui bono* question—power perspectives and individual needs—are developed in all the more detail.[18] There is a focus on "concrete people with concrete goals," who "always tried to take advantage of opportunities to improve their situation."[19]

Rummel and Voltmer distinguish three main categories of instrumentalization.[20] The first concerns "social use" in the form of bottom-up activism of the populace. Rummel especially examined this dimension in studies on *Hexenausschüsse* in the Trier and Moselle region. He argues that for ordinary people, witch-hunting could be an attractive tool to get rid of particular neighbours or to be used as a weapon against higher classes. Persecutions offered a "new means of attack," "compensation for social inferiority," and "an extraordinary seizure of power beyond their traditional means of influence."[21] Take a witch-hunt in the town of Cochem in the 1590s.[22] Under the disguise of a witch trial, Rummel discerns a peasant and artisan revolt against the ruling elite of merchants. A *Hexenausschuss* installed itself in the town hall and helped create an atmosphere of pogrom-like attacks on women from the establishment. In what was a dramatic loss of their honour, these women were publicly scolded for witchcraft and had their veils torn from their heads. Later on, witch-hunters entered their houses and dragged these women by their hair through the streets to the castle, where they were tortured.

The second category, especially prominent in Rita Voltmer's work, concerns the top-down, "stately-political use" of witch-hunting by rulers.

[16] Groß, *Hexerei in Minden*, 27–8.

[17] Rummel and Voltmer, *Hexen und Hexenverfolgung*, 98; Voltmer, "Herren und Hexen," 13, 133; Groß, *Hexerei in Minden*.

[18] Voltmer, "Von der besonderen Alchimie, aus Menschenblut Gold zu machen," 132–34.

[19] Voltmer, *Hexen. Wissen was stimmt*, 85; Rummel, "Vom Umgang mit Hexen und Hexerei," 80.

[20] Rummel and Voltmer, *Hexen und Hexenverfolgung*, 99–113.

[21] Rummel, "Das 'ungestüme Umherlaufen der Untertanen'," 154; Rummel, "So mögte auch eine darzu kommen, so mich belädiget," 223; Rummel and Voltmer, *Hexen und Hexenverfolgung*, 102.

[22] Rummel, "Vom Umgang mit Hexen und Hexerei," 97.

Among the alleged underlying purposes of witch-hunting are disciplining the population, destroying political rivals, and "demonstrating power" as compensation for internal and external weaknesses.[23] Witch trials in politically turbulent areas were used to strengthen judicial-political claims through the creation of judicial precedents. When the populace called for witch-hunting, rulers often calculated whether or not to use this powerful symbolic tool for their political ends.

Rummel and Voltmer´s third category is "use by specialists." Especially the late sixteenth- and early-seventeenth-century persecutions offered professional opportunities. For many witch-hunters, it was a "career-strategic stroke of luck" that offered new work, new income, extensive dining at other people's costs, and the conviviality of a joint mission.[24] This is also why trials increasingly targeted the rich: their money ensured further income.[25] The dynamics of persecution could give these professionals sudden and sensational power. Multifunctionalism pays little attention to the oppression of women as an underlying aim, but Rummel and Voltmer still stress that women could be targeted strategically because of their weaker societal position.[26] For instance, an attack on the wives of rich men could be a means of taking revenge on husbands who were more difficult to touch by law.

The earlier discussions of several German persecutions mainly focused on genuine panic and belief. However, their events also offer room for instrumentalist interpretations. Dietrich Flade looks like the preeminent person people might want to get rid of, if only because they owed him debts.[27] The antielitist sentiments of the Trier trials remind us of Rummel's "compensation for social inferiority." Voltmer also stresses the role of the Jesuits in Trier and how they may well have staged and manipulated the boys' confessions to enhance their role in the fight against heretics. There was not a secret master plan, but Voltmer still assumes that some of her colleagues speak too naively of the "voluntary" child confessions.[28]

As a showcase of "stately-political use," Rita Voltmer highlights the hunt of St. Maximin, the imperial abbey close to Trier. Again, Voltmer accepts that all layers of the population could be afraid of witches. Yet,

[23] Voltmer, "Herren und Hexen," 12.
[24] Rummel, "So mögte auch eine darzu kommen, so mich belädiget," 218.
[25] Rummel, *Bauern, Herren und Hexen*, 315–17.
[26] Rummel and Voltmer, *Hexen und Hexenverfolgung*, 97.
[27] Voltmer, "Zwischen Herrschaftskrise, Wirtschaftsdepression und Jesuitenpropaganda," 86.
[28] Voltmer, "Jesuiten und Kinderhexen," 208–17, 230.

while the abbot was probably "filled with fear," she thinks the hunt "was more important for him as a means of securing his rulership."[29] Internally, trials were used to discipline the population, especially defiant tenants and deceitful landlords, and to demonstrate "governmental qualities like peace-keeping and protection."[30] Externally, the imperial abbey was entangled in conflicts over judicial and political sovereignty, and witch trials were a means of signifying and solidifying its autonomy. Voltmer argues that it is not by accident that St. Maximin's trial records have been preserved carefully: authorities hoped to use them on later occasions as proof of the legality of their rule. She additionally highlights that one of the key persecutors, Claudius Musiel, enlarged his lands at the expense of executed landowners.[31]

Or take the Bamberg hunt. When discussing "use by specialists," Rummel and Voltmer mention Ernst Vasoldt as an obvious example.[32] Voltmer thinks that the vast persecutions of the prince-bishoprics also fulfilled other purposes, like "the disciplining of the villages of their rural hinterlands as well as the urban elite, who showed resistance to the witch-hunting campaign." These trials could serve "an elite disciplinary program all too effectively."[33] Others similarly interpret the attack on Bamberg's urban elite as eliminating a rival political faction.[34] Chancellor Georg Haan was a desirable target due to envy and resentment regarding his position and the wealth that could be used as income. Britta Gehm, who studied the Bamberg hunt in most detail, shows little interest in instrumentalization, but when discussing the trial of Haan, she notices that his torture was continued even after he had made confessions that legitimated a death sentence. "One cannot help but get the impression that the investigators and torturers enjoyed seeing this once high-standing and powerful man on the floor."[35] The hunt of Ellwangen has not received attention from the instrumentalists, and its two key studies by Erik Midelfort and Wolfgang Mährle very much argue for genuine beliefs. But in a later publication, Mährle seems affected by the instrumentalist school, saying that

[29] Voltmer, "Von der besonderen Alchimie, aus Menschenblut Gold zu machen," 136.

[30] Voltmer, "Die politischen Funktionen der frühneuzeitlichen Hexenverfolgungen," 98.

[31] Voltmer, "Claudius Musiel oder die Karriere eines Hexenrichters," 254.

[32] Rummel and Voltmer, *Hexen und Hexenverfolgung*, 111.

[33] Voltmer, "Witchcraft in the City," 167.

[34] Flurschütz da Cruz, *Hexenbrenner, Seelenretter*, 49; Knefelkamp, "Von der Ketzerei zur Hexerei," 30–31.

[35] Gehm, *Die Hexenverfolgung im Hochstift Bamberg*, 136.

the momentum of the persecution was "largely based on material interests" and "very quickly became a career instrument for the councillors."[36]

This model of multifunctional use should be taken more seriously than earlier attempts to uncover hidden aims, which were relatively easy to dismiss. All the idiographic historians needed to do was highlight the diversity and complexity of the phenomena. However, such objections are less effective against the German model of multifunctional instrumentalization. By presuming that accusations and trials fulfilled a different set of purposes in each instance, this model also has an idiographic twist. Moreover, Harris and some feminist authors hardly studied original historical documents, but the instrumentalists are as familiar with archival sources as any historian can hope to be. The instrumentalists also use some of the rhetorical weapons of the "genuine belief historians" against their trope of witch-hunting as an end in itself. By resorting to a "generalized" and "omnipresent, all-explanatory fear of witches," it is those "genuine belief historians" who are indicted for presenting a one-dimensional explanation.[37] The instrumentalists claim that *their* approach is the richer and more multifaceted one.

Not everyone is convinced, though, and some "genuine belief historians" have responded.[38] Notably, nobody has argued that instrumentalization *never* occurred. Dillinger accepts that Claudius Musiel in St. Maximin gained more land through trials, and Briggs considers the option of people being "sincere, however self-interested." Wolfgang Behringer offers some room for ulterior motives like revenge, political factionalism, money, or common dining.[39] Yet, what raises hackles about the instrumentalist interpretation is that it takes things very far and underestimates the power of belief. While accepting that other interests played a role, Gerd Schwerhoff finds the term "instrumentalization" too much of a modern figure of thought, suggesting that belief was used as a pretext.[40] The most outspoken critic is Johannes Dillinger, whose fundamental problem is evidence.[41]

[36] Mährle, "Fürstpropstei Ellwangen," 382, 384.

[37] Moeller, *Dass Willkür über Recht ginge*, 467; Voltmer, "Die politischen Funktionen der frühneuzeitlichen Hexenverfolgungen," 98.

[38] Tschaikner, "Nutzung oder Instrumentalisierung?"

[39] Dillinger, *Böse Leute*, 212–30, 316–17, 444; Briggs, *The Witches of Lorraine*, 374; Behringer, *Hexenverfolgung in Bayern*, 135–36; Behringer, *Witches and Witch-Hunts*, 229–30.

[40] Schwerhoff, "Hexerei, Geslecht und Regionalgeschichte."

[41] Dillinger, "Richter als Angeklagte," 167–69.

Because how does one *prove* that witch trials were used for an ulterior purpose? He writes:

> The only unquestionable evidence for the instrumentalization hypothesis would consist of an accuser's statement showing his true colours. However, this would imply an open admission of judicial abuse, which would not be in the accuser's interests and will thus not be found in the sources.[42]

The instrumentalists postulate a hidden agenda, which, by its very nature, cannot be proven in that way precisely because it was hidden. According to Dillinger, this creates a "methodologically inadmissible circular argument."[43] The attribution of hidden motives requires an enormous burden of proof, and Dillinger contends that the instrumentalists fail to deliver—he, for instance, denies that financial gain was of much importance in Trier.[44]

In addition, he argues that the instrumentalist model appears superfluous.[45] In his own "evil people paradigm" (see Sect. 6.5), the connection of witchcraft accusations to other conflicts is understood as an expression of genuine belief: witches were a part of people's perceived reality and also of a social reality in which discernibly evil individuals seemed likely suspects. Dillinger further stresses the supremacy of belief by highlighting how actions at all stages of witch trials, from shaving suspects and changing their clothing to their torture and burning, stood in a magical light.[46] Or think of the population of Elz, which in 1589 threatened their ruler to leave the town themselves if he would not persecute witches. "That people even considered such a step shows that they felt existentially threatened," he notes, illustrating how belief could override people's normal calculations.[47]

In turn, the instrumentalists regard such interpretations as too gullible. Historians today should question the rhetoric and self-legitimizations of witch-hunt proponents, not take them at face value. In Jürgen Michael Schmidt's view, there is more of a level playing field. Statement of belief

[42] Dillinger, *Hexen und Magie*, 127.
[43] Dillinger, *Böse Leute*, 230.
[44] Dillinger, 230, 341.
[45] Dillinger, *Hexen und Magie*, 127.
[46] Dillinger, "Das magische Gericht"; Dillinger, "The Political Aspects of the German Witch Hunts," 67.
[47] Dillinger, *Böse Leute*, 241.

may also be strategic, so "if we cannot retrieve the truth, this applies to both sides: the statements of the victims and their defenders should be weighed equally to those of the persecutors."[48] The differences between these two camps are not always as significant as they may first appear. Both sides accept that there was genuine belief, that instrumentalization was possible, and that witchcraft accusations and other conflicts were interrelated. Eventually, it is a question about where to put the emphasis. The lack of self-reflexive comments and autobiographical records of witch-hunters also makes historical analysis difficult.[49] It has made the debate somewhat of a stalemate, and some historians move towards an agnostic position.[50] Writing about the arrest of the Haan family in Bamberg, Günther Dippold comments that:

> In this example, we may recognize the limits of our understanding. Was the fate of Haan and his family a coldly calculated judicial murder, or was it sincerely believed that a person who stridently opposed witch-hunting had to be in league with the devil? In any case, it is not even said that the individuals who aimed to destroy the Haan family all thought the same.[51]

This debate over alleged instrumentalization is mostly a debate amongst historians, but obviously, it has interdisciplinary ramifications. The instrumentalist's claims about human behaviour and culture do not look far-fetched from an interdisciplinary point of view. Barbara Groß has justifiably linked the approach to Pierre Bourdieu, who also argued that people continually defend their hierarchical position or try to improve it, with symbolic culture being a key weapon on the battlefield. Psychologists and cognitive scientists use the concept of "motivated reasoning."[52] It assumes that genuine intentions can drive behaviour at a conscious level. Still, such intentions are crucially influenced by hidden emotions and desires that make us believe what we want to believe. So, witch-hunt proponents may well have meant what they said, but they meant it precisely *because* it served their interests. Recently, the French cognitive scientist Hugo Mercier used evolutionary psychology to draw similar conclusions.[53]

[48] Schmidt, "Ein Politisches Ausrottungsprogramm?," 149.
[49] Briggs, *Witches & Neighbours*, 196.
[50] Walinski-Kiehl, "Witch-Hunting and State Building," 256.
[51] Dippold, "Die 'Hexenpolitik' Der Bamberg Fürstbischöfe," 57.
[52] Kunda, "The Case for Motivated Reasoning"; Norman, *Mental Immunity*, 219.
[53] Mercier, *Not Born Yesterday*.

Mercier maintains that evolution has designed our brains to be naturally vigilant, protecting us against ideas that harm our interests. In Mercier's view, many seemingly senseless acts—motivated by supernatural belief systems—result from more convenient "down-to-earth considerations."[54] Amongst his examples is witch-hunting, and he argues that behind the disguise of witchcraft beliefs, "the usual self-interested motives rear their ugly heads." Mercier mainly focuses on witch persecution as a means to get rid of undesired individuals, and he renders the content of beliefs as of only marginal importance.[55]

What bearing does the model of multifunctional instrumentalization have on the proposed Darwinian scenario? The idea of multifunctionalism dovetails with a previous argument about Darwinian cultural evolution being a process in which cultural variants may survive due to different selection pressures in each instance. Both approaches thus integrate the advantages of eclecticism. Neither has it been said that blind evolution and (sub)conscious human design exclude each other as explanatory models of how cultural adaptations arise. Still, the more explanatory room there is for competent agents who (sub)consciously understood the workings of witch-hunting, the less room is left for blind evolution as a force of sociocultural design. But especially regarding the *cui bono* question, the instrumentalists have reached quite different conclusions. While none of the proposed hidden functions or aims can explain witch-hunting on its own, it is argued that we get much closer to a sufficient explanation when we throw them all together. This interpretation conflicts with the idea of witch beliefs and trials spreading at the expense of their human hosts and their societies. It would make any "selfish meme" explanation largely redundant.

However, the instrumentalist approach is not without problems. If the people accountable for witch-hunts in most cases (sub)consciously knew what they were doing, how to explain the haphazard and unpredictable course of so many trials? Why were initiators of persecution so often harmed themselves? This still suggests genuine and escalating panics without anyone being in control rather than people shrewdly pursuing their hidden goals. The idea of multifunctional instrumentalization also raises the question of why hunts were not even *more* common. For Rita Voltmer, the absence of trials is also instrumentalization because people weighed

[54] Mercier, 120.
[55] Mercier, 189, 202.

their chances and concluded that it would best suit their interests *not* to persecute.[56] But if accusations and persecutions were indeed such wonderfully effective instruments, we might expect them to have been more frequent. The correlation (see Chap. 7) between the precise content of beliefs and the way trials unfolded also strengthens the case of genuine belief. Specific convictions about sabbaths, child witches, or rich witches were not only relevant at the margins but absolutely essential. The specificity of witch beliefs still offers the most plausible explanation for the specificity of trials, while factors like greed, envy, or power play still appear somewhat too general and universal.

A closer look at the case studies reveals further problems, as some examples from Voltmer's work may illustrate. How strong is her evidence that persecutors in St. Maximin used trials for hidden purposes? In one publication, she says that trial records "prove" that disciplining defiant subjects was an underlying aim, but this statement contains no reference.[57] Her claim that persecutors cared about safeguarding the principality's judicial rights looks plausible, but does that prove this was a hidden aim? The causal connection might have been different: persecutors cared deeply about eradicating witchcraft and were thus extra determined to solidify the legal powers that made it possible. Overall, killing four to five hundred people—one-fifth of the abbey's population—looks like a bizarrely lopsided way of safeguarding judicial rights. Even if we add things like the legitimization of power or the use by specialists, it all still looks strangely out of proportion. However, if we assume that both the people and the authorities genuinely believed themselves to be involved in an apocalyptic life-or-death struggle with a sect of Satan-worshipping witches living in their midst, the scale of the persecution finally begins to look more plausible.

Or take Bamberg. Voltmer's claim that hunts in the prince-bishoprics could serve "an elite disciplinary program all too effectively" appears strange, considering all the damage trials caused. The prince-bishop and his clique even had to flee Bamberg because no money was left for defence against an approaching army. Elsewhere, Voltmer refers to Reutlingen. This southern German town experienced long trial-free periods punctuated by a few middle-sized persecutions. According to Voltmer, "recent research has found a ready explanation": generations of rulers used the

[56] Voltmer, "Hexenverfolgung und Herrschaftspraxis," 17.

[57] Voltmer, "Von der besonderen Alchimie, aus Menschenblut Gold zu machen," 136.

trials to establish their power, and persecution was thus "characterized by the influence of personal interests to gain a successful political career."[58] Voltmer's reference is a study by Thomas Fritz, but Fritz, in fact, assumes "a tremendous impact of witch belief" and a populace that recurrently demanded persecution.[59] When new and inexperienced authorities took their position, they condoned trials in a state of self-doubt and undecidedness. When the establishment then found out what incredible damage it all caused, they stopped persecution. "Through the experience and all the harm, they became smarter, never to allow trials again."[60] Only a few decades later, when this experience was forgotten, the mechanism started all over again.

Another case is Rottweil, where trials mainly targeted the poor, like vagrants and beggars. Rita Voltmer: "Obviously, this policy followed the general aims of early modern city government: to purify the city from vagrants, to discipline the poor, to demonstrate power both in the territory and in the inner city and thus, to create an omnipresent authoritative urban regime."[61] Her reference is a study by Mario Zeck, but again, her source says something else. Zeck explicitly dismisses an instrumentalist interpretation, saying that it "completely underestimates" the power of belief.[62] Instead, his argument is entirely in line with Dillinger's "evil people-paradigm." On another occasion, Voltmer mentions Oettingen as a case of stately political use, and here, Alexandra Haas is the source.[63] But like Zeck, Haas rejects instrumentalism, as she finds ideological and moral motivations more plausible. Haas acknowledges that rulers sometimes gained money or territory after a trial but sees this as a consequence rather than a cause; rulers wanted compensation for their expenses.[64]

While much of what the instrumentalists bring forward may sound plausible, Dillinger is right to address problems regarding evidence. The response about the level playing field is not wholly convincing either. Witch-hunt proponents consistently said that they acted to eradicate witchcraft. Hence, when historians postulate this motivation, they can be confident they are not making it up. But regarding the alleged

[58] Voltmer, "Witchcraft in the City," 162.

[59] Fritz, "Hexenverfolgungen in der Reichsstadt Reutlingen," 206.

[60] Fritz, 302.

[61] Voltmer, "Witchcraft in the City," 163–64.

[62] Zeck, *Im Rauch gehn Himmel geschüggt*, 70–3.

[63] Voltmer, "Herren und Hexen," 38–39.

[64] Haas, *Hexen und Herrschaftspolitik*, 98–99, 111, 218.

instrumentalist motivations, one cannot be so sure of that. When the poor were targeted, one might presume that people wanted to get rid of the poor. When the rich were targeted, one might assume that people wanted to get rid of the rich. When the middle classes were targeted, one might presume that rulers wanted to discipline the population and show them their power. One can always think of some instrumentalist goal, no matter what persecution looked like. But when explanations are that flexible and not necessarily substantiated by quotes from the sources, it becomes easy to make mistakes. It also gives the idea of multifunctional instrumentalization a ring of unfalsifiability. What findings would ***disconfirm*** their model?

Then, there is the problem of the "design" of the cumulative concept of witchcraft. The instrumentalists also admire how ingeniously the concept was adapted to serve its alleged many purposes.[65] Rummel emphasizes the elasticity of the beliefs, which created an "almost inexhaustible potential for applying it to practically every person, every behaviour and every problem."[66] The "crimen obscurum" notion further evaded the need for controllable proof. But the question staring us in the face again is: how did the concept develop these capacities? Most instrumentalists highlight competent actors who knew what they were doing, but were these actors also talented enough to design the concept that cleverly? The instrumentalists leave the issue unaddressed.

This makes the problem we posed earlier even more urgent. Did actors understand what historians understand today, namely how beliefs about witchcraft reputations, sabbaths, flights, torture, rich witches, and child witches enhanced the reproduction of trials? From the perspective of the instrumentalists, we might expect witch-hunt proponents to have understood this and elaborated upon such notions for that reason. Also, we might assume that they (sub)consciously grasped how these notions served their hidden interests. From the perspective of blind evolution, on the other hand, we might expect that the actors were not necessarily cognizant of the beliefs' enhancing effects on trials. But as the instrumentalists rightly point out, we should not only look at the narratives of proponents, as they may have misled their audiences about their true aims. We also need to look at contemporary critics. We shall now delve more deeply into what primary sources may reveal.

[65] Groß, *Hexerei in Minden*, 330; Rummel, "Vom Umgang mit Hexen und Hexerei," 208.
[66] Rummel, "So mögte auch eine darzu kommen, so mich belädiget," 225.

8.2 What Proponents Said

If anyone fits Rummel's and Voltmer's category of a witch-hunt specialist who deliberately enhanced persecution, it was Heinrich Kramer, key author of the *Malleus Maleficarum*. This Dominican cleric had already made a career in the Inquisition, and his personality matched his profession—Kramer was hard-working, belligerent, self-righteous, and unafraid of making enemies.[67] So, it was hardly surprising that the topic of diabolical witchcraft attracted his attention when trials resurged in the 1480s. Kramer began preaching against witches in Ravensburg, which helped initiate a persecution there that killed eight women. Later, he repeated his efforts in Innsbruck, but authorities ordered Kramer to leave the city this time. The bishop of Innsbruck wrote to a friend that the inquisitor had "clearly demonstrated his foolishness" by presuming "much that had not been proved."[68] This failure only made Kramer more determined and he vindictively wrote his *Malleus Maleficarum*. With the academic and inquisitor Jacob Sprenger as a possible co-author, he urged the world to eradicate diabolical witches, whose numbers were alleged to have grown enormously.

The authors assumed the apocalypse was imminent, so the work was hastily written.[69] Still, it became quite a bestseller by the standards of the time and did have some of the desired effects of putting witchcraft on the agenda as a central issue in the struggle between Satan and humankind. Historical scholars today also note that the book combined some ideas about witchcraft in a way that made the concept more inflammatory.[70] Previously, the diabolical witches' identity had remained quite abstract, but the *Malleus* linked it more effectively to people who were well-known at a popular level: village witches. Since such women could readily be found throughout Europe, it created much potential for further trials. The book also tapped more thoroughly into misogynous stereotypes and offered helpful guidance on how to persecute: Kramer and Sprenger recognized the importance of torture to extract confessions. From our

[67] Herzig, "The Bestselling Demonologist"; Broedel, *The "Malleus Maleficarum" and the Construction of Witchcraft*, 12–19.

[68] Broedel, *The "Malleus Maleficarum" and the Construction of Witchcraft*, 17.

[69] Behringer and Jerouscheck, "'Das unheilvollste Buch der Weltliteratur'?," 68.

[70] Goodare, *The European Witch-Hunt*, 49–52; Herzig, "Flies, Heretics, and the Gendering of Witchcraft"; Broedel, *The "Malleus Maleficarum" and the Construction of Witchcraft*, 59–60, 100–101.

theoretical perspective, this does not look like a blind evolutionary process silently doing its adaptive work. In Kramer, we find a flesh and blood human actor with particular aims and interests who deliberately broadcasted witchcraft notions to encourage persecution.

But was the book really that shrewdly designed? Until 1560, witch-hunting did not grow much, and a closer look at the *Malleus*' content suggests that this content was a part of the problem. One of the notions most conducive to the spread of witch-hunting was the witches' sabbath, and it plausibly would have helped Kramer's and Sprenger's efforts to discuss this incendiary notion in detail. But strikingly, the *Malleus* hardly mentioned the idea, so why this relative absence? The American historian Hans Peter Broedel presumes "calculated reservations."[71] Western Alpine ideas about witches' gatherings included both men and women with divergent social profiles, while Kramer and Sprenger wanted to focus almost exclusively on poor women. Allegedly, this made the sabbath unhelpful for their argument. Robin Briggs thinks that "the rigid scholastic format of the work somehow prevents them putting any emphasis on this," while Julian Goodare maintains that the omission was, in fact, not "a weakness in the argument, but (…) arguably a strength" as it cleared the ground for a more thorough focus on village witchcraft.[72] Christopher Mackay argues that the sabbath was "simply not relevant to the purpose of the Malleus," which was to provide a general handbook on the reality of witchcraft and how to eradicate it.[73] We might thus reconsider the relative omission as an intentional or a functional trait of the work.

Such interpretations can be questioned, though. The book was lengthy, containing hundreds of pages, so why did that not leave some room for discussing the sabbath next to other topics? Why should ideas about sabbaths and female village witches be mutually exclusive? Moreover, if the authors had calculated reasons for not making the sabbath a part of their witch image, why did they still casually refer to it occasionally? Closer analysis shows they mentioned witches' meetings when discussing something else.[74] In a discussion on witches' cannibalism of babies, for instance, they recount a case in Italian Como where a man had lost a child from its crib and "went in search of it, and when he saw a gathering of women at

[71] Broedel, *The "Malleus Maleficarum" and the Construction of Witchcraft*, 130.
[72] Briggs, *Witches & Neighbours*, 34; Goodare, *The European Witch-Hunt*, 50.
[73] Institoris, *The Hammer of Witches*, 283.
[74] Institoris, 283–84, 360.

night time, he observed that the baby was being killed and eaten while liquor was being consumed."[75] Such casual references to collective witches' meetings suggest that they had no deliberate reasons for playing this notion down. The question still stands: why did the *Malleus* not discuss witches' assemblies more prominently?

Perhaps we should consider a more parsimonious and straightforward explanation: the sabbath was not neglected deliberately, but just because the authors did not recognize its potential for boosting witch trials. Considering the late medieval German context, this is not implausible. We have already learned that the sabbath notion did not connect well with indigenous ideas about witchcraft, so it was pretty evident for Kramer and Sprenger to think of witches as acting alone or in small numbers. Other clues from the book further indicate that they did not see the potential of chains of accusations. It occasionally argues that witches lure other women into witchcraft, which further shows that Kramer and Sprenger accepted the idea of witches knowing each other.[76] However, the authors hardly capitalized on this notion for persecutory purposes. Only a few sentences passingly note that witches can denounce other witches.[77] While a long section on torture demonstrates recognition of its potential for extracting confessions, the section fails to connect torture to making suspects name accomplices.[78] One passage in this section that *does* discuss how witches knowing each other might aid persecution only offers the less auspicious advice of putting accomplices in the same cell, so that spies standing outside those cells can overhear them to learn what crimes they committed.[79] Hence, the authors of the *Malleus* plausibly neglected some of the best recipes for making witch-hunting prosper simply because they did not understand those recipes.

This apparent lack of comprehension of how essential cumulative notions could operate in combination does not seem to have been exceptional. Section 6.1 discussed how early ideas about witchcraft often remained patchy compared to the period after 1560. Like many others, Nider failed to connect sabbaths to flights, and the Strasbourg preacher

[75] Institoris, 212.
[76] Institoris, 275, 278–81.
[77] Institoris, 511, 577.
[78] Institoris, 541–58.
[79] Institoris, 558.

Geiler von Kaysersberg was ambiguous about flights and sabbaths and did not think of witchcraft in terms of a collective super crime. The mere fact that it took so long to combine various notions of witchcraft into the more robust and "virulent" concepts of the late sixteenth century indicates that early proponents did not recognize how to arrange things more effectively.

Someone who did combine some of the more inflammatory elements was the Italian count Gianfrancesco Pico della Mirandola.[80] After involvement in persecution in his principality, in 1523, he published a dialogue on witchcraft that mentioned both harmful magic, flights, and torture and even referred to a sabbath as "a very vast multitude."[81] Yet, whether this also implies much understanding of how those insights could trigger a big hunt is questionable. The dialogue was called *Strix, sive de lucificatione daemonum* (The Witch, or on the Deceptions of Demons), and its structure and content suggest that the primary purpose was to show that evil demons had been deceiving and tricking people since the beginning of antiquity, through oracles, images, and calumnies.[82] Enhancing witch-hunting does not seem to have been the book's essential aim since it did not provide much guidance on how to persecute.[83]

From 1560 onwards, cumulative notions began to stimulate trials more effectively, so we might expect more explicit wisdom on how to get trials going. But did proponents in this period indeed offer such insight? Take Lambert Daneau's *Les sorciers* from 1574. His dialogue was written in plain and accessible language and discussed ideas that influenced the rise of witch-hunting. Daneau repeatedly stressed that flights and sabbaths were actual events, noting that witches could also be found amongst people of high standing, and the young.[84] He further warns about harmful magic, and a remark about a hunt in Savoy that killed eighty people in a single village shows his interest in extensive persecution.[85] Yet again, the tract offered little practical instruction and failed to capitalize on essential incendiary elements: Daneau said nothing about the potential of torture, and neither did he discuss the potential of young witnesses. Moreover, as

[80] Mirandola, *La sorcière*; Stephens, "The Skeptical Tradition," 114–16; Behringer, "Walter Stephens. Demon Lovers," 89–94.

[81] Mirandola, *La sorcière*, 200.

[82] Mirandola, 217–18.

[83] Stephens, *Demon Lovers*, 94.

[84] Daneau, *Les Sorciers*, 48–49, 78–79, 85–90.

[85] Daneau, 7.

far as he offered concrete advice, it was not always practical. For instance, he recommended giving suspects some freedom so that judges could regularly check on them by surprise to catch them in the act of talking to demons.[86]

The lack of proof for such understanding does not imply that understanding was absent. Perhaps Daneau did comprehend how the sabbath notion generated chains of accusations, but he just did not mention it. However, that a mere discussion of witches' meetings does not imply such comprehension can be illustrated through the trial records from Rostock in 1584.[87] The documents of this German city routinely referred to sabbaths on the Blocksberg that were attended by three hundred witches who feasted and interacted with demons. This might suggest that persecutors cunningly knew how to instrumentalize the sabbath notion for furthering trials. But closer scrutiny suggests otherwise. The Austrian historian Müller observed that sabbath notions played no further role in the interrogations whatsoever. Investigators thought of harmful magic as being inflicted by solitary witches and, in some cases, by small collectives.

That early modern sources do not necessarily provide clues for a lack of understanding is shown by the most important witchcraft treatise from this period, Jean Bodin's *De la démonomanie des sorciers* from 1580. The French thinker repeatedly stressed the reality of sabbaths and flights over long distances to get there. He also went into quite some detail about what happened at assemblies and emphasized the attendance of countless witches.[88] Bodin then coupled such notions with promising persecutory practices. He repeatedly underlined that witches knowing each other from sabbaths offered the opportunity to find innumerable accomplices through painful questioning of witches and that they "need not be arrested according to the law."[89]

Bodin also commented insightfully on another promising investigatory opening, the child witches.[90] Regarding the enhancement of witch-hunting, his argument did contain a weakness: Bodin almost exclusively focused on the daughters of witches, mainly leaving out boys and children without witch parents. This limited his argument's persecutory potential.

[86] Daneau, 113.
[87] Müller, "Elaborated Concepts of Witchcraft?"
[88] Bodin, *De la démonomanie des sorciers*, 60–62, 72–73, 223–34.
[89] Bodin, 358, 376, 384, 355.
[90] Bodin, 262, 375, 396, 428.

But Bodin's discussion of *how* those daughters could be used in trials shows cunning. Girl witches, he argued, are taken to sabbaths, and "being of a tender age, they can be persuaded more easily" to tell us about "the names of the persons there, the time of occurrence, the place they travelled to, and what they did there."[91] Bodin warned that girl witches could be afraid of making confessions before several investigators, so he recommended making them feel more comfortable by letting them speak to one investigator only. The others should listen secretly behind a tapestry to write the child's admissions down. His book is full of such manifest shrewd advice on how to stimulate trials. Regarding local reputations for witchcraft, he noted that "concerning the question of witches, common gossip is nearly infallible."[92] Only the potential of flight and rich witches may not have been entirely on his radar.

In sum, somewhat anachronistically, we might say that Bodin had much sociological and psychological insight into the workings of witch-hunting. He intelligently knew what buttons to push, even though he was not professionally involved in witch trials.[93] I earlier posed the question of whether proponents understood what historians understand today, namely how cumulative notions of witchcraft enhanced the reproduction of trials. In Bodin's case, we can answer this question—with flight and rich witches being a possible exception—with a yes. Bodin did not "design" central ideas about witchcraft, as those ideas were already evolving before him. But his *Démonomanie* provides a compelling case of someone who understood how certain witchcraft notions boosted trials, propagating them with that in mind.

Bodin's book spread far and wide, so we might presume that his insights became common currency. However, other essential sources provide a mixed picture. Take Peter Binsfeld's *Tractatus* from Trier, 1589. Written while the hunts of the region were in full swing, his book reflected many innovations of those hunts. Binsfeld referred to children and people of "high standing" as possible witches and highlighted gatherings, flight, and misogynous stereotypes, and he explicitly affirmed that torture should be used to question suspects about sabbath accomplices.[94] A striking feature of the book, however, is that Binsfeld did not address one of the

[91] Bodin, 358–9.

[92] Bodin, 394.

[93] Krause, "The Will to Know and the Unknowable," 122.

[94] Binsfeld, *Tractat Von Bekantnuß der Zauberer unnd Hexen*, 123.

outstanding traits of the Trier hunts, their size. One passage tells of a suspect "mentioning one, two, or more" companions, but there is no information about sabbaths or chains of accusations being large.[95] That a person directly involved in the trials omitted this prodigious cultural "mutation" suggests that neither he nor possibly others saw the potential of what was occurring right before their eyes. We find a similar oversight in Ulm's newssheet in the wave of trials that subsequently swept Germany.[96] Its author encouraged witch-hunting and gave due attention to the size of the persecutions in Trier and other places. But he missed what made those trials so big: the author said nothing about (large) witches' gatherings.

Moreover, that the Trier hunts spread from place to place was crucially impacted by the belief that witches could know other witches over long distances. We might expect persecutors to have been well aware of the importance of flight, which stimulated this "infectious" effect. But sources suggest otherwise. Johannes Dillinger observed that interrogators in the region were not particularly interested in flight.[97] They mainly discussed the topic concerning something else that had their attention: witches' social status. Did witches fly on simple tools or luxury goods like beautiful horses and carriages? Binsfeld explicitly recounted that witches could fly over long distances very quickly, but interestingly, he also said that witches returned from sabbaths on foot.[98] Binsfeld thus mentioned a detail that threatened to undermine the reproductive advantage of the flight belief. The Trier perception of flight does not seem to have been unique. In trial records from Lorraine, Robin Briggs similarly observed that flight was not seen to enable sabbaths or long-distance contacts.[99] Interrogators discussed the topic for other reasons, for instance, to stress the abnormality of the whole affair. Julian Goodare generalizes: "Witches' flight occurred in confessions, not because the interrogators wanted to explain large gatherings at long distances, but because flight was itself an important ingredient in witchcraft beliefs."[100]

What about the general understanding of witch-hunting dynamics after Trier? This was when sabbath notions became most elaborate, with demonologists painting assemblies in full colour and detail and persecutions

[95] Binsfeld, 88.
[96] Behringer, *Hexen und Hexenprozesse in Deutschland*, 218–23.
[97] Dillinger, *Böse Leute*, 120.
[98] Binsfeld, *Tractat Von Bekantnuß der Zauberer unnd Hexen*, 59–61, 97, 123.
[99] Briggs, "Many Reasons Why," 60; Briggs, *Witches & Neighbours*, 40.
[100] Goodare, *The European Witch-Hunt*, 136–37.

being pursued more effectively. One can find many insights in the case of Henry Boguet, who hunted witches in the Franche-Comté around 1600. He spoke of an "infinity of witches," he repeatedly stressed that suspects should be asked for accomplices, and he linked flight to sabbath attendance.[101] Furthermore, Boguet devoted much attention to child witches, for example, when describing how "by means of a child an infinite number of witches was discovered." In his view, these children also deserved the death penalty, as their role in trials was "a secret judgment of god."[102]

Still, other vital sources provide indications that the proponents' shrewd understanding was not necessarily at the root of witch-hunting's success around this time. Take Nicolas Rémy's *Daemonolatreiae* (1595). The Lorraine inquisitor displayed much interest in the children of witch parents, including boys, also being witches.[103] Yet, Rémy failed to address how to use children in trials. What interested him most about the topic was the moral question of whether child witches deserved harsh punishment, which he answered affirmatively. Rémy also lengthily discussed sabbaths, which he claimed were "attended by great numbers."[104] But he added an odd detail. The book repeatedly states that witches cover their faces there to prevent recognition and accusation by their companions. This shows that Rémy saw how sabbath recognition could create chains of accusations because that was precisely what the alleged witches wanted to prevent. But remarkably, Rémy only discussed this insight in combination with the notion of face-covering, which simultaneously had the potential of undermining the reproductive dynamics of the sabbath belief entirely. This variant has also been found in trial records elsewhere, like Schwaben and Saarland.[105]

If witch-hunting's proponents understood the reproductive dynamics of the sabbath notion, we might expect them to immediately remove this dangerously unhelpful variant or advise interrogators not to be misled by such dodging. But that is not what happened. Martin Delrio also recounted that witches tried to avoid recognition by appearing at the sabbath "with their face covered by a mask, a linen cloth, or some other veil or facial representation."[106] Like Bodin's book, Delrio's *Disquisitiones* was shrewdly

[101] Boguet, *Discours exécrable des sorciers*, 24–25, 67–68, 334–35, 439–43.
[102] Boguet, 12–13, 442.
[103] Rémy, *Demonolatry*, 85, 92–98.
[104] Rémy, 47–65.
[105] Dillinger, *Böse Leute*, 206; Labouvie, *Zauberei und Hexenwerk*, 120.
[106] Rio and Maxwell-Stuart, *Investigations into Magic*, 93, 97.

written, effectively connecting cumulative beliefs to investigatory techniques. For instance, he postulated that judges would sin gravely if they did not question suspects about accomplices, and he delved into the potential of torture and common gossip.[107] But that makes it all the more striking that Delrio included the potentially undermining variant of face masking without instructions on defusing it. In some ways, the variant fitted nicely into demonological theory—it was a common trope that the devil and his witches tried to conceal their activities.[108] Yet, for reproducing the witch-hunting phenomenon, adopting this variant was a marked act of "unintelligent design" by witch-hunting's proponents. The fact that the variant did not gain more prominence was plausibly due to the hidden selective process weeding it out again.

There are further indications that proponents in the heyday of trials were not fully aware of how things worked. Regarding the Ellwangen hunt, we have already encountered peculiar confessions about a diabolical palace where executed witches lived to continue their activities.[109] That investigators recorded the notion and even allowed it to gain traction in confessions suggests that they did not recognize its undermining potential. Another interesting case is Pierre de Lancre's *Tableau de l'inconstance des mauvais anges at demons* (Picture of the Inconstancy of Evil Angels and Demons).[110] It was published in 1612 after De Lancre had supervised an extensive hunt in the Labourd region. The judge presented the sabbath in full baroque blossom, full of pomp and spectacle, and innumerable attendants. If anyone should have comprehended how such notions were to be put into persecutory action, it was De Lancre. But unlike some other works, he hardly offered instruction on persecuting. The book gives the impression that De Lancre eagerly wanted to tell the world about everything he believed to have uncovered rather than to connect the dots on how to persecute.

Or take Friedrich Förner, the inciteful preacher from Bamberg. William Bradford Smith argues that such personalities could play "a vital role in the instigation and expansion of the trials," and he sees Förner as "the principal architect of the witch hunts in Bamberg" who more than any other

[107] Machielsen, *Martin Delrio*, 282–83.
[108] Clark, *Thinking with Demons*, 22–23.
[109] Dillinger, *Kinder im Hexenprozess*, 129–31; Mährle, "O wehe der armen seelen," 435.
[110] Lancre, *On the Inconstancy of Witches*.

single individual must bear the onus.[111] Behringer describes the preacher as the "spiritus rector" of these trials and argues that the great persecutions of this period were "pre-planned ever more perfectly, and carried out ever more skilfully."[112] Förner was well-versed in demonology and witch-hunting, so we might expect him to know the tips and tricks of rousing persecution. But did the sermons of this "architect" indeed provide a blueprint, or a design, of what was coming?

The striking fact is that he did not. His sermons movingly aimed to scare people with warnings about the snares of the devil and the dangers of witchcraft, but incendiary elements like weather magic, flight, and rich witches were hardly or not addressed. There are some passages on child witches, but considering the importance of child accusations at this historical juncture, his attention was negligible.[113] More amazingly, the sabbath is almost entirely ignored. Like the *Malleus*, Förner only passingly refers to it a few times when making a point about something else.[114] Instead, Förner mainly focused on the danger of consulting village witches. The suffragan bishop mentioned that people visited such cunning women for medical cures or soothsaying, warning that this meant invoking the diabolical. People should *exclusively* seek the protection of the holy church and its sacraments. Arguably, this focus looks like instrumentalization, with Förner exploiting the witch image to solidify his own church's spiritual power at the expense of those women. But notably, this focus played no significant role in the subsequent hunt. That a well-connected person like Förner seems to have had so little understanding of what was making witch-hunting big indicates that the Bamberg hunt was less intentionally designed than is currently presumed. This finding strengthens the alternative Darwinian scenario.

Another essential source from the time is Heinrich von Schultheis's *Eine ausführliche Instruction* (An Extensive Instruction) from 1634. This Westphalian witch commissioner had been involved in the horrific hunts of the electoral state of Cologne, and his book offered guidance for

[111] Bradford Smith, *Reformation and the German Territorial State*, 183; Bradford Smith, "The Persecution of Witches," 53.

[112] Behringer is quoted in: Bradford Smith, *Reformation and the German Territorial State*, 183; Behringer, *Hexenverfolgung in Bayern*, 136.

[113] Förner, *Dämonenglaube und Zauberei im Jahre 1625*, 248, 632–33.

[114] Förner, 98, 208, 282.

inquisitors.[115] The work was written as a dialogue between a doctor—Schultheis himself—and a somewhat sceptical higher official. Schultheis addressed the importance of common gossip, encouraged anonymous accusations, and disallowed suspects a proper defence. He emphasized torture's indispensability to roll up the secret network and offered concrete advice on how to torture. The broad social profile of witches is discussed only sparsely, but Schultheis shrewdly stressed that convicted witches must bear the costs of their burnings.[116] Moreover, child witches are mentioned repeatedly.[117] "Children tend to tell us the truth," Schultheis says, and since the crime of witchcraft is so very secret, legal restrictions on the use of child witnesses should be abolished.[118]

In addition, Schultheis delved into the persecutory importance of the sabbath and endlessly stressed what this implies: that inquisitors should press suspects about whom they had seen there.[119] Schultheis sets the example in two exemplary interrogations in the book.[120] In the first, a suspect named Gretha is confronted with her bad reputation, the accusations against her, and apparently inconsistent declarations. Gretha initially refuses to plead guilty, exclaiming that she had rather never been born. But torture makes her confess her witchcraft. The questioning then moves to the sabbath. "How many of you came together there?" the doctor inquires.[121] When she gives a number of fifty, the doctor is not satisfied yet, so he asks her whether there were fewer or more. Gretha now admits that there were, in fact, two or three hundred witches present. The doctor forces her to mention names, suggesting she denounce people from other places. As he explains to her, for a witch, there is no better way of showing faith and love for Jesus Christ than through naming accomplices. It would save her soul and potentially the souls of the people she named.[122] Schultheis presents painful questioning as a means to break the devil's power over his subjects. Torturing the accused is not an injustice but a way of doing suspects a favour. Any hindrance on the path towards confes-

[115] Schultheis, *Eine Außführliche Instruction*; Grawlich, "Der Hexenkommisar Heinrich von Schultheiß."

[116] Schultheis, *Eine Außführliche Instruction*, 467.

[117] Schultheis, 2, 136, 140–41, 305–6, 337–38.

[118] Schultheis, 330.

[119] Schultheis, 30–31, 112, 269, 301, 328.

[120] Schultheis, 185–253, 280–82.

[121] Schultheis, 208.

[122] Schultheis, 191–2.

sions—like the accused retracting statements when torture is over—is then interpreted as a diabolical attempt to thwart their return to God. Eventually, Gretha sees how the devil has entrapped her and expresses her gratitude: "Let our Lord Jesus Christ reward you, my dearest Mister Doctor, for keeping your best interests at heart for a poor sinner like me."[123] So all's well that ends well, and after her burning, Gretha is expected to meet her eternal bliss.

We can hardly be surprised that witch-hunters like Schultheis were able to create the super-hunts of this time. But when looking more carefully, his work also betrays apparent limitations to the level of insight. The book focuses on extracting names of accomplices who attended crowded sabbaths, but astoundingly, the exemplary interrogations lead to only very few denunciations. The questioning of Gretha eventually delivers a mere three names, and the second questioning of a man named Thonnis does not exceed that. This tiny number also satisfies the doctor. Intriguingly, the book did not finish off the incendiary potential of large sabbaths. There is also a notable omission in his discussion of child witches. In one story Schultheis describes how three boys searched for birds' nests, with one of them, Luscus, trying to invoke the devil to find more.[124] This upset one of the other boys, and an ensuing panic in their community leads to witchcraft accusations against Luscus. Yet, while childhood fantasy and rashness thus play a role in the story, Schultheis recounts that Luscus confessed after being tortured. The book elsewhere also argues that children confess more easily when tortured.[125] Schultheis thus emphasized a particular vulnerability of children. What he omitted, however, was that children are also likely to make *voluntary* confessions and how to exploit that.

This bears upon the broader question of to what extent the role of children was instigated by knowledgeable persecutors who understood their usefulness. Behringer thinks the child accusations were "mostly undesired and unplanned and were often inexplicable."[126] Instrumentalists like Voltmer and Rainer Walz find such interpretations too naïve and presume a more deliberate use of children.[127] Yet, they do not substantiate

[123] Schultheis, 206.

[124] Schultheis, 98–101.

[125] Schultheis, 305.

[126] Behringer, "Kinderhexenprozesse," 43. Zahlreiche Einzelbeispiele zeigen, daß "meistens ungeplant und unerwünscht, oft auch ganz unerklärlich waren."

[127] Voltmer, "Jesuiten und Kinderhexen," 208, 214–17, 230; Walz, "Kinder in Hexenprozessen," 217–20.

such claims about a shrewd understanding with clear evidence from the sources. My own findings are ambiguous. Some authors understood a lot, others appear less insightful. The German pamphlet from 1626 devoted to child witches (see Sect. 7.3) also failed to mention how children could be used in trials. The author was primarily interested in the shocking question of why so many children were seduced by the devil.[128]

The increasing focus on the rich is also rife with interpretative problems. If proponents failed to address the topic, this does not automatically imply that they were incognizant of how an attack on the rich served the attention for and funding of witch persecution. Perhaps Förner and Schultheis deliberately wanted to steer trials in that direction but did not, or only sparsely, discuss the topic to avoid provoking the rich and powerful. Yet, we should certainly consider more non-intentional scenarios here. Erik Midelfort proposed that the increased targeting of the rich was just a by-product of something else: "One would expect persons under torture to agree most easily on the notorious, the eccentric, or the wealthy since they were the best-known persons in town."[129] This would also explain the disproportionate targeting of midwives and innkeepers. Another option might be that the people interrogated chose to mention members of the ruling classes out of revenge.[130] "People wanted to harm those responsible for judicial procedures," Alexandra Haas explains.[131] These suggestions do not appear implausible and can be integrated into a Darwinian model. The cultural adaptation would then look as follows: the focus on rich witches had the effect of more extensive, more spectacular trials, which in turn enhanced the focus on rich witches. What would give the adaptation a Darwinian twist is that the initial cultural variants leading to this focus occurred due to factors other than persecutors who understood these enhancing effects. In that regard, the variants' occurrence may have been a chance hit.

Let us now return to the four hypotheses concerning the evolution of the sabbath as a cultural adaptation (see Sect. 6.5). The first hypothesis—*many cultural variants regarding witchcraft appeared, most of which did not reproduce well*—had not yet been examined by witchcraft scholarship

[128] *Newer Tractat von der Verführten Kinder Zauberey.*

[129] Midelfort, *Witch Hunting in Southwestern Germany*, 172.

[130] Behringer, *Witches and Witch-Hunts*, 158; Haas, *Hexen und Herrschaftspolitik*, 87, 113–19.

[131] Haas, *Hexen und Herrschaftspolitik*, 114.

as such, and Sect. 7.5 presented many clues for its validity. The second hypothesis—*the belief in the witches' sabbath triggered larger and more spectacular persecutions*—was already well established, and especially Sects. 7.1, 7.2 and 7.3 further substantiated it. The third hypothesis—*this effect contributed to the survival of the belief in the witches' sabbath itself*—was already addressed to some degree, and Sects. 7.1, 7.2 and 7.3 offered further confirmation. The fourth hypothesis—*this enhancing effect was neither intended nor recognized as such by the actors involved; it was a chance hit*—had not yet been examined at all. Our findings here are multifaceted. There are clues for understanding, especially in Bodin, suggesting that this cultural adaptation partially evolved due to involved actors who understood its underlying mechanism. But there are also clues for significant limitations to the apparent understanding by proponents, especially regarding the enhancing effect of large sabbaths. We cannot rule out that proponents had more knowledge at subconscious levels, but as long as that does not manifest itself in the sources, it remains unprovable. Hence, intentional design does not, by a long stretch, appear sufficient an explanation for this cultural adaptation. Darwinian selection comes forward as the indispensable complementary model. This also goes for other aspects of the cumulative concept of witchcraft, especially flight. Further research is required, though.

Apart from the question of what proponents did or did not understand, there is the issue of an alleged instrumentalization of trials for ulterior purposes. Do the texts of proponents provide clues in that regard? The German instrumentalists claim that earthly purposes were perfectly conceivable at the time. The sources studied indeed offer proof for that. First, there was the widely held idea that people became witches to achieve earthly ends. What lured people in, Bodin said, was Satan's false promise to "give wealth to the poor, pleasure to the afflicted, power to the weak, beauty to the ugly, knowledge to the ignorant, honour to the despised, and the favour of the great."[132] Second, and more importantly, proponents observed that witchcraft accusations could be used for ulterior ends and warned against it. Delrio prescribed that only trustworthy people who had no anger, rancour, or hatred against the people denounced should be accepted as witnesses.[133] Others, like Bodin and Schultheis, expressed

[132] Bodin, *De la démonomanie des sorciers*, 305.
[133] Rio and Maxwell-Stuart, *Investigations into Magic*, 195.

similar caution.[134] Throughout the era, it was routine practice in trials to make witnesses swear an oath not to abuse procedures.[135] Overall, proponents presented themselves as reasonable and ethical people who wanted evidence to be weighed carefully and cared about not putting the innocent at risk. Delrio and Schultheis proclaimed that it is better to let ten guilty witches go than to harm a single innocent person.[136] The latter also clarified that his daunting task of saving the witches' souls had rapidly made his hair turn grey.[137]

The German instrumentalists interpret these pious self-representations cynically.[138] Rummel: "Of course, there were strategies of legitimization, in which interest-based actions were camouflaged and concealed under apparently genuine motives."[139] But he sees such virtuous talk as a way of ideologically sanctioning the hidden benefits. That sources show that instrumentalization of accusations and trials by proponents was perfectly conceivable offers some support for this scenario. Delrio's and Schultheis's words of caution might also be taken with a pinch of salt, as their proposals were, in the end, blatantly incautious. Yet, we should still be careful here. Ulterior motives were thinkable, but does that also make them plausible?

Take Jean Bodin, our best example of a knowledgeable agent. He was not involved in trials, and his anti-witchcraft fanaticism seems to have hindered his career. That makes it rather challenging to say what his ulterior motives might have been. Virginia Krause identifies different dimensions of his zeal.[140] She depicts Bodin as a man with deeply held religious convictions who thought that an angel accompanied him and communicated with him through a system of signs. Bodin had visions and dreams in which he believed he could hear God's voice and receive instructions. The French thinker considered himself a prophet in the tradition of the Hebrew prophets, so plausibly, it is also in this light that he saw his writing on

[134] Bodin, *De la démonomanie des sorciers*, 394; Schultheis, *Eine Außführliche Instruction*, 10, 143–48, 152–53.

[135] Behringer, *Hexen und Hexenprozesse in Deutschland*, 216, 227; Rublack, *The Astronomer & the Witch*, 216; Moeller, *Dass Willkür über Recht ginge*, 71.

[136] Machielsen, *Martin Delrio*, 283; Schultheis, *Eine Außführliche Instruction*, iv, 102, 108–9; Rio and Maxwell-Stuart, *Investigations into Magic*, 199–201, 215–16.

[137] Schultheis, *Eine Außführliche Instruction*, 445.

[138] Voltmer, "Von der besonderen Alchimie, aus Menschenblut Gold zu machen," 131.

[139] Rummel, "So mögte auch eine darzu kommen, so mich belädiget," 206.

[140] Krause, "The Will to Know and the Unknowable."

witchcraft. "The *Démonomanie* was not the result of a strictly reasonable enterprise: it was perhaps first and foremost the fruit of Bodin's religious vocation and mystical experiences," Krause writes.[141] Profound, genuine belief takes centre stage here.

There are more indications of instrumentalization in the case of an inquisitor like Schultheis. Rita Voltmer claims that he probably made financial gains and that his function as witch commissioner played a role in his ennoblement. "He made a social, if not financial, profit from such activities."[142] Tanja Grawlich says that the sparsity of sources makes it impossible to determine whether he enriched himself but thinks it conceivable, as other regional commissioners made gains.[143] Yet, even here, we should be careful.[144] The allegation lacks proof, and Schultheis's activities and devotion still seem extreme, even considering earthly advantages. Why not presume that his avowed purposes were genuine? Maybe he really feared the consequences of not acting against witches, and he honestly thought himself to be saving witches' souls. Perhaps the stress of this grim task really did turn his hair grey. Proponents consistently expressed such convictions, and within the contemporary system of witch beliefs, they had some logical force.[145] Seen from the reproductive interests of the witch-hunting phenomenon itself, this belief may strike us as yet another of the phenomenon's ingenious "tricks" to motivate people and further its own "selfish" spread.

Fortunately, our empirical options have not been exhausted yet. What makes the current debate on instrumentalization so tricky is that proponents of trials possibly concealed their real motivations. Contemporary opponents of trials, however, were probably less secretive in that regard. The German instrumentalists rightly point to them as an invaluable source of information. What were those critics' views on this matter? To what extent did they claim that trials were used successfully for ulterior aims? And did the critics recognize how notions like the sabbath, flight, or the child witch engendered trials? We now take a closer look.

[141] Krause, 126.
[142] Voltmer, "Schultheiss, Heinrich von (ca 1580-ca 1646)."
[143] Grawlich, "Der Hexenkommisar Heinrich von Schultheiß," 318–20.
[144] Decker, "Der Hexen-Richter Dr. Heinrich von Schultheiß," 1054–55.
[145] See, for instance: Haas, *Hexen und Herrschaftspolitik*, 100–102; Bodin, *De la démonomanie des sorciers*, 254.

8.3 What Opponents Said

When the Württemberg preachers Matthaeus Alber and Wilhelm Bidenbach in 1562 responded to the new surge of witch trials, they warned about the false and groundless talk of witches "that arose from malice and resentment and is enhanced and strengthened by all too eager credence, and which may overcome even the most innocent of one's adversaries through cunning and deceit."[146] These Lutheran pastors were far from unique in pointing out malevolent motives behind the trials. In 1635, Johann Meyfart identified malice, hatred, envy, and resentment as crucial causes, creating an atmosphere in which misfortune was quickly interpreted as witchcraft.[147] This was what such gossip could sound like: "He became rich very soon, so he's a witch. That young lady is beautiful, so that's why she's a witch. That man gives a lot of alms, so he's a witch, and she has good harvests, so that's why she's a witch."[148]

A few years earlier, Friedrich Spee observed that such sentiments were especially prominent in his German fatherland. The Jesuit priest argued that people in other countries could endure it when certain people became rich, but in Germany, "there were others, who had not been as lucky, who would put their heads together, and pretend that this could only have happened because of witchcraft."[149] Germans had more difficulty accepting other people's happiness, so that was why witch trials so unusually ravaged these lands.

Various passages quoted in 8.1 have already indicated that contemporary critics suspected witch-hunt proponents of harbouring ulterior motives. Many more examples can be added to the list. In an area bordering on the Palatine, the princedom of Mainz initiated a witch persecution that killed many serfs there. The Palatine objected, explicitly accusing Mainz of a political campaign: the serfs tended to side with the Palatine, revealing the underlying motive for their extermination.[150] In a report on witch-hunts in the Basque Country, the Spanish Jesuit Hernando de Solarte in 1611 recounted the story of a boy who had told him that three women summoned him to accuse another woman of luring him into witchcraft. Also, they instructed the boy which sabbath companions he

[146] Alberum and Bidenbach, *Ein Summa etlicher Predigten von Hagel und Unholden.*
[147] Meyfart, *Christliche Erinnerung*, 98, 186–90.
[148] Meyfart, 187.
[149] Spee, *Cautio Criminalis*, 3.
[150] Schmidt, *Glaube und Skepsis*, 327–28.

should mention by name. In return, the women promised him four *reales*, a shirt, and other goods. "Their real motive for bribing the boy was their personal hatred of these people," De Solarte explained.[151] Allegedly, the women had made similar requests to other boys.

Friedrich Spee spoke about a case of a city official who was blamed for misconduct but then diverted attention to his accusers by initiating charges of witchcraft against them. "What easier way these days to take revenge?" he commented.[152] Spee further mentioned "greedy and vicious judges" who were after the money of the people they convicted, while Johann Weyer spoke of "bloodthirsty judges."[153] Persecutors in Ellwangen, Eichsätt, Bavaria, and Cologne had to defend themselves against charges of financial abuse, and as Cornelius Loos said about the witch-hunt system: "Isn't it obvious how writers and judges try to strengthen and defend the machinery, all of them, when they describe vacuous confessions"?[154] Anyone rendering the interpretation of witch-hunting as a means to oppress and stereotype women as entirely anachronistic should take note of what the German preacher C. Beerman said in a 1593 sermon:

> That we have so many witch burnings, these days is not least due to innumerable authors who write so very dreadfully of women and condemn virtually all of them for their evil, poisonous, and diabolical nature but subsequently pride themselves that incomparably more women are burned as witches than men, who are by nature better, and not so poisonous, cunning and devious. The public, which listens to such talk, then says that justice is being done, as these women are infernal and treacherous like the devils.[155]

Contemporary critics also understood much of how witch-hunting worked, like the role of torture in extracting false confessions. Take people who had to undergo this horrific ordeal.[156] In a secret letter, smuggled out of a prison in Nördlingen in 1590, a woman named Rebecca Lemp wrote to her husband: "Oh, you, my chosen sweetheart, they have tortured me,

[151] Henningsen, *The Salazar Documents*, 154.

[152] Spee, *Cautio Criminalis*, 129.

[153] Spee, 10, 36, 300; Weyer, *Witches, Devils, and Doctors in the Renaissance*, 550.

[154] Durrant, *Witchcraft, Gender, and Society in Early Modern Germany*, 41; Behringer, *Hexenverfolgung in Bayern*, 303–5; Midelfort, *Witch Hunting in Southwestern Germany*, 110–11; Eerden, "Cornelius Loos und die 'magia falsa,'" 150.

[155] Schwerhoff, "Hexerei, Geslecht und Regionalgeschichte," 330.

[156] Briggs, *The Witches of Lorraine*, 43.

although I'm as innocent as God in heaven."[157] Lemp implored him to send her something with which she could kill herself, as she feared that torture would destroy her body and make her confess to things dangerous to her soul. Many authors of printed critiques dwelt upon the horrors of the torture chambers.[158] A jurist in Schleswig stated that "also the innocent, due to the pain, are forced to lie."[159] In his extensive assessment of the witch trials, *Von Zauberern, Hexen und Unholden* (On Magicians, Witches, and Sorcerers), the German jurist Johann Gödelmann in 1592 explicitly targeted Bodin's notion of the *crimen exceptum* as a source of false testimonies.[160] This professor from Rostock argued that witchcraft verdicts should always be based on material proof or eyewitness accounts. This, of course, made the alleged crime of witchcraft almost impossible to prove, but Gödelmann probably chose this line of attack because he understood where witch-hunting could be crucially tackled. Friedrich Spee acknowledged that some robust people might endure torture and that the practice sometimes produced genuine confessions.[161] But overall, he thought it a source of innumerable lies and noted that not only the judges, inquisitors, and commissioners but also the assistant torturers had studied the art of interrogation and could apply it with much agility.[162]

The chain mechanism of alleged witches naming ever more companions was also understood. Alber and Bidenbach observed how, through confessions extracted by torture, certain women "not only falsely doom themselves but also others."[163] Or take Johann Meyfart's vivid description of witch-hunting's propensity for escalation: "Once a witchcraft trial has started, it tends not to come to a halt, but to move quickly from person to person, from sex to sex, from village to village, from town to town. Monthly, weekly, daily, hourly, the number of indicated witches multiplies."[164] The more alleged witches one kills, the more alleged

[157] Grasmück, Lorenz, and Schmidt, *Hexen Und Hexenverfolgung Im Deutschen Südwesten*, 188.

[158] Dillinger, *Böse Leute*, 366; Witekind, *Christlich Bedencken und Erinnerung von Zauberey*, 165–66.

[159] Behringer, *Hexen und Hexenprozesse in Deutschland*, 127.

[160] Gödelmann, *Von Zäuberern, Hexen und Unholden*.

[161] See, for instance: Spee, *Cautio Criminalis*, 60–61, 93–101, 183.

[162] Spee, 70.

[163] Alberum and Bidenbach, *Ein Summa etlicher Predigten von Hagel und Unholden*.

[164] Meyfart, *Christliche Erinnerung*, 68.

witches one gets, critics remarked.[165] Some also explicitly identified the inflammatory potential of the sabbath belief. In his reflections on the mechanism of suspects naming accomplices, a Lutheran theologian from Lemgo, Jodocus Hocker, in 1568, said, "On what else is this most of the time based, but those fantastical dances, of which they [the suspects] recount, to the greater honour and glory of themselves, that they saw at those dances the wives of the mayor, the councillor, and the pious citizen."[166] In an attack on witch-hunting from 1585, the *Christlich bedencken und erinnerung von Zaubery* (Christian Thoughts and Recollections Concerning Witchcraft), the Calvinist professor Hermann Witekind similarly observed that as a result of the idea of witches' dances, "many an innocent woman, through malicious talk and slander, has been killed."[167]

There were more contemporary insights about the workings of witch-hunting that preceded modern insights. Cornelius Loos pointed to long winter evenings, where peasants sat around the stove, telling each other fantastical stories. In such tales, he recognized a crucial ingredient of subsequent concepts of witchcraft.[168] Meyfart knew how official accusations arose from common gossip: "If one blows the trumpet, it resonates through the entire forest; if one accuses Anna, it resounds through the whole village, hamlet, or town."[169] Spee addressed how the public reading of confessions during executions caused the spread of specific ideas and names, explaining why accusations and confessions about witchcraft could be pretty similar: various people had heard the same things during those readings.[170] Contemporary critics also observed that the witchcraft doctrine was a hermetically sealed system in which everything counted as a confirmation. Weyer spoke of "an inextricable labyrinth from which no way of escape, no end, can be seen."[171] Spee commented that if suspects had lived indecent lives, this was considered evidence of their evil nature. But if they had lived seemingly decent lives, this was also considered

[165] Witekind, *Christlich Bedencken und Erinnerung von Zauberey*, 136, 212.

[166] Hocker, *Der Teufel selbst*, 297.

[167] Witekind, *Christlich Bedencken und Erinnerung von Zauberey*, 136.

[168] Eerden, "Cornelius Loos und die 'magia falsa,'" 150; Eerden, "Der Teufelspakt bei Petrus Binsfeld und Cornelius Loos," 61.

[169] Meyfart, *Christliche Erinnerung*, 190.

[170] Spee, *Cautio Criminalis*, 98–101.

[171] Weyer, *Witches, Devils, and Doctors in the Renaissance*, 522.

evidence, as it showed their dishonesty. Spee further comprehended another essential mechanism: people were afraid to criticize trials due to the risk of self-incrimination.[172]

In sum, contemporary critics' texts confirm what the German instrumentalists maintain. In early-modern discourse, it was already assumed that witchcraft accusations and trials were used for a wide variety of ulterior purposes. The discourse does not *prove* that multifunctional instrumentalization was real, but that instrumentalization was so clearly conceivable makes it more likely. Contemporary opponents apparently also understood many aspects of how witch-hunting functioned. As such, it does not *prove* that the persecutors possessed the same insights; perhaps critics saw things persecutors failed to see. However, it reveals that particular insights were available then, making it more likely that proponents shared such awareness. Hence, these empirical findings fail to confirm the proposed Darwinian scenario. All this does not suggest incognizant actors, subjected to the trickery of a culturally evolved "design without designer" that spread at the expense of its human hosts.

This, however, is also the moment to begin stressing limitations to early-modern critics' understanding. One is the issue of how the notion of flight contributed to the persecutions' spread. Critics repeatedly contested the reality of flight and often linked this notion to belief in sabbath attendance. But what is not identified explicitly—at least not in the sources consulted—is how the notion of flight enabled sabbaths and chains of accusations over longer distances. Then, there is the impact of child accusations. None of the consulted texts analysed this in much detail—Bodin's sharp insights on the topic were neither adopted nor contested. The bishop of Pamplona did remark about the witch panic in the Basque Country that "while there may be some element of truth in it, much is fiction, delusion, and calumny which has been caused by the children and ignorant people who talked about these matters and have picked up the terms from what they have heard and from the news" and Meyfart passingly noted that we could neither build on nor trust child confessions.[173] But overall, the topic appears remarkably absent. Even Spee, who wrote his book at a time when child accusations played a significant role, did not delve into the subject.

[172] Spee, *Cautio Criminalis*, 121, 128.

[173] Henningsen, *The Salazar Documents*, 192; Meyfart, *Christliche Erinnerung*, 220.

As pointed out, critics of witch-hunting seem to have had a better understanding of the mechanism of sabbath accusations. But there are also indications to the contrary. Alber and Bidenbach did not address the sabbath, even though the notion played a crucial role in nearby Wiesensteig. Johann Weyer did discuss and contest the idea, but he failed to highlight how it created chains of accusations. In one instance where he referred to this connection, he did so in quite an intriguing manner. Weyer approvingly quoted the Italian physician Girolamo Cardano, who argued that sabbaths are completely fraudulent because otherwise, "one of them [the suspects] could expose a hundred of others or even more if as many were present (…). But they accuse only those who are known to them by sight or by rumour."[174] This passage suggests an insight into the sabbath's potential for creating large chains of accusations. But remarkably, Cardano and Weyer apparently argued that persecutors were *not* aware of that, as they did *not* press suspects to name many accomplices. On closer scrutiny, Hocker's quote also contains a notable detail: he did not think that the names of highly placed women appeared on the initiative of interrogators but because suspects mentioned them to their own greater glory. He presents persecutors in a rather incognizant and passive role. Spee's *Cautio criminalis* also fails to explicitly point out that the idea of witches' gatherings triggered chains of accusations.

Furthermore, essential observations must be made regarding how contemporary critics mostly answered the *cui bono* question and what they thought *primarily* drove persecutions forward. True, they often suggested instrumentalization, but there was much more going on. In that respect, let us again look at the insightful contemporary comments on how the sabbath notion created havoc. Here, for instance, is Witekind's quote above in its fuller context:

> Do these witches see each other at such dances, and do they know each other? Answer: if they see someone, they do not see a human or physical being, but a spectre, and as a result of that, many an innocent woman, through malicious talk and slander, has been killed. And I presume that the people who pretend to know where these dances occurred and who denounce the people allegedly present some people are closely aligned to

[174] Weyer, *Witches, Devils, and Doctors in the Renaissance*, 504.

> the devil, and who serve him in his deceit and murder, and who get orders and payment from him, which makes them worse and more punishable than the witches.[175]

According to Witekind, the critical force behind the sabbath notion and the human misery it created was not the persecutors but the devil. *He* manufactured the spectres and instigated people to make false accusations. "So clever and swift is Satan," Witekind said, that he created the imagination of the witches' dance to make innocent people die.[176]

Similarly, Jodocus Hocker's passage quoted above continues as follows: "As proven above, these dances are nothing but diabolical spectres and deception, on which no reasonable judgment can be based."[177] The proof Hocker refers to, among other things, contains the following:

> This is the truth: this sly fox, the devil, infatuates the poor women in their sleep to the extent that he creates such imaginations and illusions in them that they honestly think that they go to wonderful meals, music, dances, and handsome young lads, with whom they briefly have intercourse, all of which is vain fantasy, deceit, and trickery.[178]

The idea of witch beliefs being a diabolical manufacture was central to Hocker's account.[179] His book *Der Teufel selbst* (The Devil Himself) provided an overview of Satan's trickery. Johan Meyfart similarly saw the devil as the evil genius behind witch trials, including the notion of the sabbath. Satan influenced people's imaginations so that those people believed in innocent people attending witches' dances. Thus, nobody would be safe anymore.[180]

This idea of the devil using witch trials to ruin humankind could be found throughout the critics' works.[181] It was a common thread in the sceptical thinking in Württemberg, of which Alber and Bidenbach were representatives. But the argument was also used pervasively in the Calvinist

[175] Witekind, *Christlich Bedencken und Erinnerung von Zauberey*, 136.
[176] Witekind, 256.
[177] Hocker, *Der Teufel selbst*, 297.
[178] Hocker, 277.
[179] Hocker, 261–316.
[180] Meyfart, *Christliche Erinnerung*, 230–35.
[181] Clark, *Thinking with Demons*, 204–8; Duni, "Doubting Witchcraft"; Behringer, *Hexenverfolgung in Bayern*, 277–78.

Palatine, another centre of resistance and scepticism.[182] For instance, while judicial advice from 1581 did not dismiss witch trials entirely, it did stress that sabbaths were devilish fantasies and that judges should be careful "not to be deceived and led astray by such diabolical spectres."[183] When Leonhard Schug, a theologian, reflected on the mass hunts in Trier, he highlighted that the Bible proves that the devil is a deceiver who wants innocent people harmed. That was why Satan created the sabbath illusion.[184] In the conflict with Mainz, the Palatine authorities did not solely interpret its persecution as a shrewd political operation. They additionally argued that alleged witches' dances were diabolical dreams, and the prince of the Palatine warned that witch trials "offered the devil an excellent opportunity to play and shed innocent blood."[185] Regarding flights and sabbaths, Gödelmann stated, "In my simplicity, I hold them for nothing but diabolical spectres, treachery, and fantasy."[186] Let us also read Weyer's statement about the labyrinth (mentioned above) in its fuller context:

> Assuredly, in matters admitting of little certitude, the less cautious would not then be slipping from one single error into a thousand ones, as though trapped in an inextricable labyrinth from which no way of escape, no end, can be seen. From long experience, that crafty old weaver knows how to weave such webs skilfully.[187]

Weyer thought that Satan, not the witch-hunters, had contrived the labyrinth.

In the early modern context, there was nothing special about this line of argumentation; interpreting witchcraft beliefs as devilish trickery was an orthodox position going back to the Middle Ages.[188] Ulrich Molitor's late fifteenth-century critique typically dismissed belief in visible sabbath attendants as a diabolical imagination.[189] Erik Midelfort rightly describes Molitor's views as "not the first glimmerings of an enlightened mentality

[182] Schmidt, *Glaube und Skepsis.*
[183] Schmidt, 185–93.
[184] Schmidt, 264–65.
[185] Schmidt, 361.
[186] Gödelmann, *Von Zäuberern, Hexen und Unholden*, 273.
[187] Weyer, *Witches, Devils, and Doctors in the Renaissance*, 522.
[188] Duni, "Doubting Witchcraft," 205–10.
[189] Kieckhefer, *Hazards of the Dark Arts*, 142–44.

but the echoes of a well-known and respected medieval tradition."[190] Invoking the devil's forces was a more common way of explaining why humans caused each other pain and suffering. Wick's news collection contained a story about a man who came home drunk and killed his pregnant wife with a knife. The illustration shows how the devil stood smiling at the door opening to enjoy the spectacle he had helped create.[191] The religious persecution and warfare of the sixteenth century were interpreted along similar lines.[192]

The idea of the devil as the arch-contriver does not play an equally important role in all critical texts. It was not a central theme in Spee's book. Still, Spee used the idea in various ways, for instance, when he argued that people responsible for trials could be witches themselves.[193] He depicted the devil as a "jack-of-all-trades" who can play with people's senses to the extent that those people no longer see the difference between black and white, truth and falsehood.[194] This especially enabled Satan to create imaginations in old women's minds. Spee further thought of persecutors who were themselves eventually burned as witches as the devil's way of enlarging his realm.[195] Or take Alonso de Salazar, a highly sceptical member of the Spanish Inquisition, who examined the witch trials in the Basque Country. Overall, de Salazar's writing was calm and matter-of-fact, and he carefully dissected the inconsistencies and improbabilities of witch beliefs. But on various occasions, de Salazar refers to Satan as the perfidious genius behind it, saying about one local panic that: "Immediately and without any effort, the Devil leaves the village in an uproar, and those unjustly incriminated exposed to condemnation and other afflictions."[196]

This betrays an apparent ambiguity in the sceptical argument. On the one hand, these critics—especially Protestant ones—said that witch-hunt proponents overestimated Satan's power at the expense of divine power.[197] Yet, as soon as it came to the devil's abilities to create illusions, the same critics magnified the devil's power, turning him into the essential architect

[190] Midelfort, *Witch Hunting in Southwestern Germany*, 24.
[191] Wick, *Die Wickiana*, 66.
[192] Waite, "Sixteenth-Century Religious Reform and the Witch-Hunts," 493.
[193] Spee, *Cautio Criminalis*, 37.
[194] Spee, 179.
[195] Spee, 26.
[196] Henningsen, *The Salazar Documents*, 280, 351, 404.
[197] Weyer, *Witches, Devils, and Doctors in the Renaissance*, 85.

of witch persecution.[198] Here again, their argument became similar to that of witch-hunt proponents. The devil was God's minister of vengeance, his executioner. It was with divine permission that through the devil and then through the horrors of witch trials, humans were punished for their sins. One implication of witch persecution as a devil's tool was that the people responsible for trials had allowed themselves to become his marionettes. In a message to rulers, Johann Meyfart urged them to see how Satan was "amused and exalted about your Belialian tyrannies, and cheered about your grunting, scolding, and snorting, which is music to his ears."[199] Cornelius Loos thought that Satan was so dismayed by the fact that many Catholics had escaped the abyss of Protestantism that he now used superstitious witch belief as an alternative trap.[200] In places where Catholics had once lived in harmony, Satan incited the discord of witch trials, and the persecutors had been foolish enough to let themselves be exploited for that purpose. The imaginary danger of witchcraft distracted Catholics from real threats, like Protestant heresies.

Satan exerted his influence by interfering in many of the trials' details. He made people spit up nails or speak in strange voices so that others would interpret that as witchcraft.[201] Gödelmann considered the water ordeal a diabolical invention.[202] Johann Meyart presented torture as Satan's intelligent creation. Being a master of lies, the devil loved all the falsehoods that painful questioning elicited.[203] Another diabolical trick, as we have seen, was to make people honestly believe that they were diabolical witches. The devil better understood bodily humours than any physicist, so he could move the humours to alter people's state of mind.[204] Being "a subtle and devious spirit, an artist skilled through long experience," Meyfart said, Satan knew how to beguile the senses and confuse people's thoughts.[205] Demented and melancholic old women with unstable minds who lacked proper Christian faith were particularly

[198] Eerden, "Cornelius Loos und die 'magia falsa'" 156–58.

[199] Meyfart, *Christliche Erinnerung*, 255.

[200] Franz, "Antonius Hovaeus, Cornelius Loos und Friedrich Spee," 127.

[201] Pleier, *Malleus Judicum*, 17–18; Weyer, *Witches, Devils, and Doctors in the Renaissance*, 290-1–301.

[202] Gödelmann, *Von Zäuberern, Hexen und Unholden*, 319–38.

[203] Meyfart, *Christliche Erinnerung*, 49–54, 136–53.

[204] Hocker, *Der Teufel selbst*, 278; Pleier, *Malleus Judicum*, 54.

[205] Meyfart, *Christliche Erinnerung*, 61, 226.

vulnerable.[206] Gödelmann argued that when such women sat down "in a state of great sorrow and melancholia, they might easily be captured by the devil's swift rhetoric."[207] Satan further had an ingenious way of making gullible women think that they performed real magic: the devil knew when divinely ordained hailstorms were on their way, so he quickly told these women to make a storm.[208] When the hailstorm appeared, these women mistakenly thought that they had magically caused it. The underlying diabolical rationale: due to their genuine belief, such women more easily confessed, thereby giving rise to persecution.

In consequence, contemporary critics did not necessarily consider alleged witches innocent. In a sceptical treatise from 1583, Paulus Frisius portrayed witch trials as a big mistake built upon diabolical deception. But still, he spoke of women who themselves believed to be witches as hell whores who deserved nothing but death.[209] Hocker, Molitor, or Alber and Bidembach similarly argued that the mere attempt to be a witch already warranted the death penalty.[210] Others were more lenient. Pleier maintained that a lack of proper faith did not deserve the death penalty. After all, many rulers also let Jews live in their midst, and eventually, it was up to God to decide.[211] Especially Witekind showed compassion, proposing to reconvert the poor wretches.[212] Men should be friendlier to women, as it would make them less sad and melancholic, leaving the devil fewer opportunities to play with their minds.

Genuine belief was also often thought to motivate the actors responsible for persecution, and again, the devil was potentially involved. What educated contemporaries regarded as the simple-minded common rabble was said to be full of delusional and fantastical stories and beliefs.[213] Hocker warned judges never to listen to the "reckless and capricious mob" as that

[206] Witekind, *Christlich Bedencken und Erinnerung von Zauberey*, 20–22, 95.

[207] Gödelmann, *Von Zäuberern, Hexen und Unholden*, 278–9.

[208] Alberum and Bidenbach, *Ein Summa etlicher Predigten von Hagel und Unholden*; Witekind, *Christlich Bedencken und Erinnerung von Zauberey*, 103.

[209] Frisium, *Deß Teuffels Nebelkappen*, Dritten punct.

[210] Alberum and Bidenbach, *Ein Summa etlicher Predigten von Hagel und Unholden*; Ahrent-Schulte, "Lauter falscher Wahn und starke Einbildung," 95; Kieckhefer, *Hazards of the Dark Arts*, 151.

[211] Pleier, *Malleus Judicum*, 57–59.

[212] Witekind, *Christlich Bedencken und Erinnerung von Zauberey*, 164, 186, 202–6, 215, 239–44.

[213] Alberum and Bidenbach, *Ein Summa etlicher Predigten von Hagel und Unholden*; Spee, *Cautio Criminalis*, 2.

would offer the devil further opportunities to wreak havoc on the innocent.[214] As mentioned above, Spee described judges of witch trials as "greedy and vicious," but he more often thought them clumsy, careless, and ignorant.[215] Spee called superstition and unreasonableness key causes of trials. In analysing how suggestive questioning created apparently consistent confessions, he observed that this made judges and commissioners honestly think that they had "seized the truth with both hands."[216] Meyfart argued that Satan could even trick wise and soft-hearted people. He knew how to arouse latent evil inclinations, and in the case of witch persecutors, this resulted in a diabolical frenzy. These people had become so "deluded and manipulated by the devil these days that governments can no longer hire them."[217] Within the contemporary framework, there was not necessarily a contradiction between evil ulterior motives and false beliefs manufactured by the devil. Satan used and enhanced people's evil inclinations to motivate them to witch-hunting and so to bring about human misery.[218]

Attributing a holy zeal to persecutors was nothing special, as the role of witch-hunting in a broader debate about the secular use of religion illustrates. That religion could be instrumentalized for social or political ends was a widely discussed topic, especially associated with the work of Niccolò Machiavelli.[219] His plea for secular use of religion outraged many contemporaries. Machiavelli disregarded the veracity of religious doctrines, and that was playing with religion. Christian doctrines should be endorsed not because they are useful but because they are true. An accusation of Machiavellianism, or political motives, became a severe allegation in early modern discourse. The trope entered the debate over witch trials, but it was not so much the witch-hunt supporters who were associated with Machiavellianism. Instead, it was their opponents who were framed as "politici," as "cold political Christians" who placed earthly concerns above Christian ideals.[220] The trials' advocates, on the other hand, were more

[214] Hocker, *Der Teufel selbst*, 296.
[215] Spee, *Cautio Criminalis*, 300.
[216] Spee, 2, 70.
[217] Meyfart, *Christliche Erinnerung*, 132, 257.
[218] Hocker, *Der Teufel selbst*, 296.
[219] Clark, "Feigned dieties, pretended conferences, imaginary apparitions."
[220] Behringer, *Hexenverfolgung in Bayern*, 249–51, 310; Clark, "Feigned dieties, pretended conferences, imaginary apparitions," 280–81.

likely to be put in the camp of the "zelanti," the zealots, who pursued holy goals irrespective of secular consequences.[221]

Contemporaries also had good reasons for not interpreting hunts in primarily Machiavellian terms, as they already saw how persecution often *harmed* secular interests. Spee stressed that chains of accusations, once set in motion, became difficult to stop and eventually even killed people who engendered allegations in the first place.[222] "It is only when they have seen and learned that they were denounced and then imprisoned that they finally open their eyes and bemoan their land, but only when it is too late." Regarding authorities accountable for trials, he wrote that "through this burning, they devastate their lands more than any war has ever done, and yet they achieved nothing with it, which should make us weep bloody tears."[223] Meyfart noted that persecutions were detrimental to rulers' interests, warning them about the diabolical dynamics that witch trials brought about:

> You will find hatred of rabble against rabble (…). You will find the envy of the poor against the rich, of craftsmen against scholars, of house owners against officials, of the evil against the noble. You will find fury, anger, and quarrels in alleys and streets, calumny and slander in courtyards and castles, and unrest and riots in churches and schools.[224]

The devil also stimulated the escalating dynamics of trials. Witekind argued that the horrifying prison conditions brought female suspects to an even worse state of confusion, making them more susceptible to the devil's pursuits.[225] Confessions thus inflated even further. When Satan observed that his attempts at creating witch trials were successful, he doubled his efforts. But when he noted that the burning stopped, he lost confidence that people would fall for his tricks, so he gave up.[226] Critics warned that the negative consequences of persecution extended into the afterlife.[227] Witekind urged people to listen to their hearts and see how

[221] Behringer, *Witches and Witch-Hunts*, 174–76.
[222] Spee, *Cautio Criminalis*, 37, 80, 112–13.
[223] Spee, 98, 7.
[224] Meyfart, *Christliche Erinnerung*, 57.
[225] Witekind, *Christlich Bedencken und Erinnerung von Zauberey*, 148.
[226] Witekind, 214.
[227] Hocker, *Der Teufel selbst*, 297; Pleier, *Malleus Judicum*, 61–70; Spee, *Cautio Criminalis*, 49.

irresponsible and unchristian witch-hunts were. "Killing persons is a big thing," he said, so doing that without proper evidence was a grave sin, sure to provoke God's wrath.[228] Meyfart reminded the people responsible for trials that the end of times was near and that their unspeakable crimes severely jeopardized their salvation.[229] His message was clear: do not let yourself be fooled by the devil.

That both sides in the debate used remarkably similar supernatural arguments could make things puzzling and arbitrary. The French literary scholar Marianne Closson remarks: "Great confusion reigned: how could one know the true nature of diabolical illusion? The lack of a firm answer to this question fuelled the fear of being oneself a victim of the devil's deceptions."[230] Whether people considered something real or illusory could eventually depend on what seemed most convenient or plausible. The "trump card of demonology" is what Dillinger calls the idea of diabolical deception. It could be invoked to neutralize any ostensible phenomenon that appeared unfitting.[231] We might think of the debate as occurring between two "bodies" of thought possessing similar "immune systems" that could disarm any attack quickly.

This arbitrariness was a strength but also a weakness for both camps. Proponents argued that witches were seemingly in bed at night while the real people feasted at the sabbath. But critics remarked, why not presume that the real people were in bed while the ones feasting were illusory?[232] On the other hand, a sceptic like Weyer poked fun at apparently crazy supernatural beliefs, but as Bodin retorted: "Weyer, who wants to discuss metaphysics as a physicist, confesses in a thousand instances in his book that the demons, who are intelligent spirits, enter people's bodies."[233] Yet, despite this level playing field, we may still appreciate the moral and intellectual validity of one of the critics' key arguments: in case of uncertainty, one should not burn people.

[228] Witekind, *Christlich Bedencken und Erinnerung von Zauberey*, 251.

[229] Meyfart, *Christliche Erinnerung*, 266.

[230] Closson, *L'imaginaire démoniaque en France (1550–1650)*, 32.

[231] Dillinger, *Hexen und Magie*, 45.

[232] Gödelmann, *Von Zäuberern, Hexen und Unholden*, 209; Witekind, *Christlich Bedencken und Erinnerung von Zauberey*, 170.

[233] Bodin, *De la démonomanie des sorciers*, 473. "Wier, qui veut dispiuter de la Metaphysique en phisicien, confesse en mille endroits de ses livres que les Diables, qui sont formes intelligibles, entrent au corps des hommes."

One difficulty in interpreting such claims about the supernatural is the extent to which people meant what they said. Stuart Clark is amongst those most determined about proponents and opponents sharing the same demonological worldview. Still, even he considers the option that Spee was merely strategic or ironic in stating that witch-hunters were themselves witches.[234] Regarding De Salazar, the expert Gustav Henningsen thinks that he was not really interested in demonology and only invoked diabolical delusions to take the witch-hunters down on their own turf.[235] But again, we should be careful when reading things in texts that the texts do not explicitly say. Take De Salazar. The expert Lu Ann Homza more recently argued that his viewpoints were entirely traditional, demonstrating obedience towards authority.[236] De Salazar's work contains no assertions that demons do not exist, and he advocated a fierce inquisition of real heresy. No valid reasons suggest that De Salazar was a proto-enlightened sceptic. Religious stridency can also be found amongst other critics, like Loos and Spee, who favoured the persecution of Protestants, or Witekind, who vehemently denounced dancing as a diabolical sin.[237] Witch-hunts proponent Jean Bodin thought he received help from a spirit, but interestingly, so did Weyer.[238] It was evident and plausible to think that demonic or angelic interferences were real. So, the burden of proof today remains on those who claim that early modern critics did not mean what they said.

All this bears upon the instrumentalization debate. On the one hand, the German instrumentalists were right to point to contemporary opponents of witchcraft who, too, assumed that witchcraft accusations and trials were used for various hidden ends. While these findings do not *prove* the instrumentalist interpretation, they offer an indication. However, sources also reveal that this was a limited part of the story. Contemporary suspicions about ulterior motives dwindle compared to the supernatural dimension of diabolical trickery. Much of what contemporary critics said is more in line with what the "genuine belief historians" maintain today. The people responsible for witch-hunting were not so much thought of as carefully weighing their chances on how to use trials to their own hidden

[234] Clark, *Thinking with Demons*, 207.

[235] Henningsen, *The Salazar Documents*, 94–95.

[236] Homza, "An Expert Lawyer and Reluctant Demonologist."

[237] Voltmer, "Demonology and Anti-Demonology," 160; Schmidt, *Glaube und Skepsis*, 234; Pleier, *Malleus Judicum*, 44, 110; Witekind, *Christlich Bedencken und Erinnerung von Zauberey*, 148.

[238] Valente, "'Against the Devil, the Subtle and Cunning Enemy,'" 104–5.

advantage strategically. At the time, they were already described as zealots, full of sincere belief, while the trials' consequences were already considered harmful, even for these people themselves. The instrumentalists are cherry-picking the sources: they do not, or hardly, focus on the supernatural aspects so crucial to critics of the time. The instrumentalist model does not sufficiently address the impact of magical and religious thought.

However, in a different way, the early-modern critics *did* see witch-hunting as a massive case of instrumentalization. But it was not the accusers or persecutors who were primarily thought to instrumentalize witch trials. The actors in the witch-hunting events were trapped and exploited by something beyond themselves. To their own detriment, they operated as the puppets of an invisible puppeteer. Nor were these actors the ones to shrewdly think through the incendiary details that made witch persecutions so big, like torture or the sabbath. The alleged creator and beneficiary behind witch-hunting's intelligent design was Satan. This "subtle and devious spirit," "sly fox," "crafty old weaver," and "jack-of-all-trades" manufactured and instrumentalized trials to *his* benefit. Satan shrewdly knew how to use aspects of the environment, like people's gullibility, unstable old women, the potential of torture, evil inclinations, or the randomness of chain accusations, to set people against each other and to bring about humanity's ruin and damnation. *This* was how the critics primarily answered the *cui bono* question.

These early-modern claims are neither very much in line with the German instrumentalist model nor with that of the genuine belief historians. Yet, in some striking regards, they do bear an eerie resemblance to the proposals made here. This book likewise argues that the witch-hunt system was a design, albeit not one designed by human actors. Furthermore, it is assumed that a hidden beneficiary was involved, with its own purposes. This "agency" tricked people into making them do things that significantly harmed these people's neighbours, their communities, and, in many cases, even these people themselves. But the hidden "agency" was not Satan, and neither was the ruin of humankind its ultimate goal. Instead, it was the cultural phenomenon of witch-hunting that benefitted, and its underlying purpose was blind reproduction. The analyses of contemporary critics cannot *prove* that these Darwinian proposals are warranted, but they offer substantial indications. Many of the empirical findings presented here are surprisingly congruent with the idea of a virus-like cultural entity that spread at the expense of its human hosts. This "viral" analogy will now be spelled out in more detail.

8.4 "Viral" Hunts?

"As the Alps represented the product of tectonic activity, so too was this a region of cultural tectonics, where the interaction of cultures was reflected in this remarkable conjunction of writing."[239] This is how Richard Kieckhefer describes the appearance of the cumulative concept of witchcraft in a linguistically splintered area. Jürgen Michael Schmidt notes how a fifteenth-century trial in Heidelberg turned the city into "a small epicentre with a certain power of its own," while Katrin Moeller remarks how criticisms of trials created "processes of erosion."[240] Rita Voltmer and Karl Weisenstein speak of a "spark" of accusations that made witch-hunting disperse from place to place, and Schmidt states that, in 1446, the witch-hunting moved from Switzerland straight to Heidelberg "like a bolt of lightning."[241] Another comparison regularly used is that between the simultaneous occurrence of trials and the phenomenon of waves.[242]

What should we make of such comparisons to physical phenomena? Are they mere stylistic metaphors without scientific import, or do they reveal more profound similarities between witch-hunting and physical nature? The question lies beyond the scope of this book. Yet, one analogy that is more common than any *does* matter here: epidemic disease. Already in 1675, the French philosopher Nicolas Malebranche used witch beliefs as an example of "contagious communication and powerful imaginations," arguing that people's thirst for scary stories led to the accumulation of tall tales about witchcraft.[243] More recently, Wolfgang Behringer observed that "The implementation of witchcraft persecutions spread contagiously, but any politically coordinated effort with that direct intent was conspicuously lacking."[244] Elsewhere, he writes that Petrus Canisius, a supporter of exorcisms and witch-hunting, "presumably infected a whole generation of young Jesuits with these ideas."[245] Voltmer speaks of lists of accomplices that could turn "endemic" trials into "epidemic" ones, while Johannes Dillinger notes that "At the end of the 1620s, as before in the 1590s, the

[239] Richard Kieckhefer, "The First Wave of Trials," 168.

[240] Schmidt, *Glaube und Skepsis*, 48; Moeller, *Dass Willkür über Recht ginge*, 482.

[241] Voltmer and Weisenstein, *Das Hexenregister des Claudius Musiel*, 33; Schmidt, *Glaube und Skepsis*, 11, 48.

[242] Behringer, *Hexenverfolgung in Bayern*, 128.

[243] Malebranche, *Recherche de la vérité*, I: 278–338.

[244] Behringer, "Witchcraft Studies in Austria, Germany and Switzerland," 84.

[245] Behringer, "Canisius, St. Peter (1521–1597)."

population of the Dreiherrischen became 'infected' by the witch-hunts of the electoral state of Trier."[246] The English historian Diarmaid MacCulloch explores a further intriguing analogy: "The chronology and geography are as episodic and developmental as the gradual spread of a slowly incubating disease."[247]

The similarities between pathogens and the witch-hunting phenomenon are indeed striking. Sudden outbursts of witch persecution resemble "outbreaks" of contagious disease. Through the spread of fear and the naming of accomplices, witch trials in one place could then "infect" other places, leaving many casualties in their wake. Incubation means that people can carry a pathogen and infect others without outwardly showing disease symptoms. This looks vaguely similar to the relationship between the cumulative concept of witchcraft and the practice of witch-hunting: people often carried the concept in their minds without directly instigating trials. When the prince-provost Von Westerstetten left Ellwangen to take a position as prince-bishop of Eichstätt, it was not immediately visible from the outside that within his brain cells resided an extremely lethal "strain" of witch beliefs that later helped trigger the hunt of Eichstätt that killed hundreds. Epidemiology is also familiar with the concept of "superspreaders," unusually contagious people who contaminate many others. We might think of Petrus Canisius as a case in point. Canisius became seriously "contaminated" around 1562, and through his preaching and teaching, he then "infected" innumerable others with this potentially lethal cultural "virus."

Coronaviruses are known for their little spikes, which enable them to ingeniously attach themselves to human cell receptors; the proteins on the cell's surface that function as openings.[248] The cumulative concept of witchcraft did not literally contain spikes, but it did possess ingenious methods of attaching itself to people's thought worlds. By connecting to deeply rooted fears and interpersonal conflicts, the concept entered people's minds and made them see things in a different light. Unusual events became acts of witchcraft, and distrusted individuals became diabolical witches. We noticed that viral and bacterial pathogens "manipulate" or "hijack" the behaviour of their carriers for the pathogen's own reproductive purposes. Think of the rabies virus and how it turns dogs into restless

[246] Voltmer, "Herren und Hexen," 16; Dillinger, *Böse Leute*, 382.

[247] MacCulloch, *Reformation*, 563.

[248] Saplakoglu, "Here's Why the New Coronavirus Is so Good at Infecting Human Cells."

wanderers with foaming mouths, eager to bite. Asking what function this behaviour has for the dog will not get us anywhere. But when we ask what function it has for the rabies virus, it all makes perfect functional sense. Along similar lines, we might interpret the cumulative concept of witchcraft as a cultural "virus" that entered people's minds and "manipulated" their behaviour: people started panicking, accusing others, torturing and killing suspects, and spreading accusations over long distances. The lesson of epidemiology is that asking what function it had for the actors involved does not necessarily give us more insight. But when we ask what function it had for reproducing the witch-hunting phenomenon itself, it all makes perfect functional sense.

Are such comparisons merely permissible as stylistic metaphors, or do they warrant serious scientific consideration? The comparison between socio-cultural phenomena and infectious disease is an old one. During the Reformation, opposing factions eagerly framed each other's religions as "plagues," and after the hypothesis of infectious disease turned into a scientific fact, the comparison became even more prevalent.[249] The French precursor of social psychology, Gustave Le Bon, studied the behaviour of mobs and thought that it could be dictated by forces outside of people's personal interests: "Within crowds, ideas, sentiments, emotions, and beliefs possess a contagious power that is as intense as the power of microbes."[250] In popular culture, the analogy remains pervasive to this day—just think of "epidemics of violence," "computer viruses," or social media content "going viral." The comparison also keeps popping up in academic literature.[251]

Much academic scholarship, however, has grown sceptical of the comparison.[252] One of the reasons is moral misgivings; the analogy has proven susceptible to political and ideological abuse. The Nazis were obsessed with biological comparisons and often marked the Jews as "bacilli," "germ carriers," or a "racial tuberculosis" that threatened the Aryan body.[253] Since disease should be eradicated, this dehumanizing language helped remove moral considerations and made people more prone to committing genocide. Strikingly, the moral hazards of the pathogenic comparison also

[249] Mercier, *Not Born Yesterday*, 96–97.

[250] Bon, *Psychologie des foules*, 113.

[251] Hatfield et al., *Emotional Contagion*; Christakis and Fowler, *Connected*; Gladwell, *The Tipping Point*.

[252] Davis, "Contagion as Metaphor."

[253] Lang, *Genocide*, 25; Sontag, *Illness as Metaphor*, 84–85.

became apparent during witch-hunts.[254] Friedrich Förner described witchcraft as "a plague of the soul" and wanted to ensure that "the healthy part will not be impaired and the sick cattle will not infect the entire sheepfold."[255] Such language did not stimulate caution and mildness on the persecutors' part.

In addition to moral reservations, there are scientific misgivings. The comparison is often considered a catchy but deceptive metaphor, and many objections raised against meme theory are also generally used against this idea of socio-cultural "epidemics." The differences between biological and cultural processes are thought to be too substantial for this analogy to take hold. Zachary Warren and Séamus Power note, "Ideas, beliefs, and behavior are normatively regulated and made up of meanings and symbols, whereas diseases are made of biological agents." Microbes and viruses "cannot make meaning out of their social world the way persons do and do not negotiate and choose to participate in 'contagion' the way humans do."[256] The comparison is thought to reify culture and discharge humans from their responsibilities. The cognitive scientist Hugo Mercier (see 7.1) raises evolutionary objections: "We evolved neither to send nor to receive pathogens – indeed, a good chunk of our evolution is devoted to avoiding the effects of pathogens."[257] By contrast, we *did* evolve to receive signals from other humans, and our natural "vigilance" shields us from harmful signals.

The many ideological mishandlings of the comparison should indeed alarm us, especially when humans rather than forms of culture are likened to pathogens. Still, moral hazards should not prevent us from further exploring the analogy, as this would imply a moralistic fallacy. Having said that, good moral reasons do exist for examining the parallel in more detail. If cultural phenomena can evolve pathogen-like traits, would it not be in our best interest to learn more about this? Our understanding of contagious diseases has saved innumerable lives. Might a better understanding of conceivably similar cultural patterns enhance human well-being, too? More insight into culture's ways may help us defuse some of culture's harmful tricks. Or as Ted Cloak put it: "If we *are* the slaves of some of

[254] Rügge, *Die Hexenverfolgung in der Stadt Osnabrück*, 112.
[255] Förner, *Dämonenglaube und Zauberei im Jahre 1625*, 131–32.
[256] Warren and Power, "It's Contagious," 361–66.
[257] Mercier, *Not Born Yesterday*, 106.

'our' cultural traits, isn't it time we knew it?"[258] In this light, it is laudable that the philosopher Andy Norman is pioneering the science of "mental immunity" against alleged mind parasites.[259]

We might better not literally describe cultural phenomena as viruses, bacteria, or epidemics. Those words were developed for biological things, and the differences with cultural phenomena are innumerable. This is why such biological terms are continually put between inverted commas here when applied to culture. Yet, what may still be interpreted in a more literal sense and what does not require inverted commas are some of the similarities among the underlying processes. Alex Mesoudi justifiably states that what happens in cultural processes is not "a bit like" Darwinian selection but, to a certain extent, *is* Darwinian selection.[260] The reality of the equivalence is also plausibly why the outcomes of biological and cultural processes can be markedly similar in some regards.

That analogies regarding epidemic disease can have real scientific value becomes apparent in Wolfgang Behringer's following comment: "Witch beliefs are not confined to functionally explainable features. There is a fascinating surplus of fantastical elements, fairy-like attributes, which cannot be explained by mere functionalism, such as the ability to fly through the air."[261] Behringer thought of Durkheimian functionalism here, and from that angle, the belief in flight indeed appears superfluous. But this is why Daniel Dennett stressed that we need to cast our nets out widely, as functional beneficiaries can be elusive. The analogy of the "selfish" cultural "pathogen" can make us ask a new question: how does the witch-hunting phenomenon's own virus-like reproduction benefit? From that perspective, the flight belief's functionality suddenly stares us in the face: it enhanced the sabbath notion, made witchcraft more dangerous, and helped trigger chains of accusations over long distances.

Importantly, analogies about cultural "pathogens" do not make meaning or symbols irrelevant, let alone imply that humans cannot make decisions. The lesson is different: human decision-making is significantly impacted by the culture we absorb, and not all functional properties of symbolic and meaningful culture were designed by human hand. Paraphrasing Clifford Geertz, we might say that humans are animals

[258] Cloak, "Is a Cultural Ethology Possible?," 178.
[259] Norman, *Mental Immunity*.
[260] Mesoudi, *Cultural Evolution*, 2011, viii–ix.
[261] Behringer, *Witches and Witch-Hunts*, 2.

suspended in webs of significance that they have not only spun themselves. To a certain extent, it was blind Darwinian evolution that did the spinning. The pathogenic analogy teaches us that these webs are not necessarily spun to serve human interests. An essential purpose involved is the "egotistical" reproduction of those webs themselves.

To many humanistic scholars and social scientists, this may sound like reification at its worst, with culture operating as an agency with its own interests. Human culture obviously cannot exist without us and only obtains meaning when we give it meaning. But the analogy to viruses is especially enlightening here, as viruses cannot exist without a host either.[262] Cellular organisms survive due to the ability to manufacture proteins and metabolisms of their own, but viruses cannot do that. "Viruses are just matter with informational content," the virologist Michael Cordingley explains.[263] To persist, viral DNA or RNA must reside inside a cellular organism and use the organism's machinery for the viruses' own reproductive purposes. Cordingley: "Viruses are obligate parasites of cells, but act egotistically and without regard to the success of their host's genome: they are *egotistical, independently evolving infectious information*."[264] This is where the viral analogy is helpful: culture also consists of tangible matter with informational content—in this case, located in brains or artefacts—which cannot persist without a population of (human) hosts. The viral analogy shows that none of this implies that culture cannot have "egotistical" purposes of its own. Reification means turning something abstract into something real. But culture *is* real, so its reification does not have to be a sin. It may actually be a virtue.

Mercier is right that our minds possess evolved mechanisms that defend our interests. Yet, though our bodies also contain evolved immune systems that protect our interests, pathogens often make us ill—they evolved ways of evading immune systems. Why should cultural "pathogens" not be able to elude human inborn vigilance? A crucial trick of the cumulative concept of witchcraft was that it appealed to some of our human vigilance for certain recurrent dangers. Take fear or anger; emotions evolved to protect humans against real threats. Yet, during a witch-hunt, these defensive mental features were directed at an imaginary threat and no longer served the purposes for which they originally evolved. In this case, a

[262] Cordingley, *Viruses*.

[263] Cordingley, 2.

[264] Cordingley, 18.

cultural phenomenon *exploited* those traits to its own benefit. Some cognitive scholars are also more open-minded about the idea of witch-hunting subverting people's interests. Nora Parren argues that biologically adaptive cognitive traits can stimulate witch persecution but that "the end result may actually be anti-social and have negative consequences for everyone."[265]

As I said, such ideas will raise many eyebrows among historians. However, as with Darwinian cultural evolution in general, it is not difficult to reconcile the pathogenic analogy with things that historians traditionally hold dear. Take historical contingencies. That a new coronavirus was able to conquer the world in 2020 was not due only to its inherent properties but also to a stroke of luck. The strain appeared early in the well-connected Chinese city of Wuhan, after which people who had passed through quickly took the virus across the globe. Things could have ended differently if the pathogen had appeared in a provincial backwater. In this light, we might wonder what would have happened to the Alpine witchcraft mythology if the great ecumenical council of 1431–1437 had not occurred in its vicinity. Without this "superspread event" of Basel, the Alpine concept might have remained stuck in its local context. Evolutionary success in a "viral" model is not predetermined but depends on contingent conditions.

A further parallel is the circumstances in which epidemic diseases are liable to strike. Take the example of cholera.[266] The bacterium was long endemic in the Bay of Bengal, where it caused seasonal epidemics. But in the nineteenth century, new reproductive opportunities appeared. The bacterium spreads through the faecal-oral route and thrives in poor sanitary conditions. When humans drink contaminated water, it enters the body, causing heavy and potentially lethal diarrhoea. Through this route, the bacteria re-enter the water to be drunk by other humans, and the cycle starts again. The Industrial Era, with its globalized trade and polluted slums full of malnourished people, provided ideal circumstances for the bacterium. Cholera happened to be pre-adapted to, and then further adapted to, the conditions of a new era, becoming a worldwide evolution-

[265] Parren, "The (Possible) Cognitive Naturalness of Witchcraft Beliefs," 411; Koning, "Witchcraft Beliefs and Witch Hunts."

[266] Snowden, *Epidemics and Society*, 233–68; Crawford, *How Microbes Shaped Our History*, 130–38.

ary success. As a historical expert writes about epidemic diseases: "They exploit the features of a society that are social, economic, political, and environmental."[267]

Similarities to the spread of the witch-hunting phenomenon come to mind. Here, we might think of a reproductive cycle in which beliefs caused trials, and trials caused beliefs. The phenomenon also revealed ingenious abilities to exploit social, economic, political, and environmental circumstances. By 1560, witch-hunting had already existed for some time, but like cholera, it happened to be pre-adapted and able to further adapt to the conditions of a new era. The climatic disasters, economic hardship, religious fervour, apocalyptic anxieties, and politico-religious division of the decades around 1600 provided ideal conditions for this cultural "pathogen." Especially Germany, with its political fragmentation, vulnerable economies, ecclesiastical states, ardent clerics, and zealous rulers, was susceptible to "outbreaks" of witch-hunting.

The comparison to viruses or bacteria can remind us that Dennett's *cui bono* question does not have to be addressed unequivocally. We learned that the relationship between organisms and the microbes or viruses residing inside them can be ambiguous. Just think again of Ed Yong's statement about "a continuum of lifestyles, between 'bad' parasites and 'good' mutualism, with allies who can disappoint us and enemies who rally to our side." If applied to our cultural case study, this would suggest that the relationship between the witch-hunting phenomenon and its human hosts may also have been ambiguous. The key focus here is how the phenomenon harmed its human hosts, but this does not rule out that hosts in certain cases benefitted. I have already pointed out that the German instrumentalists have brought forward a relevant perspective that has some grounding in the sources, and we need to remain open-minded about cases where people did (sub)consciously use trials to their own advantage. Hence, the Darwinian approach can help us address the *cui bono* question in a multifaceted manner, integrating the strengths of other approaches while adding an unexplored layer of explanation.

The functions we attribute to specific elements of the witch-hunt system can also be multifaceted. The example of the elephant's trunk already showed how things in a Darwinian world can have multiple functions simultaneously. Regarding the sabbath notion, the focus here mainly lies on its enhancing effect on trials through the creation of chains of

[267] Snowden, *Epidemics and Society*, 234.

accusations, but we should also consider additional functions, that is, other effects that enhanced the notion's survival. Stuart Clark argues that the sabbath notion, in this "age of cognitive extremism," with people thinking in binary opposites, made a good fit.[268] This is why it "was able to survive for so long as something invariant and intelligible – something eminently thinkable."[269] In a Darwinian framework, we can translate this insight as the sabbath notion being well adapted to its intellectual environment. In addition, we might think of entertainment and pornography as functions.[270] The extravagant and sexually loaded sabbath descriptions of authors like Delrio and De Lancre must have been exciting to read. For such authors, as well as for their readers, it possibly fulfilled a subconscious function of indulging in pornographic interests under the self-deceptive veneer of a pious cause. De Lancre considered what happens at the sabbath "so dirty that it is shameful to talk about," but this did not prevent him from discussing it in outstanding colour and detail.[271]

Another insightful evolutionary notion is the "evolutionary trade-off." It concerns the idea that an increase in the potency of one adaptive trait causes a decrease in the potency of other adaptive traits.[272] For instance, running fast offers animals the advantage of escaping predators or catching prey. But the benefits come at significant costs. Running fast demands much energy, creates risks of falls and slippages, and requires longer legs prone to bone fractures, while enhanced manoeuvrability reduces physical strength.[273] From this angle, we might also consider the evolutionary costs and benefits of some aspects of the witch-hunting phenomenon. The sabbath notion was an adaptive trait, but we should also consider a reproductive downside: reduced credibility. Witches' gatherings already appeared fantastical to many contemporaries, so the notion simultaneously hindered the spread of witch-hunting.[274] Regarding evolutionary trade-offs, we might furthermore wonder why the expensive practice of burning witches at stakes did not give way to burning witches in ovens, as happened in Zeil

[268] Clark, *Thinking with Demons*, 1–148.
[269] Clark, 134.
[270] Goodare, *The European Witch-Hunt*, 139; Machielsen, *Martin Delrio*, 221.
[271] Lancre, *On the Inconstancy of Witches*, 199.
[272] Flatt and Heyland, *Mechanisms of Life History Evolution*, 141–42.
[273] Wynn et al., "Running Faster Causes Disaster."
[274] Briggs, *Witches & Neighbours*, 34–36, 51.

and some other places.[275] Plausibly, in the economy of culture, stake burnings were worth their costs as their public display fulfilled a pivotal function of "infecting" the audience more deeply with the witchcraft doctrine.[276]

We could also think in terms of evolutionary trade-offs regarding the social categories that were targeted. After its initial phase in the Alps, witch beliefs and trials long focused especially on women, who were often relatively poor. For the reproduction of the witch-hunting phenomenon, this offered the advantage of tapping into misogynous stereotypes, while people in power did not have to feel threatened about being targeted themselves. Witch-hunting should not be understood as "women hunting," but this type of persecution still crucially depended on and benefitted from its patriarchal and misogynous context. However, this persecutory focus also created evolutionary disadvantages. It reduced the short-term reproductive potential, as half the population, men, remained largely unassailable. Confiscations also provided little funding for further trials, and the witchcraft doctrine made itself vulnerable to specific criticisms: sceptics poked fun at witch trials often being based on confessions of women, who purportedly made credulous and unreliable witnesses.[277] And why would a figure as imposing as Satan have to rely on such insignificant creatures?[278]

This weak spot was resolved by the late sixteenth and early seventeenth centuries, as almost all social groups were targeted. Chapter 7 showed how this opened up new reproductive vistas for the witch-hunting phenomenon. But in their turn, these trends also had their costs. A crucial effect of indiscriminate persecution was that more groups in society began to feel threatened by trials and to become more determined in their resistance.[279] Earlier on, we learned that the persecutions of Trier and Ellwangen enhanced trials in other places but hindered further hunting in Trier and Ellwangen themselves. Once people had seen with their own eyes what horror such trials caused and learned how their networks had been attacked, they became more hesitant about starting a witch-hunt again.

[275] Behringer, *Witches and Witch-Hunts*, 154; Voltmer, "Hexenbrenner Und Hexenbischöfe," 215.

[276] Schormann, *Hexenprozesse in Deutschland*, 33.

[277] Krause, *Witchcraft, Demonology, and Confession in Early Modern France*, 126; Kieckhefer, *Hazards of the Dark Arts*, 118.

[278] Weyer, *Witches, Devils, and Doctors in the Renaissance*, 563.

[279] Midelfort, *Witch Hunting in Southwestern Germany*, 121–63, 178–92.

Such effects were also notable after more modest panics.[280] Overall, indiscriminate panics were often followed by increased caution and resistance.[281]

It is probably not by accident that in places where trials continued steadily, persecutions were more restricted in their social profiles and less overwhelming in their scale. In a comparative study of sabbath notions in Lorraine and Trier, the German historian Elisabeth Biesel noted that Lorraine sabbaths, in contrast to Trier's extravagant sabbaths, were imagined as festival-like village events attended by around six to thirty witches.[282] Apart from the content of Lorraine's beliefs, Robin Briggs also points to Lorraine's judicial peculiarities that prevented trials from spiralling out of control.[283] They led to "a curious equilibrium" in which no mounting resistance to persecution came about, and trials eventually produced one of the highest death tolls in Europe.[284] Rottenburg provides another relevant case. Its panics of the late sixteenth century failed to bring witch-hunting to an end, and Erik Midelfort links this to its relatively narrow range of victims: "the stereotyped identification of witches as old women suffered so little shock," he argues, that support for persecution remained substantial enough for trials to continue. Only around 1600, when high officials came under attack, did Rottenberg's witch-hunts decline.[285] In sum, in the places where they occurred, indiscriminate panics offered short-term reproductive advantages but long-term reproductive disadvantages.

The observation that large "outbreaks" of witch-hunting provoked continued resistance has given rise to one of the most common epidemiological analogies of witchcraft historiography: immunization. Wolfgang Mährle observes that after the persecution of 1611–1618, Ellwangen reached a "high level of immunization," and regarding the southwestern German hunts of the late 1620s, Erik Midelfort notes that the "lesson

[280] Behringer, *Hexenverfolgung in Bayern.*

[281] Levack, "The Decline and End of Witchcraft Prosecutions," 432; Midelfort, *Witch Hunting in Southwestern Germany*, 121–63.

[282] Biesel, "Les descriptions du sabbat."

[283] Briggs, *The Witches of Lorraine.*

[284] Briggs, 386.

[285] Midelfort, *Witch Hunting in Southwestern Germany*, 90–94, 179.

learned in those panics may have provided immunity from witch trials."[286] Epidemiologists also know that a population's immunity can be temporary: when the immunized generation has died, people become vulnerable again, and diseases strike anew.[287] This recurrently happened in the history of the plague. It may remind us of what Thomas Fritz observed in Reutlingen (see 8.1): after a few generations, the horrid experience of earlier hunts was forgotten, so trials occurred anew. Yet, overall, "immunization" often seems to have had durable effects, as panics did not frequently occur in the same place more than once or twice.[288] Wolfgang Behringer observes that local historians often note that witch persecution in their region was only an exception, making them atypical. But Behringer argues that the atypical was, in fact, typical: "As a general rule we can conclude that the burning of witches was an exception – everywhere."[289] He explains: "Normally, people are not interested in burning their neighbours, and governments do not wish to wipe out their tax-payers."

In Sect. 6.5, it was stressed that this finding of panics not often occurring in the same place more than once or twice provides a counterargument to the idea that the witch-hunting phenomenon was a product of cultural group selection. After all, witch trials should have occurred more frequently within single groups for any such selective process to take hold. Witch burnings being the exception everywhere also creates further problems for the instrumentalization approach. If trials were as multifunctional as presumed, "immunizing" effects should have been unlikely. If people (sub)consciously learned how witch-hunts served their ulterior interests, we should expect them to become eager to re-use the tool. But historical records reveal something else: many people were overtaken by the witch-hunting phenomenon and, in retrospect, realized that their interests had been harmed and something had gone awry. This provides solid clues for a "selfish meme"-like model. It seems that we are dealing here with a cultural "pathogen" that moved from place to place, leaving many casualties and "immunized" survivors in its wake. The intriguing parallel of "immunization" may also help us address our last question: why did this well-adapted phenomenon eventually cease?

[286] Mährle, "Ellwangen - Hexenverfolgungen"; Midelfort, *Witch Hunting in Southwestern Germany*, 194.

[287] Krause and Trappe, *Die Reise unserer Gene*, 197–200.

[288] Roper, *Witch Craze*, 11; Goodare, *The European Witch-Hunt*, 245.

[289] Behringer, *Witches and Witch-Hunts*, 148–49.

8.5 The End of Witch-Hunting

The critical question in this book is why witch-hunting became so big, not why it eventually ended. Explaining its rise is also the more challenging theoretical problem: that people started doing something that harmed their interests is more puzzling than that they stopped doing it. Still, the decline of witch-hunting is an exciting topic in its own right and may also help us further reflect on the Darwinian proposals made here. After all, if the witch-hunting phenomenon was so well adapted to enhance its own spread, why did it falter after 1630 to come to a definitive end in the eighteenth century?

The most obvious direct cause of witch-hunting's decline after 1630 was a European development external to the phenomenon: the Thirty Years' War. We learned that its first phase presented no hindrance to witch-hunting. However, the early 1630s signified a tipping point, as the impact of the Swedish army on the Bamberg hunt has already illustrated. In those years, the war turned into one of the most brutal and devastating conflicts in European history. Many German and foreign armies traversed the lands, pillaging and brutalizing the population. Famines and epidemics raged, and in some regions, up to sixty to eighty percent of the population perished.[290] Hardship typically enhanced witch trials, but in this case, the deprivation was so extreme that it left little capacity for witch persecution.

As such, there was nothing special about witch-hunting receding for a while. Earlier waves also came and went. What especially warrants attention is that trials did not return to earlier levels after 1648.[291] Persecution resurged in the 1650s and 1660s, and some waves of trials continued after that. Yet, the overall trend was downward.[292] One explanatory strategy of historians is to consider broad changes to European history that were not necessarily related to the witch-hunting phenomenon but still negatively impacted it. From a Darwinian angle, we might frame that as changes to the "environment" that made witch beliefs and trials less well adapted.

One factor was the climate, which stabilized somewhat.[293] Weather conditions remained extreme compared with earlier times, but people had

[290] Labouvie, *Zauberei und Hexenwerk*, 251–52; Midelfort, *Witch Hunting in Southwestern Germany*, 75; Decker, "Die Hexenverfolgungen im Herzogtum Westfalen," 335.

[291] Goodare, *The European Witch-Hunt*, 321.

[292] Dillinger, *Hexen und Magie*, 137–52.

[293] Matthews and Briffa, "The 'Little Ice Age,'" 21.

become more used to them. Severe cold and frost no longer seemed to require a particular explanation. After 1650, the strains of overpopulation and contracting resources also lessened.[294] The dramatic population loss in the Thirty Years' War left more resources for subsequent generations, and while life remained harsh for many, socio-economic conditions did improve a bit. One can imagine people to have felt slightly less at the mercy of contingent events. This left fewer opportunities for the witchcraft concept to stoke hatred.

Another significant development was the consolidation of state power and shifts in judicial practices.[295] Witch panics had especially flourished in politically fragmented areas where sloppy judicial procedures were not hindered by external interference. After 1650, such conditions became more unusual. Centralized states grew more powerful and developed increasingly professional judicial procedures that relied less on torture. States were also more secular in their orientation, with religious zealotry giving way to raison d'état.[296] Religion remained pivotal in politics and people's personal lives, and persecution of religious deviance remained common, but the ideal of the godly state did lose some of its vigour.[297] The horrors of religious intolerance and warfare had disillusioned many people, making them wary of religious fanaticism and more lenient towards religious pluralism.

In religious thought, the providential tradition that explained misfortune as direct divine punishment also remained a competitor to witchcraft as an explanation.[298] But a different competitor posed the most lethal threat: natural causation.[299] The judicial confrontation in 1620 between the accusers of Katharina Kepler and the defence by her son, Johannes Kepler, foreshadowed a conflict of worldviews that would eventually bring witch-hunting to its knees. Kepler did not think in terms of a ubiquitous devil in a decaying world, but being an exponent of the Scientific Revolution, he was more interested in natural causes and human

[294] Levack, *The Witch-Hunt in Early Modern Europe*, 249–51; Schilling, *Die neue Zeit*, 285–344.

[295] Dillinger, *Hexen und Magie*, 144–48.

[296] Levack, *The Witch-Hunt in Early Modern Europe*, 245–49; Goodare, *The European Witch-Hunt*, 329–34.

[297] Kaplan, *Divided by Faith*.

[298] Bever, "Witchcraft Prosecutions and the Decline of Magic," 276.

[299] Levack, *The Witch-Hunt in Early Modern Europe*, 240–45.

progress.[300] We should not overestimate the importance of this revolution for witch-hunting's decline, as trials were already on the retreat before the new science began to have much of an impact. Yet, when elites began to adopt rationalized views in the seventeenth century, they were less likely to favour supernatural explanations over natural ones.

Lastly, changes to ideas about women deserve consideration.[301] In contrast to stereotypes about sexually rapacious creatures who easily allied themselves with the devil, from the mid-seventeenth century onwards, stereotypes came to focus on the alleged weakness and delicacy of females. "Women were increasingly seen as frail rather than threatening, needing protection, not control," Bernard Capp writes.[302] Women came to be thought of as less interested in, or even resistant to, sex. Such stereotypes also enhanced patriarchy, as they presented females as unfit to rule. However, the idea of women being involved in diabolical intercourse and rampant feasting at the witches' sabbath did not connect well with these new categories.

In addition to specific *external* changes, we must address *internal* mechanisms within the witch-hunting phenomenon. We already encountered mechanisms of "immunization" and in certain ways, such self-limiting effects were already present from the beginning.[303] In Arras, people had become wary of witch-hunting after the panic of 1459–1460. Jurists in Rothenburg ob der Tauber wrote negatively about what happened in Wiesensteig in 1562, interpreting the event as a warning of the unpleasant consequences of the judicially questionable witch trials.[304] The scale of the Wiesensteig hunt enhanced further persecution but hindered it at the same time. Yet, in this period, such consequences did not yet compensate for the enhancing ones. The wave of 1630, however, seems to have been a turning point. Bamberg was a crucial case in this shift, at least in Germany.[305] The fact that many higher authorities spoke out against it while the Bamberg government was subjected to enormous external pressure discouraged trials elsewhere. Persecutors in the electoral state of Cologne were alarmed by the *Reichshofrat's* verdict, becoming more hesitant in their persecutions. The scale of the Bamberg hunt was so

[300] Rublack, *The Astronomer & the Witch*; Behringer, *Witches and Witch-Hunts*, 137, 184.
[301] Bever, *The Realities of Witchcraft*, 409.
[302] Capp, "Gender and Family," 45.
[303] Goodare, *The European Witch-Hunt*, 245–46.
[304] Rowlands, *Witchcraft Narratives in Germany*, 25–26.
[305] Gehm, *Die Hexenverfolgung im Hochstift Bamberg*, 180–81, 227.

overwhelming, its judicial procedures so irresponsible, and the carnage amongst higher classes so brutal that it stimulated wariness about witch-hunting amongst German establishments.

Throughout Germany, the enormity of the persecutory wave around 1630 stimulated what Erik Midelfort has called "the crisis of confidence."[306] Contemporaries, including people who had encouraged persecution, widely observed that the whole system seemed to have developed dynamics of its own that nobody could control and that invited social chaos and anarchy.[307] This helped bring about a shift in public opinion, with lawyers, preachers, and rulers becoming worried about public safety and innocent people being killed. "By an agonizing process, towns and finally whole regions came to reject the procedure of torture and denunciation," Midelfort notes.[308] Significantly, in his view, this new German resistance did not stem from any early Enlightenment thought, nor was it related to a decreasing belief in the devil. It was mainly the escalating *practices* of trials that made people cautious.

The new scale of persecution also stimulated more well-thought-out and compelling criticisms.[309] Johann Weyer, in 1563, directed much of his criticisms against the *belief* in witchcraft, but as we saw earlier, this made his plea somewhat arbitrary; he primarily used the same demonological reasoning as his opponents. This plausibly made his book less effective than he hoped.[310] Shrewdly, Friedrich Spee, in 1631, shifted his attention from witch beliefs to the practices of witch trials. In what Behringer calls "a Ciceronian masterpiece," Spee, at the beginning of his *Cautio criminalis*, accepts that witches exist and trials could be warranted.[311] With some false modesty, he presents his sole aim to ensure that no innocent person will be convicted. But his subsequent attack on the procedures of witch trials becomes so very devastating that little of those procedures remain. Spee eventually contests that diabolical witchcraft exists at all.[312]

Stuart Clark sees Spee's argument as part of an effective shift in criticism that increasingly focused on procedures instead of beliefs. "It was

[306] Midelfort, *Witch Hunting in Southwestern Germany*, 121–63.

[307] Briggs, *Witches & Neighbours*, 51, 360; Dillinger, *Hexen und Magie*, 144–4; Labouvie, *Zauberei und Hexenwerk*, 195, 255.

[308] Midelfort, *Witch Hunting in Southwestern Germany*, 191.

[309] Clark, "Glaube und Skepsis in der deutschen Hexenliteratur."

[310] Behringer, *Witches and Witch-Hunts*, 171–72.

[311] Spee von Langenfeld, *Cautio criminalis*, 217.

[312] Spee von Langenfeld, 287, 407.

much more difficult for critics to distance themselves intellectually from orthodox demonology than to pick holes in particular trial procedures and investigate techniques on largely technical grounds."[313] Especially torture could be criticized in a non-arbitrary manner within the boundaries of acceptable contemporary discourse. Spee's book is widely thought to have decisively weakened the witch-hunting phenomenon in Germany.[314] Within a Darwinian framework, we might interpret this as a case of culture-culture coevolution, in which the increased "virulence" and adaptedness of witch beliefs and trials provoked more well-adapted "immunizing" responses. This does not imply that everything about this process was Darwinian—the shrewd new adaptedness of Spee's argument looks more like a product of intentional design.

The decline of witch trials may also help us further compare the accounts of the German instrumentalists with those of the "genuine belief historians." The instrumentalists say relatively little about the decline, which seems related to their framework not offering a plausible scenario. After all, why would people stop persecuting witches if it provided them with such an excellent multifunctional tool? It is unlikely that people after 1630 were no longer interested in money, power, envy, or revenge. The instrumentalists have made some proposals, however. Moeller and Rummel argue that the underlying interests of witch-hunters just became too apparent at some point.[315] But this argument raises more questions than it answers. If people increasingly understood how witchcraft accusations and trials were instrumentalized for hidden ends, then why did they not decide to use it for their *own* ends? Did people after 1630 become more morally conscious? Moeller and Rummel also provide no evidence that people after 1630 gained more understanding of instrumentalization.

Another instrumentalist explanation for why witch-hunting ended is that it increasingly lost many of its functions and came to affect too many groups in society too negatively.[316] Rita Voltmer argues that the mass persecutions became so obviously destructive that people came to reject them. This, again, raises more questions than it answers. If witch beliefs and trials were created by knowledgeable agents who (sub)consciously

[313] Clark, *Thinking with Demons*, 208.

[314] Robisheaux, "The German Witch Trials," 190.

[315] Rummel, "Das 'ungestüme Umherlaufen der Untertanen'," 160–61; Moeller, *Dass Willkür über Recht ginge*, 481.

[316] Labouvie, *Zauberei und Hexenwerk*, 265; Voltmer, *Hexen. Wissen was stimmt*, 88.

knew what they were doing, then why did persecutions become dysfunctional in the first place? Voltmer's statement about mass persecutions being dysfunctional also challenges her other claim about them serving "an elite disciplinary program all too effectively." Again, the "genuine belief historians" account appears more plausible. Midelfort's scenario of a "crisis of confidence" presumes that genuine belief in the guilt of convicts crucially drove persecutions forward. When this genuine belief began to falter, trials also faltered. In addition, the self-limiting effects of panics indicate that persecutions were often dysfunctional and harmful to the actors involved.

These mechanisms allow us to draw further parallels with the evolution of contagious diseases. Among epidemiologists, it is well-known that pathogens can limit their reproduction by being too lethal.[317] Take Ebola, one of the deadliest diseases we know. The disease expert Johannes Krause writes that "Ebola kills extremely quickly, so the virus has very little time to jump from one infected person to another. Ebola outbreaks thus burn down rapidly, meaning that the virus does not have enough time to spread to more distant populations." This differs from the seasonal flu, of which "new forms migrate from Southeast Asia to the whole world almost every year."[318] Notably, the killing of hosts is not what pathogens are necessarily after—their sole purpose is survival. That pathogens still often kill can be a by-product of disease symptoms enhancing a pathogen's spread. We saw that the cholera bacterium often causes lethal diarrhoea, but the underlying rationale is not the killing but entering the water again.[319]

Pathogens survive in a world full of evolutionary dilemmas and trade-offs.[320] When highly virulent, viruses can cause rapidly progressive disease and are more readily transmitted. While reduced virulence may lengthen the life of hosts, under certain conditions, it can come at the price of lower levels of replication and fewer successful transmissions. But in their turn, the aggressive "hit-and-run" epidemics are at risk of overplaying their hand by being too deadly and triggering more vigorous responses of immune systems.[321] After their initial successes, such diseases often burn themselves up and evolve into milder forms. This, for instance, happened

[317] Modrow, *Viren*, 46; Crawford, *How Microbes Shaped Our History*, 16, 24.
[318] Krause and Trappe, *Die Reise unserer Gene*, 176.
[319] Krause and Trappe, 176–81.
[320] Cordingley, *Viruses*, 195; Crawford, *How Microbes Shaped Our History*, 24.
[321] Cordingley, *Viruses*, 117–19.

to syphilis.[322] When arriving in late-fifteenth-century Europe, syphilis was an extreme killer that slaughtered around sixteen million people within fifty years. However, due to self-limiting effects, these strains gave way to the less fearsome ones we know today.

These mechanisms loosely remind us of what happened in the evolution of witch persecution. The ultimate goal of this cultural phenomenon was not necessarily to kill people, but self-reproduction. The cumulative concept of witchcraft constantly needed to enter new minds again, and we might interpret deadly and spectacular hunts as the concept's way of achieving that end. Yet, its reproductive "strategy" of creating indiscriminate panics also made the witch-hunting phenomenon overplay its hand. Such comparisons should not be taken too literally, but the end of witch persecution does reveal striking parallels with overly deadly pathogens burning themselves up. This observation raises a follow-up question: why did evolutionary processes in the history of witch trials create such self-limiting aspects in the first place? Why did this cultural phenomenon not "decide" to stick to a more restricted profile of victims, thus safeguarding a more durable survival? Darwinian theory provides an answer: evolutionary processes do not consider the long term. As Richard Dawkins said, "Genes have no foresight. They do not plan ahead."[323] If a specific new variant enhances short-term reproductive success, it will reproduce, irrespective of long-term effects. The evolutionary mechanisms that were set in motion by the hunts of Wiesensteig, Trier, Ellwangen, and Bamberg could not foresee how, after 1630, those same mechanisms would help put the "R-number" of this cultural "virus" below 1.0.

At the same time, we should not underestimate the capacities of this cultural "virus" to respond to challenges. When real viruses are confronted with new defences of immune systems, they often evolve ways of evading them. In our cultural case study, we might thus expect processes of culture-culture coevolution, with the witch-hunting phenomenon evolving ways of escaping its new hindrances. Due to critics like Spee, torture came to be applied less frequently in witchcraft cases, thereby removing one of witch-hunting's essential reproductive tricks. If the phenomenon was to survive, it needed to find a way around that. And that is precisely what happened.

[322] Krause and Trappe, *Die Reise unserer Gene*, 221; Crawford, *How Microbes Shaped Our History*, 63–67.

[323] Dawkins, *The Selfish Gene*, 22.

Witch beliefs and trials, more than ever, came to focus on the one category of people who often confessed without torture: children.[324] From 1630 onwards, witch trials almost exclusively started with accusations from children and also targeted them more prominently as victims. The category of rich men, on the other hand, receded. This had the downside that less funding for persecution could be harnessed, making trials smaller again. But at this historical juncture, that offered the reproductive advantage of not irritating elites too much. "The mitigation of witch trials thus simultaneously ensured their perpetuation," Wolfgang Behringer writes.[325]

The late phase of witch-hunting also further shows mechanisms of "immunization." Late trials often occurred in places that had shown restraint during the peak period from 1560 to 1630.[326] We earlier saw that Lutheran Esslingen kept its head relatively cool in 1562, but one century later, it experienced a hunt that led to dozens of killings, most of whom were men.[327] Other traditionally cautious towns like Rothenberg, Nurnberg, and Salzburg also only now experienced their most dramatic trials, while former hotspots remained largely unaffected. The Lutheran town of Calw, southwestern Germany, illustrates the fact that the witchcraft concept still could cause much havoc.[328] In 1683, an eleven-year-old boy accused an older woman of teaching him demonic witchcraft while also naming children whom he allegedly saw at the sabbath. These children were pressed to tell what they knew, which provoked fanciful testimonies. Panic ensued, and parents guarded their children at night to prevent witches from approaching them. The town was rife with hatred and suspicion, with accusations becoming rampant. A few alleged witches were executed, and a mob lynched one woman. Higher authorities, however, responded differently. Calw fell under Württemberg's rule, and an investigatory committee was installed that advised against further trials. In 1684, all talk about witchcraft was prohibited, and a company of forty soldiers restored order.

Moeller's and Rummel's instrumentalist viewpoint might lead us to expect that an escalation was avoided because critics understood how

[324] Dillinger, *Hexen und Magie*, 116–18; Behringer, *Hexenverfolgung in Bayern*, 332–99.
[325] Behringer, *Hexenverfolgung in Bayern*, 341.
[326] Roper, *Witch Craze*, 133; Goodare, *The European Witch-Hunt*, 318–21.
[327] Midelfort, *Witch Hunting in Southwestern Germany*, 154–57.
[328] Midelfort, 158–63; Dillinger, *Kinder im Hexenprozess*, 117–23.

accusers and persecutors instrumentalized the panic. But this was not the case. The religious and political authorities interpreted the fears as genuine and considered the panic harmful to these people and the community. They still presumed it was the devil who cunningly benefitted from it all. *He* created the dreams in children's minds and the subsequent delusions to ruin the town and drive innocent people to their deaths. In response, preachers urged the populace not to fear witchcraft but to be all the more on the alert for Satan's trickery and deception. Even these relatively late critiques still fit nicely with the idea of the witch-hunting phenomenon being a "design" that overcame people and spread at those people's own expense.

The pattern of late witch trials mostly occurring in regions and towns that had not been "immunized" yet can be seen throughout the continent.[329] When West-Central Europe became more resistant, witch-hunting moved towards the largely untainted outskirts of the Western world. Having remained relatively unaffected so far, Poland and Hungary only experienced their historic peaks of witch-hunting in the late seventeenth and eighteenth centuries. Other persecutions occurred in places like Sweden and New England; think of the infamous Salem trials. Yet, such events on the fringes of the Western world were amongst the last significant throes of European witch-hunting. In the eighteenth century, the phenomenon lingered in some pockets, but its end was inescapable. Due to their obscure, zealous, and intolerant character, witch-hunting became the ideal target for enlightened thinkers.[330] Proponents of persecution were attacked ever more openly and severely. Calls for trials remained prominent amongst peasants, but for the reproduction of witch-hunting, that was perhaps more a curse than a blessing. The dismissal of witch beliefs became an elitist way of distinguishing oneself from the apparently backward populace.[331] Moreover, witch-hunting required "a mood of high seriousness," as Goodare puts it, and in that regard, a lethal danger was on the rise: witch beliefs were increasingly ridiculed and satirized.[332] As Stuart Clark has argued, witch beliefs lost their allies in the world of

[329] Goodare, *The European Witch-Hunt*, 318–21; Behringer, *Witches and Witch-Hunts*, 138–40, 185–90.

[330] Dillinger, *Hexen und Magie*, 143; Behringer, *Witches and Witch-Hunts*, 135, 185–90.

[331] Levack, *The Witch-Hunt in Early Modern Europe*, 240–45.

[332] Goodare, *The European Witch-Hunt*, 346; Bever, "Witchcraft Prosecutions and the Decline of Magic," 281–82.

thought, just as those allies were losing theirs.[333] Country after country repealed its anti-witchcraft laws. The phenomenon suffered a fate that it shares with almost all living species: it failed to adapt to new circumstances and went extinct.[334]

Or did it? Witch trials indeed ended, but one could argue that the concept of witchcraft adapted to the new age and is still vivid among us. The new evolutionary future that lay ahead already became apparent in 1697, when the French writer Charles Perrault included the witch figure in his collection of fairy tales, the *Histoires ou contes du temps passé* (Stories and Tales of Times Gone By). From then on, witches appeared more prominently in fictional stories, often read to children.[335] Especially through the tales of the Grimm brothers and later the films of Disney, the witch turned into an iconic figure in fairy tales. Some elements of the early modern witch were preserved, like flight, cannibalism, evilness, and their predominantly female identity. But the new "environment" of fairy tales also produced new selection pressures. The adoration of the devil retreated, and a typical new trope became the witch who lived solitarily in the forest, like the crone in the story about Hansel and Gretel. In the early modern period, that variant would have been disastrous for the spread of witch-hunting, hindering chains of accusations. But in the modern fairy tale environment, it fitted nicely to create a sense of mystery and horror that frightened children.

Again, the developments are vaguely reminiscent of the evolution of pathogens. Some illnesses once ravaged adult populations but evolved into milder forms as childhood diseases.[336] Might we interpret the fairy tale witch as the cultural equivalent of a childhood disease? The question lies beyond the scope of this research. The topic here is the evolution of the witch-hunting phenomenon. Marked by the infamous trial of Anna Göldi in Glarus 1782, that phenomenon went extinct. So, it is also here that our story ends.

[333] Clark, *Thinking with Demons*, 686.
[334] Levack, *The Witch-Hunt in Early Modern Europe*, 252–56.
[335] Goodare, *The European Witch-Hunt*, 346–47; Roper, *Witch Craze*, 251.
[336] McNeill, *Plagues and Peoples*, 50–51, 140–41.

CHAPTER 9

Conclusion

"To come very near to a true theory, and to grasp its precise application, are two very different things, as the history of science teaches us. Everything of importance has been said before by somebody who did not discover it."[1] If there is one theory that confirms these observations by the philosopher Alfred North Whitehead, it is Darwin's theory of evolution by natural selection. We saw that the ingredients and outcomes of the evolutionary process had already been widely recognized before Darwin. Naturalists were perfectly aware that traits of living organisms are ingeniously adapted to environments. They also saw that organisms vary, that there is a reproduction of traits, and that most organisms quickly vanish in a relentless struggle for life. Hence, it is not entirely surprising that his theory long encountered scepticism. Was his mechanism not just a re-description of things that people knew all along or a tautology that merely stated that what survives, survives, because it survives? However, what made Darwin's discovery of natural selection so important was not the recognition of its various ingredients but the understanding of the counter-intuitive effect of these ingredients *in combination*. Immanuel Kant professed that the mechanistic model of science would never explain the purposefulness of living nature. But that was precisely what Darwin did. His theory explains how a mindless and purposeless process can produce purposeful outcomes in the form of design without designer. This insight has helped unite the

[1] Quoted in: Merton, *Social Theory and Social Structure*, 1.

S. Hofhuis, *The Evolutionary History of Witch-hunting*,
https://doi.org/10.1007/978-3-031-66936-1_9

disciplines of the life sciences into a more consistent framework, and it set a fruitful research agenda that continues to this day.

These observations are important when assessing the future of Darwinian theory in the social sciences and the humanities. We started with Daniel Dennett's prediction of Darwinism being a "universal acid" that eats through any barrier it encounters. The focus here has been on qualitative history. It raised my main question: can Darwinism also renew qualitative history? Or does the theory face an impenetrable barrier after all, and will it fail to succeed, like many earlier science-oriented approaches, in turning history into an integral part of the sciences? My answer has been that Darwinian theory does hold the promise of revitalizing qualitative historical scholarship. This book supports what some Darwinians claim. The theory can unite existing disciplines into a more consistent framework while producing new insights and research questions. Yet, it has also been argued that this transformative power is not so much built upon seeing things that nobody ever saw before, nor that "soft" qualitative methodologies will finally be subservient to "hard" science. Instead, the prospects of Darwinism mainly concern the recombination of ideas, tools, and empirical findings that were already available. Rather than making qualitative methodologies like hermeneutics or thick description compliant, it will give them a whole new relevance.

How can Darwinism exactly renew the study of history? The programmes of evolutionary psychology and gene-culture coevolution shed new light on the evolved biological propensities of human behaviour and the intriguing interactions between genes and culture, but they also have limitations. Cultural phenomena possess semi-autonomous dynamics of their own, which cannot be reduced to biological traits. Yet, through the idea of Darwinian *cultural* evolution, it is precisely the semi-autonomous "emergent" properties of culture to which Darwinism can contribute insights. Two crucial ideas take centre stage. First, the selection and retention process of accidentally adaptive cultural variants leads to cultural forms of "design without designer." Second, those designs may be well-built to benefit no one except the reproduction of those cultural forms themselves. Like "selfish" genes or pathogens, some cultural phenomena may spread at the expense of their human hosts and societies.

Some sceptics consider cultural Darwinism old wine in new bottles, and indeed, the upcoming field of cultural evolution makes claims similar to older non-Darwinian approaches. Numerous social scientists and scholars of the humanities argued that socio-cultural phenomena possess functional traits that seem unintended and unrecognized by the actors involved. Many also observed that forms of culture vary, that they are reproduced, that some forms survive better than others, and that there is such a thing as unintended consequences. Yet, what had not been adequately assessed is the counter-intuitive outcome of these ingredients *in combination*. Critics of old-school functionalism pointed out that the movement failed to explain how "latent functions" arise, and that is precisely what cultural Darwinism has to offer. Beliefs and practices may survive because of initially accidental effects that were neither intended nor recognized as such. This non-tautological idea sets a clear research agenda: can we find such hidden functions? Do specific cultural variants persist because of effects, unperceived by the involved actors?

The question of who or what benefits from cultural phenomena also has old precedents. *Cui bono*? The functionality of cultural forms has traditionally been sought in three directions: the maintenance of social systems, power structures, and the fulfilment of individual needs. These approaches all have their merits, and the field of cultural evolution shows that the advantages of these perspectives can potentially be integrated into a more consistent overarching Darwinian framework. The survival of a particular cultural variant can be due to multiple factors simultaneously and be different in each instance. This implies that the beneficiaries of a specific cultural form can be multifaceted and subject to change. Yet, the theory of Darwinian cultural evolution also has something genuinely new in store for us, in the idea of culture reproducing "selfishly." Selfish gene theory in biology taught us that biological adaptations are best understood when we look at them from the perspective of the reproductive interests of the hereditary information—located on genes—behind those adaptations. Authors like Dawkins, Dennett, and Blackmore have brought forward that we may look at cultural information similarly. Cultural variants that make people spread those variants will be most successful. Like pathogens, their reproduction may even continue when human carriers and their societies are significantly harmed.

What the cultural Darwinians have not sufficiently addressed so far, is the importance of qualitative historical tools. Being a strongly quantitative field, cultural evolution has little time for things like thick description, hermeneutics, and *verstehen*, that is, the careful interpretation of how actors themselves perceive things and the contextual meaning of symbols. This is a loss since those tools are essential if we want to put the Darwinian hypotheses to the test. The claim that actors often do not comprehend the functionality of many cultural assets requires detailed consideration of what actors did or did not understand. The hypothesis that cultural phenomena spread "selfishly" also requires careful attention to how involved actors perceived the benefits of their cultural items. To see how this integration can be implemented, we delved into one case study: European witch persecutions.

Regarding this case study, Whitehead's dictum proves applicable again. Darwinian theory can produce new insights about these infamous trials, but this newness is mainly built upon a rearrangement and recombination of insights and empirical findings familiar to witchcraft scholarship. One essential non-Darwinian insight is that the early modern witch-hunt system has the semblance of an intelligent design. Contemporary critics of trials, like Johann Weyer or Johann Meyfart argued that the devil was the evil genius behind it. A "subtle and devious spirit," a "sly fox," a "crafty old weaver," and a "jack-of-all-trades," the devil knew how to activate people's evil inclinations and lure them into a labyrinth of witch beliefs from which there was hardly any escape. Satan's ultimate goal was to make people suffer from the horrors of witch-hunting; innocent people were tortured and burned while persecutors ruined their salvation. Modern scholars thought that the witch-hunt system contained "built-in structures" that were "fabricated," "braided," "designed," and "perfected" by the actors responsible for persecutions. Purportedly, it was the ruling classes, men, inquisitors, city elites, or witch-hunt professionals who designed the system to rob or oppress either the poor, women, the ordinary populace, the rich, or traditional peasant culture.

These commentators had good reasons for assuming an intelligent manufacture. Cumulative witchcraft beliefs indeed became remarkably well adapted to scaring the Europeans of the time. This was an era of widespread fears of diabolical evil, subversive anti-Christian communities, harmful magic of "village witches," uncontrollable female power and sexuality, as well as the hazards of illness and ecological disaster. Gradually, a cumulative concept of witchcraft ingeniously came to arrange these anxieties into something even more frightening. The devil allegedly made a pact with people, women in particular, through sexual intercourse, and then formed anti-Christian communities who met each other at perverse nightly gatherings called witches' sabbaths. To enable their attendance, Satan offered his witches the possibility of a nightly flight. Also, he gave them magical powers to harm ordinary Christians, like causing illness or extreme weather conditions that ruined harvests. Due to the secrecy of witchcraft activities, it was challenging to find out who the witches were, so torture was used to make people confess and tell who else they had seen at the witches' sabbath. An additional option was to resort to the witness statements of alleged child witches, who were less hesitant in making confessions.

In combination, the elements of this "cumulative concept of witchcraft" developed a striking propensity for creating witch-hunts. When something terrible happened under suspicious circumstances, like an unusual frost that ruined a harvest, people might think of diabolical witchcraft. If one alleged witch had then been found, like a local "village witch" or a child full of fanciful testimonies, a chain of accusations could start. Torture then potentially produced confessions about an ever-widening circle of alleged sabbath accomplices. Chains of accusations could also spread geographically since the belief in the nightly flight implied that witches knew sabbath accomplices who lived far away. In combination, this amalgam of ideas could mobilize people's deepest fears and create the dynamics of an intensifying and "contagious" persecution.

Over the past decades, many historians have concluded that there do not seem to have been shrewd designers involved who created this apparent "intelligent design." Trials occurred haphazardly, while it was highly diverse who took the initiative and became the victims. True, females were pointedly overrepresented amongst the victims—misogyny and the unequal power balance between the sexes were an essential part of the picture. However, a significant minority of victims still consisted of men, which makes the oppression of women an unlikely ultimate aim. Moreover, the effects of witch-hunts were often damaging for all and everyone. Persecutions were expensive, they disrupted communities, and in quite a few cases, even the people who perpetrated trials were burned as witches themselves. Many experts today thus conclude that witch-hunting was not a means to a hidden end. Driven forward by a genuine belief in the danger of witchcraft, it was mostly an end in itself. But if this is so, how did the concept of witchcraft become so well adapted to create witch trials? If the actors involved were so often harmed, why did the phenomenon spread?

To explain this, Darwinian cultural evolution is indispensable. The hypothesis developed here is as follows: many cultural variants regarding witchcraft appeared, most of which did not reproduce well. However, occasionally, variants appeared or were recombined, which accidentally created more persistent and "infectious" witch trials. The size of those persecutions made those variants receive wide attention and enter many new minds, stimulating similar trials elsewhere. It was through repeated rounds of selection and accumulation that the beliefs about witchcraft evolved into an ever more "virulent" concept. Whether the actors involved understood how particular ideas or practices enhanced the spread of witch-hunting or whether they or their societies benefitted from it does not necessarily matter. The "designer" was the blind process of Darwinian evolution, while the beneficiary involved was the witch-hunting phenomenon itself. This hypothesis lends itself to entirely new empirical scrutiny. Did innumerable variants regarding witchcraft indeed appear, only to lose the "struggle for survival"? Did some of the crucial notions about witchcraft, like the sabbath, nightly flight, and the child witch, indeed create larger witch-hunts and then reproduce because of that effect? Did the actors who introduced and maintained those variants understand their enhancing effect and broadcast those elements for that reason?

To examine the proposed scenario, I especially delved into German trials between 1560 and 1630. This was the time and place where witch-hunting experienced its most significant growth. The persecutions of Wiesensteig in 1562, the Trier region around 1590, and the Ellwangen hunt from 1611–1618 show that it was primarily due to accidental causes, like previous panics about Anabaptists, weak rulership, or weather disasters, as well as some particular cultural variants, like child accusations, an attack on rich witches, or beliefs in huge sabbaths, that these hunts became unusually large. But it was precisely because of their size that those persecutions received the most attention, with their cultural variants thus becoming more prevalent. Historical sources also show a "struggle for life" between elements of culture and a wide range of less successful variants. Paraphrasing Darwin, we might say that the cultural evolutionary process was daily and hourly scrutinizing every variation, even the slightest, rejecting which was bad, preserving and adding up all that was good, silently and insensibly working, whenever and wherever opportunity offered, at the improvement of the witchcraft concept.

Yet, a thorough engagement with primary sources, like pamphlets, sermons, treatises, chronicles, and witch-hunt manuals, also shows that we should not routinely assume that involved actors were unaware of how certain aspects of witch-hunting worked. The most striking case is the witch-hunt supporter Jean Bodin. He showed much understanding of how cultural elements like the witches' sabbath, the child witch, or torture could be used in trials to enforce many confessions, broadcasting them for that reason. The work of other contemporaries—proponents and opponents of trials alike—further indicates that many functional features of the witch-hunt system were already understood. Theorists of cultural evolution should thus not too quickly presuppose incognizant actors; if possible, one should always try to examine what people comprehended or not.

Neither should we too easily assume that there were no human beneficiaries involved. The model of multifunctional instrumentalization, as developed by several German historians, can be of value here. Instead of older and somewhat cynical approaches that postulated just one or a few underlying aims, these historians argue that witchcraft accusations and trials were instrumentalized for a different set of hidden purposes in each instance. Examples include getting rid of enemies, career opportunities, or enhancing power. This interpretation corresponds to the sources to some extent, as people at the time also presumed that such instrumentalization was possible. Proponents of trials said it was a risk, while opponents said it happened. It seems plausible that instrumentalization indeed played a role.

However, this research also produced many results that support the idea of blind evolution, with the witch-hunting phenomenon spreading in a pathogen-like way. One finding is that there were notable limitations to what contemporaries seem to have understood about what made cumulative notions of witchcraft so inflammatory. Influential supporters such as Kramer, Binsfeld, or Förner do not seem to have grasped how the idea of large witches' sabbaths could be used to stimulate large chains of accusations. The contemporary understanding of the incendiary role of flight or child witches also appears limited at best. Overall, human intention and recognition seem insufficient to explain the adaptedness of the witch-hunting phenomenon, making Darwinian selection an indispensable additional explanation.

The model of multifunctional instrumentalization also has its flaws. Supporters of trials consistently said that they hunted witches to eradicate witches. So, how do we prove that there were different motivations at play? Critics at the time regularly maintained the trials were used for ulterior purposes, but they generally paid far more attention to the idea of Satan being the crucial designer and beneficiary of it all. Contemporary opponents argued that the people responsible for witch-hunts were exploited by something beyond themselves. Satan operated as the invisible puppeteer, with the witch-hunters being his marionettes. Persecutors had been stupid enough to let themselves be tricked by the devil, who then made them do things that not only harmed others but also those persecutors' interests. While this contemporary interpretation is not in line with the German instrumentalist model, it does bear a striking semblance to the Darwinian proposal made here. But instead of the perceived sinister agency of Satan, the "agency" of a cultural "pathogen" was involved, ingeniously adapted to survive for another day. It was not the devil but the virus-like cultural phenomenon that operated as the hidden "puppeteer."

Historians have already compared witch-hunts to contagious diseases innumerable times. There were "outbreaks" of witch-hunting, with towns "infecting" other towns and many casualties as a result. Yet, the comparison has only been used as a figure of speech. But why not take it more seriously? Many aspects of the witch-hunting phenomenon seemed to defy all functionalist analysis. Even after we throw together numerous known functionalist perspectives, like multifunctionalists, much of the phenomenon remains mysterious. Witchcraft beliefs and trials often appear to have been *dys*functional for all and everyone and to have contained innumerable superfluous details. Yet, as soon as we look at it from the perspective of the virus-like reproductive interests of the witch-hunting phenomenon itself, all these aspects suddenly make functional sense. The cumulative notions of witchcraft were all wonderfully adapted to ensure the phenomenon's own "selfish" spread. The socially transmitted information that we call culture, which is physically located in human brains and artefacts, can use human carriers for its own reproductive purposes. Hence, through sincere belief in the righteousness of what they were doing, people responsible for witch trials appear to have let themselves be overtaken by a nasty cultural "parasite."

This approach will appear outlandish to many historians, but it can be reconciled with much of traditional qualitative history. Several German persecutions have been presented here in the form of a chronological narrative. Doing this was important because it helped us comprehend how witchcraft concepts developed, how trials occurred, and how actors experienced things and acted upon those experiences. Emile Durkheim once dismissively wrote that the historians' focus on a "series of accidents" hindered scientific progress. But for this Darwinian approach, such accidents and events remain crucial. It is in innumerable small-scale and quite accidental incidences that the creative source of Darwinian cultural evolution must be sought. This further illustrates how Darwinism is a nomothetic and idiographic theory, which simultaneously studies the general and the particular. Concepts of witchcraft and practices of trials were continually changing and adapting to specific circumstances, with the witch-hunting phenomenon lacking any timeless "essence." Neither does the theory imply determinism. I especially traced how the concept of witchcraft became so well adapted that it enabled the huge persecutions of the late 1620s, as, for instance, occurred in Bamberg. But nothing of that is to say that it *had* to evolve like that. Had the climate not deteriorated, had religious strife not been as intense, or had Germany not remained as politically fragmented, the phenomenon would probably not have grown so massive. Contingency and complexity remain essential.

The empirical questions regarding witch trials raised here deserve more qualitative scrutiny. Can we uncover further examples of "failed" cultural variants? Do additional sources tell us more about what people comprehended about the witch-hunt system's functional features? How did other contemporary critics not studied here address who benefitted? Moreover, one challenging source material type has mostly been assessed here through the accounts of other historians, namely trial records. But surely, they are bound to contain further clues regarding the critical questions asked. Plausibly, they may also provide additional hints on how the mostly female victims of persecution experienced what happened to them and what role they played. Of course, we hear their voices through the writings of interrogators, but the records will still likely provide us with additional insights.

Moreover, this qualitative Darwinian approach and the idea of "selfish" culture can be applied to further case studies. We might think of other topics of early modern history, like the fanaticism of the religious wars, anti-Semitism, apocalyptic anxieties, or the self-destructiveness of certain Anabaptist sects. Another obvious area of future research is non-Western forms of witch-hunting. We can also think of current conspiracy theories or the widespread practice of overwork. Are many people so immersed in overwork because they and their societies benefit from it? Or is this cultural practice especially well-built for self-reproduction, irrespective of its mental and social consequences, like burn-outs? And what to think of woke ideology? Does it solely serve the emancipation of disadvantaged groups, or is it also well adapted to set people against each other and spread through the discord and publicity it engenders?

This is also where the theory carries the promise of enhancing human welfare. Considering the pervasiveness of biological pathogens, it is worth the effort to further probe whether cultural equivalents exist and, if so, how they work. Perhaps it could help us disarm some of culture's nasty tricks. If the study of cultural "pathogens" would emulate only a fraction of the successes in our medical fight against biological pathogens, the efforts would already be worth their weight in gold. At the same time, we must proceed carefully. Like most microorganisms, most forms of culture are probably neutral or beneficial for their human carriers. When adopting the selfish culture hypothesis, we risk becoming overly paranoid about our culture, seeing harmful cultural "pathogens" under every rock. It might make us eradicate beneficial forms of culture. In that regard, let us always keep the merits of other viewpoints on the *cui bono* question in mind. Hopefully, the selfish culture hypothesis will eventually sharpen our understanding of how culture works and enable us to decide better what forms of culture to remove or cherish.

But most of all, the aim here was a scientific one. It has been argued that, despite innumerable nuances, Darwinian theory has features of a "universal acid" that carries the promise of revitalizing qualitative historical scholarship. Most essentially, it can help us realize that many social scientists and scholars of the humanities have overestimated human capabilities. Our socio-cultural structures and webs of meaning were not only spun by ourselves. To a large extent, it was blind Darwinian evolution that did the creative work. More ominously, these hidden evolutionary processes do not necessarily work out well for us. At least in some cases, culture may be adapted to exploit our vulnerabilities and turn us into instruments for its own propagation. The case study of the European witch persecutions has shown how this theory provides a whole new perspective in studying history. Experts have been looking for the beneficiaries of this horrific phenomenon in every nook and cranny but failed to find them in substantial numbers. This was because one crucial beneficiary has been overlooked all the time: the witch-hunting phenomenon itself.

References

Ahrent-Schulte, Ingrid. "'Lauter falscher Wahn und starke Einbildung': Der Lemgoer Pfarrer Jodokus Hocker und seine Rezeption der Hexenverfolgung." In *Realität und Mythos: Hexenverfolgung und Rezeptionsgeschichte*, by Katrin Moeller and Burghart Schmidt, 88–101. Hamburg: Dobu Verlag, 2003.

Akey, Joshua M. "Constructing Genomic Maps of Positive Selection in Humans: Where Do We Go from Here?" *Genome Research* 19, no. 5 (2009): 711–22.

Alberum, Matheum, and Wilhelmum Bidenbach. *Ein Summa etlicher Predigten von Hagel und Unholden*. Tübingen, 1562.

Allen Orr, H. "A Troublesome Inheritance: Genes, Race and Human History by Nicholas Wade." *New York Review of Books* 61, no. 10 (2014): 18–21.

———. "Darwin and Darwinism: The (Alleged) Social Implications of The Origin of Species." *Genetics* 183, no. 3 (2009): 767–72.

Ammon, Otto. *Die Gesellschaftsordnung und ihre natürlichen Grundlagen*. Jena: G. Fischer, 1900.

Ankersmit, Frank. "Representatie Als Cognitief Instrument." *Algemeen Nederlands Tijdschrift Voor Wijsbegeerte* 103, no. 4 (2011).

Anonymous of Arras. "A History of the State, Case, and Condition of the Waldensian Heretics (Witches) (1460)." In *The Arras Witch Treatises*, translated by Andrew Colin Gow, Robert B. Desjardins, and François V. Pageau, 19–80. University Park, Pennsylvania: The Pennsylvania State University Press, 2016.

Appleby, Joyce O., Lynn A. Hunt, and Margaret C Jacob. *Telling the Truth about History*. New York: Norton, 1995.

S. Hofhuis, *The Evolutionary History of Witch-hunting*,
https://doi.org/10.1007/978-3-031-66936-1

Arnold, Matthew. *Culture and Anarchy: An Essay in Political and Social Criticism.* London: Smith, Elder, 1869.

Aunger, Robert. *The Electric Meme: A New Theory of How We Think.* New York: Simon and Schuster, 2013.

Aya, Rod. "Reign of Error; Or, the Case against Cognitive Cultural Relativism." *Irish Journal of Sociology* 13, no. 1 (May 1, 2004): 29–43.

Ayala, Francisco J. "Darwin's Greatest Discovery: Design without Designer." *Proceedings of the National Academy of Sciences* 104, no. 1 (May 15, 2007): 8567–73.

Bacon, Francis. *The Advancement of Learning and the New Atlantis.* Oxford: Benediction classics, 2008.

———. *The Works of Francis Bacon.* Vol. 4: Translations of the Philosophical Works 1. Cambridge: Cambridge University Press, 2011.

Baer, Charles F. "Mutation." In *The Princeton Guide to Evolution*, by Jonathan B. Losos, 315–20. Princeton NJ: Princeton University Press, 2014.

Baghramian, Maria, and Annalisa Coliva. *Relativism.* 1st edition. London: Routledge, 2019.

Bailey, Michael. "The Medieval Concept of the Witches' Sabbath." *Exemplaria* 8, no. 2 (January 1, 1996): 419–39.

Bailey, Michael D. *Battling Demons: Witchcraft, Heresy, and Reform in the Late Middle Ages.* 1st edition. University Park, Pennsylvania: The Pennsylvania State University Press, 2002.

Bailey, Michael D., and Edward Peters. "A Sabbat of Demonologists: Basel, 1431–1440." *The Historian* 65, no. 6 (December 1, 2003): 1375–96.

Bailey, Nathan W. "Evolution of Apparently Nonadaptive Behavior." In *The Princeton Guide to Evolution*, by Jonathan B. Losos, 710–17. Princeton NJ: Princeton University Press, 2014.

Banerjee, Konika, and Paul Bloom. "Why Did This Happen to Me? Religious Believers' and Non-Believers' Teleological Reasoning about Life Events." *Cognition* 133, no. 1 (October 2014): 277–303.

Barkow, Jerome H. *Darwin, Sex, and Status: Biological Approaches to Mind and Culture.* Toronto: University of Toronto Press, 1989.

———. "Introduction: Sometimes the Bus Does Wait." In *Missing the Revolution: Darwinism for Social Scientists*, by Jerome H. Barkow, 3–60. Oxford: Oxford University Press, 2006.

Barry, Jonathan, and Owen Davies. *Palgrave Advances in Witchcraft Studies.* New York: Palgrave Macmillan, 2005.

Barstow, Anne Llewellyn. *Witchcraze: A New History of the European Witch Hunts.* London: HarperCollins, 1994.

Bauer, S. Wise. *The Story of Science: From the Writings of Aristotle to the Big Bang Theory.* New York: W. W. Norton & Company, 2015.

Beck, Naomi, and Ulrich Witt. "Austrian Economics and the Evolutionary Paradigm*." *Studies in Logic, Grammar and Rhetoric* 57, no. 1 (March 1, 2019): 205–25.

Becker, Thomas P. "Hexenverfolgung in Kurköln: Kritische Anmerkungen zu Gerhard Schormanns „Krieg gegen die Hexen"." *Annalen des Historischen Vereins für den Niederrhein* 195 (1992): 204–14.

Behringer, Wolfgang. "Canisius, St. Peter (1521–1597)." In *Encyclopedia of Witchcraft: The Western Tradition*, I:161–62. Santa Barbara: ABC-CLIO, 2006a.

———. "Climatic Change and Witch-Hunting: The Impact of the Little Ice Age on Mentalities." *Climatic Change* 43, no. 1 (1999): 335–51.

———. "Das 'Reichskhündig Exempel' von Trier: Zur Paradigmatischen Rolle Einer Hexenverfolgung in Deutschland." In *Hexenglaube Und Hexenprozesse Im Raum Rhein-Mosel-Saar*, by Günther Franz and Franz Irsigler, 435–48. Trier: Spee Buchverlag, 1995.

———. *Hexen und Hexenprozesse in Deutschland.* München: Deutscher Taschenbuch, 2000.

———. *Hexenverfolgung in Bayern: Volksmagie, Glaubenseifer und Staatsräson in der frühen Neuzeit.* Munich: Oldenbourg, 1997.

———. "Hexenverfolgungen im Spiegel zeitgenössischer Publizistik: Die 'Erweytterte Unholden Zeyttung' von 1590." *Oberbayerisches Archiv* 109:2 (1984): 339–60.

———. "Kinderhexenprozesse: Zur Rolle von Kindern in Der Geschichte Der Hexenverfolgung." *Zeitschrift Für Historische Forschung* 16, no. 1 (1989): 31–47.

———. *Kulturgeschichte des Klimas: von der Eiszeit bis zur globalen Erwärmung.* München: Beck, 2007.

———. "Späte Hexenprozesse—ein Pfahl im Fleisch der Aufklärung." In *Späte Hexenprozesse: der Umgang der Aufklärung mit dem Irrationalen*, edited by Sönke Lorenz, Dieter R Bauer, and Wolfgang Behringer. Bielefeld: Verlag für Regionalgeschichte, 2016.

———. "Walter Stephens. Demon Lovers: Witchcraft, Sex, and the Crisis of Belief." *American Historical Review* 108 (2003): 1207.

———. "Westerstetten, Johann Christoph von (1563–1637)." In *Encyclopedia of Witchcraft: The Western Tradition*, IV:1191–93. Santa Barbara: ABC-CLIO, 2006b.

———. "Witchcraft and the Media." In *Ideas and Cultural Margins in Early Modern Germany: Essays in Honor of H.C. Erik Midelfort*, edited by Marjorie Elizabeth Plummer and Robin Barnes, 217–38. Farnham: Ashgate, 2009.

———. "Witchcraft Studies in Austria, Germany and Switzerland." In *Witchcraft in Early Modern Europe: Studies in Culture and Belief*, by Jonathan Barry, Marianne Hester, and Gareth Roberts, 64–95. Cambridge: Cambridge University Press, 1996.

———. *Witches and Witch-Hunts: A Global History*. Cambridge: Polity Press, 2008.

Behringer, Wolfgang, and Günther. Jerouscheck. "'Das unheilvollste Buch der Weltliteratur'? Zur Entstehungs- und Wirkungsgeschichte des Malleus Maleficarum und zu den Anfängen der Hexenverfolgung." In *Der Hexenhammer: Kommentierte Neuübersetzung*, 98. Munich: Deutscher Taschenbuch Verlag, 2001.

Behringer, Wolfgang, and Claudia Opitz-Belakhal. "Hexenkinder—Kinderbanden—Straßenkinder: Eine Einführung in das Thema." In *Hexenkinder—Kinderbanden—Straßenkinder*, 1–46. Bielefeld: Verlag für Regionalgeschichte, 2016.

Beinhocker, Eric D. *The Origin of Wealth: Evolution, Complexity, and the Radical Remaking of Economics*. Boston: Harvard business school, 2006.

Beiser, Frederick C. *The German Historicist Tradition*. Oxford: Oxford University Press, 2011.

Benedict, Ruth. *Patterns of Culture*. London: Routledge & Kegan Paul, 1935.

Bentley, Michael. *Modern Historiography: An Introduction*. London: Routledge, 1999.

Benton, Ted. "Social Causes and Natural Relations." In *Alas Poor Darwin: Arguments against Evolutionary Psychology*, by Hilary Rose and Steven Rose, 206–24. London: Vintage, 2000.

Ben-Yehuda, Nachman. "The European Witch Craze of the 14th to 17th Centuries: A Sociologist's Perspective." *American Journal of Sociology* 88, no. 6 (1980): 1–31.

Berger, Peter L, and Thomas Luckmann. *The Social Construction of Reality: A Treatise in the Sociology of Knowledge*. Harmondsworth: Penguin, 1966.

Berlin, Isaiah. "The Concept of Scientific History." *History and Theory* 1, no. 1 (1960): 1–31.

Berry, Stephan. "The Laws of History." In *A Companion to the Philosophy of History and Historiography*, 162–71. Malden, Mass.: John Wiley & Sons, Ltd, 2009.

Bever, Edward. *The Realities of Witchcraft and Popular Magic in Early Modern Europe: Culture, Cognition, and Everyday Life*. New York: Palgrave Macmillan, 2008.

———. "Witchcraft, Female Aggression, and Power in the Early Modern Community." *Journal of Social History* 35, no. 4 (2002): 955–88.

———. "Witchcraft Prosecutions and the Decline of Magic." *The Journal of Interdisciplinary History* 40, no. 2 (2009): 263–93.

Biesel, Elisabeth. "Les desciptions du sabbat dans les confessions des inculpés lorrains et trévirois." by Nicole Jacques-Chaquin and Maxime Préaud, 183–97. Grenoble: Jerôme Millon, 1993.

Binsfeld, Peter. *Tractat Von Bekantnuß der Zauberer unnd Hexen: Ob und wie viel denselben zu glauben*. Trier: Bock, 1590.

Blackmore, Susan. *The Meme Machine*. Oxford: Oxford University Press, 2000.
Blauert, Andreas. *Frühe Hexenverfolgungen: Ketzer-, Zauberei- und Hexenprozesse des 15. Jahrhunderts*. Hamburg: Junius Verlag, 1989.
Blécourt, Willem de. "Sabbath Stories: Towards a New History of Witches' Assemblies." In *The Oxford Handbook of Witchcraft in Early Modern Europe and Colonial America*, by Brian P. Levack, 84–100. Oxford: Oxford University Press, 2013.
———. "The Flying Witch: Its Resonance in the Sixteenth-Century Netherlands." *Magic, Ritual, and Witchcraft* 11, no. 1 (June 16, 2016): 73–93.
———. "The Making of the Female Witch: Reflections on Witchcraft and Gender in the Early Modern Period." *Gender & History* 12, no. 2 (2000): 287–309.
Bloch, Marc. *Apologie pour l'histoire ou Métier d'historien*. Paris: Armand Colin, 1967.
———. *La société féodale*. Paris: Albin Michel, 1949.
Bloch, Maurice. "A Well-Disposed Social Anthropologist's Problems with Memes." In *Darwinizing Culture: The Status of Memetics as a Science*, by Robert Aunger, 189–204. Oxford: Oxford University Press, 2001.
Blute, Marion. *Darwinian Sociocultural Evolution: Solutions to Dilemmas in Cultural and Social Theory*. Cambridge: Cambridge University Press, 2010.
———. "History versus Science: The Evolutionary Solution." *The Canadian Journal of Sociology / Cahiers Canadiens de Sociologie* 22, no. 3 (1997): 345–64.
Boas, Franz. *The Mind of Primitive Man*. New York: Macmillan, 1931.
Bodin, Jean. *De la démonomanie des sorciers*. Travaux d'humanisme et Renaissance, N° DLIX. Genève: Librairie Droz SA, 2016.
Boghossian, Paul Artin. *Fear of Knowledge: Against Relativism and Constructivism*. Oxford: Clarendon Press, 2006.
Boguet, Henri. *Discours exécrable des sorciers*. Rouan: Romain de Beavuais, 1606.
Bon, Gustave Le. *Psychologie des foules*. Paris: Félix Alcan, 1895.
Boudry, Maarten, and Steije Hofhuis. "Parasites of the Mind. Why Cultural Theorists Need the Meme's Eye View." *Cognitive Systems Research* 52 (December 1, 2018): 155–67.
Bourdieu, Pierre. *La distinction: critique sociale du jugement*. Le sens commun. Paris: Éditions de Minuit, 1979.
———. *Le Sens Pratique*. Paris: Editions de Minuit, 1980.
Bourke, Andrew F. G. "The Validity and Value of Inclusive Fitness Theory." *Proceedings of the Royal Society B: Biological Sciences* 278, no. 1723 (November 22, 2011): 3313–20.
Bowersock, Glenn W. *Martyrdom and Rome*. Cambridge: Cambridge University Press, 2002.
Bowler, Peter J. *Evolution: The History of an Idea*. Berkeley: University of California Press, 2003.

———. *The Eclipse of Darwinism: Anti-Darwinian Evolution Theories in the Decades Around 1900.* Baltimore: John Hopkins University Press, 1992.
Bowles, Samuel. "Genetically Capitalist?" *Science* 318, no. 5849 (2007): 394–96.
Boyd, Robert, and Peter J Richerson. *Culture and the Evolutionary Process.* Chicago: University of Chicago Press, 1985.
Bradford Smith, William. *Reformation and the German Territorial State: Upper Franconia, 1300–1630.* Rochester: University of Rochester Press, 2008.
———. "The Persecution of Witches and the Discourse on Toleration in Early Modern Germany." In *Topographies of Tolerance and Intolerance: Responses to Religious Pluralism in Reformation Europe*, by Marjorie Elizabeth Plummer and Victoria Christman, 50–77. Leiden: Brill, 2018.
Braudel, Fernand. *Ecrits sur l'histoire.* Paris: Flammarion, 1969.
Brian Ribeiro. "Montaigne on Witches and the Authority of Religion in the Public Sphere." *Philosophy and Literature* 33, no. 2 (2009): 235–51.
Briggs, Robin. "Many Reasons Why: Witchcraft and the Problem of Multiple Explanation." In *Witchcraft in Early Modern Europe: Studies in Culture and Belief*, by Jonathan Barry, Marianne Hester, and Gareth Roberts, 49–63. Cambridge: Cambridge University Press, 1996a.
———. *The Witches of Lorraine.* Oxford: Oxford University Press, 2007.
———. *Witches & Neighbours: The Social and Cultural Context of European Witchcraft.* London: HarperCollins, 1996b.
Broadie, Sarah. "The Ancient Greeks." In *The Oxford Handbook of Causation*, by Helen Beebee, Christopher Hitchcock, and Peter Menzies, 21–39. Oxford: Oxford University Press, 2009.
Broedel, Hans Peter. "Fifteenth-Century Witch Beliefs." In *The Oxford Handbook of Witchcraft in Early Modern Europe and Colonial America*, edited by Brian P. Levack, 32–50. Oxford: Oxford University Press, 2013.
———. *The "Malleus Maleficarum" and the Construction of Witchcraft: Theology and Popular Belief.* Manchester: Manchester University Press, 2003.
Brown, Callum G. *Postmodernism for Historians.* Harlow: Pearson Longman, 2005.
Brown, Donald E. *Hierarchy, History and Human Nature: The Social Origins of Historical Consciousness.* Tucson: University of Arizona Press, 1988.
———. "Human Nature and History." *History and Theory* 38, no. 4 (1999): 138–57.
Browne, Janet. "Charles Darwin and ideology: Rethinking the Darwinian revolution." *Metode Science Studies Journal*, no. 7 (June 20, 2017): 61–67.
Bruggeman, Jeroen. *A Sociology of Humankind: How We Are Formed by Culture, Cooperation, and Conflict.* Taylor & Francis, 2024.
Bryant, Joseph M. "An Evolutionary Social Science? A Skeptic's Brief, Theoretical and Substantive." *Philosophy of the Social Sciences* 34, no. 4 (December 2004): 451–92.
Burke, Peter. *History and Social Theory.* 2nd ed. Cambridge: Polity Press, 2005.

———. *The French Historical Revolution: The Annales School, 1929–2014*. Stanford: Stanford University Press, 2015.
———. *What Is Cultural History?* Malden: Polity, 2008.
Buskes, Christiaan Jozef Joannes. *Evolutionair denken: de invloed van Darwin op ons wereldbeeld*. Amsterdam: Uitgeverij Nieuwezijds, 2006.
Buss, David M. *Evolutionary Psychology: The New Science of the Mind*. 4th ed. Harlow: Pearson, 2014.
Butler, Christopher. *Postmodernism: A Very Short Introduction*. Oxford: Oxford University Press, 2002.
Butler, Judith. *Undoing Gender*. New York: Routledge, 2004.
Callow, John. *Embracing the Darkness: A Cultural History of Witchcraft*. London: I.B. Tauris, 2018.
Cameron, Elissa Z., and Johan T. du Toit. "Winning by a Neck: Tall Giraffes Avoid Competing with Shorter Browsers." *The American Naturalist* 169, no. 1 (January 1, 2007): 130–35.
Campbell, Donald T. "On the Conflicts between Biological and Social Evolution and between Psychology and Moral Tradition." *American Psychologist* 30, no. 12 (1975): 1103–26.
———. "Variation and Selective Retention in Socio-Cultural Evolution." In *Social Change in Developing Areas: A Reinterpretation of Evolutionary Theory*, by Herbert R. Barringer, George Irving Blanksten, and Raymond Right Mack, 19–49. Cambridge, Mass.: Schenkman Publishing Company, 1965.
Capp, Bernard. "Gender and Family." In *The Early Modern European World 1500–1800: An Introduction to Early Modern History*, by Beat Kümin, 37–47, Second edition. Oxon: Routledge, 2014.
Carneiro, Robert L. *Evolutionism in Cultural Anthropology: A Critical History*. Boulder: Westview Press, 2003.
Carr, Edward H. *What Is History?* London: Macmillan & Co. LTD, 1961.
Carrasco, David. "Sacrifice/Human Sacrifice in Religious Traditions." In *The Oxford Handbook of Religion and Violence*, by Mark Juergensmeyer, Margo Kitts, and Micheal Jerryson, 209–25. Oxford: Oxford University Press, 2013.
Cavalli-Sforza, L. L, and M. W Feldman. *Cultural Transmission and Evolution: A Quantitative Approach*. Princeton: Princeton University Press, 1981.
Charlesworth, Deborah, Nicholas H. Barton, and Brian Charlesworth. "The Sources of Adaptive Variation." *Proceedings of the Royal Society B: Biological Sciences* 284, no. 1855 (May 31, 2017): 20162864.
Chomsky, Noam. "Noam Chomsky on Post-Modernism," July 13, 2021. http://bactra.org/chomsky-on-postmodernism.html.
———. "'Rationality/Science.'" *Z Papers* 1, no. 4 (1992): 52–57.
Christakis, Nicholas A., and James H. Fowler. *Connected: The Surprising Power of Our Social Networks and How They Shape Our Lives*. New York: Little, Brown, 2009.

Christian, David. *Maps of Time: An Introduction to Big History*. Berkeley: University of California Press, 2003.

———. *Origin Story: A Big History of Everything*. London: Allen Lane, The Penguin Press, 2018.

Clark, Gregory. *A Farewell to Alms: A Brief Economic History of the World*. Princeton, NJ: Princeton University Press, 2010a.

Clark, Stuart. "'Feigned deities, pretended conferences, imaginary apparitions': Scepticism in state theory and witchcraft theory." In *Hexenprozess und Staatsbildung*, by Johannes Dillinger, Jürgen Michael Schmidt, and Dieter R Bauer, 265–84. Bielefeld: Verlag für Regionalgeschichte, 2008.

———. "Glaube und Skepsis in der deutschen Hexenliteratur von Johann Weyer bis Friedrich von Spee." In *Vom Unfug des Hexen-Processes: Gegner der Hexenverfolgung von Johann Weyer bis Friedrich Spee*, by Hartmut Lehmann and Otto Ulbricht, 15–34. Wiesbaden: Otto Harrassowitz, 1992.

———. "One-Tier History." *Magic, Ritual, and Witchcraft* 5, no. 1 (June 5, 2010b): 84–90.

———. *Thinking with Demons: The Idea of Witchcraft in Early Modern Europe*. Oxford: Oxford University Press, 1997.

Cloak, F. T. "Is a Cultural Ethology Possible?" *Human Ecology* 3, no. 3 (July 1, 1975): 161–82.

Closson, Marianne. *L'imaginaire démoniaque en France (1550–1650): genèse de la littérature fantastique*. Travaux d'Humanisme et Renaissance, no. 341. Genève: Droz, 2000.

Cochran, Gregory, and Henry Harpending. *The 10,000 Year Explosion: How Civilization Accelerated Human Evolution*. New York: Basic Group, 2009.

Cohn, Norman. *Europe's Inner Demons: An Enquiry Inspired by the Great Witch-Hunt*. London: Chatto-Heinemann, 1975.

Collingwood, R. G. *The Idea of History*. Oxford: Oxford University Press, 1994.

Collins, Randall. *Interaction Ritual Chains*. 2nd printing. Princeton: Princeton University Press, 2004.

———. "Reassessments of Sociological History: The Empirical Validity of the Conflict Tradition." *Theory and Society* 1, no. 2 (June 1, 1974): 147–78.

Cook, David. "Martyrdom in Islam." In *The Oxford Handbook of Religion and Violence*, by Mark Juergensmeyer, Margo Kitts, and Micheal Jerryson, 226–41. Oxford: Oxford University Press, 2013.

Cordingley, Michael G. *Viruses: Agents of Evolutionary Invention*. Cambridge, Massachusetts: Harvard University Press, 2017.

Cosmides, Leda, and John Tooby. "Evolutionary Psychology: A Primer." Center for Evolutionary Psychology, 1997. https://www.cep.ucsb.edu/primer.html.

Cosmides, Leda, John Tooby, and Jerome H. Barkow. "Introduction: Evolutionary Psychology and Conceptual Integration." In *The Adapted Mind: Evolutionary*

Psychology and the Generation of Culture, by Jerome H. Barkow, Leda Cosmides, and John Tooby, 3–15. Oxford: Oxford University Press, 1992.

Coyne, Jerry A. "Genes First: Review of The Social Conquest of Earth by E. O. Wilson." *Times Literary Supplement*, January 2, 2013, 4731 edition.

———. "Once Again: Misguided Calls for a Thorough Revamping of Evolutionary Biology." *Why Evolution Is True* (blog), November 23, 2016. https://whyevolutionistrue.wordpress.com/2016/11/23/once-again-a-call-for-a-revision-of-evolutionary-biology/.

———. "The Demise of Group Selection." *Why Evolution Is True* (blog), June 24, 2012. https://whyevolutionistrue.wordpress.com/2012/06/24/the-demise-of-group-selection/.

Crawford, Dorothy H. *How Microbes Shaped Our History*. Oxford: Oxford University Press, 2007.

Currie, Elliott P. "Crimes without Criminals: Witchcraft and Its Control in Renaissance Europe." *Law & Society Review* 3, no. 1 (1968): 7–32.

Dahrendorf, Ralf. "Out of Utopia: Toward a Reorientation of Sociological Analysis." *American Journal of Sociology American Journal of Sociology* 64, no. 2 (1958): 115–27.

Daly, Martin, and Margo Wilson. *Homicide: Foundations of Human Behavior*. New York: Routledge, 2017.

Daneau, Lambert. *Les Sorciers: Dialogue tres utile et necessaire pour ce temps*. Bourgais, 1574.

Darwin, Charles. *The Descent of Man and Selection in Relation to Sex*. London: Penguin Books, 2008.

———. *The Origin of Species by Means of Natural Selection; Or, The Preservation of Favoured Races in the Struggle for Life*. London: John Murray, 1875.

Darwin, Charles Robert. *The Origin of Species: Complete and Fully Illustrated*. Edited by Patricia Horan. New York: Gramercy Books, 1979.

Davies, Owen. "The World of Popular Magic." In *The Oxford Illustrated History of Witchcraft and Magic*, edited by Owen Davies, 167–94. Oxford: Oxford University Press, 2017.

Davies, Owen, and Jonathan Barry. "Introduction." In *Palgrave Advances in Witchcraft Historiography*, edited by Owen Davies and Jonathan Barry, 1–10. Basingstoke: Palgrave Macmillan, 2007.

Davis, Cynthia J. "Contagion as Metaphor." *American Literary History* 14, no. 4 (2002): 828–36.

Davis, Kingsley. *Human Society*. Berkeley: The Macmillan Company, 1949.

Davis, Kingsley, and Wilbert E. Moore. "Some Principles of Stratification." In *Readings in Social Theory: The Classical Tradition to Post-Modernism*, by James Farganis, 183–91, 4th ed. New York: McGraw-Hill, 2004.

Dawkins, Richard. "Postmodernism Disrobed." *Nature* 394, no. 6689 (July 1998): 141–43.

———. *The Ancestor's Tale: A Pilgrimage to the Dawn of Life*. London: Phoenix, 2005.

———. *The Blind Watchmaker*. London: Penguin Books, 1988.

———. *The Extended Phenotype: The Long Reach of the Gene*. Oxford: Oxford University Press, 1989.

———. *The God Delusion*. London: Bantam Press, 2006a.

———. *The Greatest Show on Earth: The Evidence for Evolution*. New York: Simon and Schuster, 2009.

———. *The Selfish Gene*. Oxford: Oxford University Press, 2006b.

———. "Universal Darwinism." In *Evolution from Molecules to Man*, by D. S Bendall, 403–28. Cambridge: Cambridge University Press, 1983.

Decker, Rainer. "Der Hexen-Richter Dr. Heinrich von Schultheiß (ca. 1580–1646) aus Scharmede." In *750 Jahre Stadt Salzkotten: Geschichte einer westfälischen Stadt*, edited by Detlef Grothmann, 1045–61. Paderborn: Bonifatius, 1996.

———. "Die Hexenverfolgungen im Herzogtum Westfalen: Forschungsstand, Quellenlage und Zielsetzung." *Westfälische Zeitschrift—Zeitschrift für vaterländische Geschichte und Altertumskunde* 131/132 (February 1981): 339–86.

———. "Die Hexenverfolgungen im Hochstift Paderborn: Forschungsstand, Quellenlage und Zielsetzung." *Westfälische Zeitschrift—Zeitschrift für vaterländische Geschichte und Altertumskunde* 128 (1978): 315–56.

Delumeau, Jean. *La Peur en Occident: Une cité assiégée (XIVe-XVIIe siècle)*. Paris: Fayard, 1978.

Dennett, Daniel C. *Breaking the Spell: Religion as a Natural Phenomenon*. London: Penguin UK, 2007.

———. *Darwin's Dangerous Idea: Evolution and the Meanings of Life*. New York: Simon & Schuster, 1995.

———. "Foreword." In *Darwinizing Culture: The Status of Memetics as a Science*, by Robert Aunger. Oxford: Oxford University Press, 2001.

———. *From Bacteria to Bach and Back: The Evolution of Minds*. London: Allen Lane, 2017.

Depew, David J., and Bruce H. Weber. *Darwinism Evolving: Systems Dynamics and the Genealogy of Natural Selection*. Cambridge, Mass.: MIT Press, 1995.

Desmond, Adrian, and James Moore. *Darwin's Sacred Cause: How a Hatred of Slavery Shaped Darwin's Views on Human Evolution*. Boston: Houghton Mifflin Harcourt, 2014.

Dijksterhuis, Ap, and Loran F Nordgren. "A Theory of Unconscious Thought." *Perspectives on Psychological Science* 1, no. 2 (2016): 95–109.

Dijstelbloem, Huub. "De democratie anders: politieke vernieuwing volgens Dewey en Latour." 2007.

Dillinger, Johannes. *"Böse Leute": Hexenverfolgungen in Schwäbisch-Österreich und Kurtrier im Vergleich*. Trier: Spee, 1999.

———. "Das magische Gericht. Religion, Magie und Ideologie." In *Hexenprozesse und Gerichtspraxis*, by Herbert Eiden and Rita Voltmer, 545. Trier: Spee Buchverlag, 2002.

———. "Friedrich Spee und Adam Tanner: Zwei Gegner der Hexenprozesse aus dem Jesuitenorden." *Spee Jahrbücher* 7 (2000): 30–58.

———. *Hexen und Magie: 2., aktualisierte Auflage*. Frankfurt: Campus Verlag, 2018.

———. "Hexenverfolgungen in der Grafschaft Hohenberg." In *Zum Feuer verdammt: Die Hexenverfolgungen in der Grafschaft Hohenberg, der Reichsstadt Reutlingen und der Fürstprostei Ellwangen*, by Johannes Dillinger, Thomas Fritz, and Wolfgang Mährle, 1–162. Stuttgart: Franz Steiner Verlag, 1998a.

———. *Kinder im Hexenprozess: Magie und Kindheit in der Frühen Neuzeit*. Stuttgart: Franz Steiner Verlag, 2013.

———. "Richter als Angeklagte: Hexenprozesse gegen herrschaftliche Amtsträger in Kurtrier und Schwäbisch-Österreich." In *Vergleichende Perspektiven: Perspektiven des Vergleichs*, by Helga Schnabel-Schulte, 123–70. Mainz: Verlag Philip von Zabern, 1998b.

———. "The Political Aspects of the German Witch Hunts." *Magic, Ritual, and Witchcraft* 4, no. 1 (June 5, 2009): 62–81.

Dippold, Günther. "Die 'Hexenpolitik' Der Bamberg Fürstbischöfe." In *"Hexenprozesse Und Hexenverfolgung" Im Hochstift Bamberg: Eine Vorlaüfige Bilanz*, by Ulrieke Siebenhaar, 33–61. Bamberg: Verlags-UG, 2013.

———. "Hexereiprozesse Im Hochstift Bamberg." In *Hexenwahn in Franken*, by Margus Mergenthaler and Margarete Klein-Pfeuffer, 224–39. Dettelbach: Verlag J.H. Röll, 2014.

Dobzhansky, Theodosius. "Nothing in Biology Makes Sense except in the Light of Evolution." *The American Biology Teacher* 35, no. 3 (March 1, 1973): 125–29.

Doolittle, W. Ford. "Uprooting the Tree of Life." *Scientific American* 282, no. 2 (2000): 90–95.

Downing, Lisa. *The Cambridge Introduction to Michel Foucault*. Cambridge: Cambridge University Press, 2008.

Dreytwein, Dionysius. *Dionysius Dreytweins Esslingische Chronik (1548–1564)*. 221. Tübingen: Literarischen Vereins Stuttgart, 1901.

Duhr, Bernhard. *Geschichte der Jesuiten in den Ländern Deutscher Zunge*. Vol. 2:1. Freiburg im Breisgau: Herder, 1907.

Dülmen, Richard van. *Theater des Schreckens: Gerichtspraxis und Strafrituale in der frühen Neuzeit*. München: Beck, 1985.

Duncan, John Shute. *Botanical Theology: Or, Evidences of the Existence and Attributes of the Deity, Collected from the Appearances of Nature*. Oxford: J. Vincent, 1826.

Duni, Matteo. "Doubting Witchcraft: Theologians, Jurists, Inquisitors during the Fifteenth and Sixteenth Centuries." *Studies in Church History* 52 (June 2016): 203–31.

Durham, William H. *Coevolution: Genes, Culture, and Human Diversity.* Stanford: Stanford University Press, 2000.

Durkheim, Émile. *Cours de science sociale. Leçon d'ouverture.* Paris: A. Colin, 1888.

———. *Les formes élémentaires de la vie religieuse: le système totémique en Australie.* Paris: Presses universitaires de France, 1968.

———. *Les Règles de La Méthode Sociologique.* Paris: Quadrige / Puf, 1981.

Durrant, Jonathan Bryan. *Witchcraft, Gender, and Society in Early Modern Germany.* Leiden: Brill, 2007.

Dussen, W. J. van der. *Filosofie van de geschiedenis: Een inleiding.* Muiderberg: D. Coutinho, 1986.

Eerden, P.C. van der. "Cornelius Loos und die 'magia falsa.'" In *Von Unfug des Hexen-Processes: Gegner der Hexenverfolgungen von Johann Weyer bis Friedrich Spee*, by Hartmut Lehmann and Otto Ulbricht, 139–60. Wiesbaden: Wolfenbütteler Forschungen, 1992.

———. "Der Teufelspakt bei Petrus Binsfeld und Cornelius Loos." In *Hexenglaube und Hexenprozesse im Raum Rhein-Model-Saar*, by Günther Franz and Franz Irsigler, 51–72. Trier: Spee Buchverlag, 1995.

Ehrenreich, Barbara, and Deirdre English. *Witches, Midwives and Nurses: A History of Women Healers.* London: Compendium, 1974.

Ein Warhafftige und gründtliche Beschreibung, Auß demm Bistumm Würtz- und Bamberg, 1626.

Eire, Carlos M.N. *Reformations: The Early Modern World, 1450–1650.* New Haven: Yale University Press, 2018.

Elias, Norbert. *Was ist Soziologie?* Grundfragen der Soziologie; Bd. 1. München: Juventa Verlag, 1970.

Elster, Jon. *Explaining Technical Change.* Cambridge; New York: Cambridge University Press, 1983.

Erdi, Peter. *Complexity Explained.* Springer Science & Business Media, 2007.

Evans, Richard J. *In Defense of History.* New York: W.W. Norton, 2000.

Evans-Pritchard, Edward Evan. *Witchcraft, Oracles and Magic among the Azande.* Oxford: Clarendon Press, 1950.

Farganis, James. *Readings in Social Theory: The Classical Tradition to Post-Modernism.* New York: McGraw-Hill, 2004.

Febvre, Lucien. *Le problème de l'incroyance au 16e siècle: La religion de Rabelais.* Paris: A. Michel, 1942.

Ferguson, Niall. "Introduction." In *Virtual History: Alternatives and Counterfactuals.*, 1–90. London: Macmillan Publishers Ltd, 1997.

———. *The Ascent of Money: A Financial History of the World.* New York: Penguin Books, 2009.

Flatt, Thomas, and Andreas Heyland. *Mechanisms of Life History Evolution: The Genetics and Physiology of Life History Traits and Trade-Offs*. Oxford: Oxford University Press, 2011.

Flock, Tristan. "What Good Is Evolutionary Psychology?" *Quillette* (blog), October 17, 2018. https://quillette.com/2018/10/17/what-good-is-evolutionary-psychology/.

Flurschütz da Cruz, Andreas. *Hexenbrenner, Seelenretter: Fürstbischof Julius Echter von Mespelbrunn (1573–1617) und die Hexenverfolgungen im Hochstift Würzburg*. Hexenforschung 16. Bielefeld: Verlag für Regionalgeschichte, 2017.

Förner, Friedrich. *Dämonenglaube und Zauberei im Jahre 1625: die "Hexenpredigten" des Bamberger Bischofs Friedrich Förner*. Translated by Burghard Schmanck. Nordhausen: Traugott Bautz Verlag, 2015.

Fort, Charles. *The Book of the Damned*. Book Tree, 2006.

Foucault, Michel. *Surveiller et Punir: Naissance de La Prison*. Paris: Gallimard, 1975.

Fracchia, Joseph, and Richard C Lewontin. "Does Culture Evolve?" *History and Theory* 38, no. 4 (1999): 52–78.

Franc, Martin le. "Le Champion des Dames, livre IV, vers 17377–18200." In *L'imaginaire du sabbat: Edition critique des textes les plus anciens (1430 c.–1440 c.)*, by Martine Ostorero, Agostino Paravicini Bagliani, and Kathrin Utz Tremp, 451–82. Lausanne: Cahier Lausannois d'histoire médiévale, 1999.

Frank, Robert H. *The Darwin Economy*. Princeton: Princeton University Press, 2011.

Franz, Günther. "Antonius Hovaeus, Cornelius Loos und Friedrich Spee—drei Gegner der Hexenprozesse in Echternach und Trier." In *Hexenprozesse und deren Gegner im Trierisch-Lothringischen Raum*, 117–42. Weimar: Verlag Rita Dadder, 1997.

———. "Hexenprozesse in Der Stadt Trier Und Deren Umgebung: Gerichtsbarkei von St. Maximum, St. Paulin Und St. Matthias." In *Hexenglaube Und Hexenprozesse Im Raum Rhein-Mosel-Saar*, by Günther Franz and Franz Irsigler, 333–54. Trier: Spee Buchverlag, 1995.

Freeman, Derek. *Margaret Mead and Samoa: The Making and Unmaking of an Anthropological Myth*. Harmondsworth: Penguin, 1984.

Frisium, Paul. *Deß Teuffels Nebelkappen, Das ist: Kurtzer Begriff, den gantzen Handel von der Zauberey belangend*. Frankfurt am Main, 1583.

Fritz, Thomas. "Hexenverfolgungen in der Reichsstadt Reutlingen." In *Zum Feuer verdammt: Die Hexenverfolgungen in der Grafschaft Hohenberg, der Reichsstadt Reutlingen und der Fürstpropstei Ellwangen*, by Johannes Dillinger, Thomas Fritz, and Wolfgang Mährle, 163–324. Stuttgart: Franz Steiner Verlag, 1998.

Fründ, Hans. "Rapport sur la chasse aux sorciers et aux sorcières menée dés 1428 dans le diocèse de Sion." In *L'imaginaire du sabbat: Edition critique des textes les plus anciens (1430 c.–1440 c.)*, by Martine Ostorero, Agostino Paravicini

Bagliani, and Kathrin Utz Tremp, 30–45. translated by Katherine Chène. Lausanne: Cahier Lausannois d'histoire médiévale, 1999.

Fuller, Steve. *The Philosophy of Science and Technology Studies*. 1st edition. New York: Routledge, 2006.

Füssel, Stephan, and Hans Joachim Kreutzer. Historia von *D. Johann Fausten*. Stuttgart: Reclam, 1988.

Futuyma, Douglas J. "Evolutionary Biology Today and the Call for an Extended Synthesis." *Interface Focus* 7, no. 5 (October 6, 2017): 20160145.

Gaddis, John Lewis. *The Landscape of History: How Historians Map the Past*. Oxford: Oxford University Press, 2002.

Gaskill, Malcolm. "The Pursuit of Reality: Recent Research into the History of Witchcraft." *The Historical Journal* 51, no. 4 (2008): 1069–88.

Geertz, Clifford. *The Interpretation of Cultures: Selected Essays*. New York: Basic Books, 1973.

Geertz, Hildred. "An Anthropology of Religion and Magic, I." *The Journal of Interdisciplinary History* 6, no. 1 (1975): 71–89.

Gehm, Britta. *Die Hexenverfolgung im Hochstift Bamberg und das Eingreifen des Reichshofrates zu ihrer Beendigung*. Jena: [S.l] [s.n], 1999.

Gershman, Boris. "Witchcraft Beliefs and the Erosion of Social Capital: Evidence from Sub-Saharan Africa and Beyond." *Journal of Development Economics* 120 (May 1, 2016): 182–208.

Geschiere, Peter. *The Modernity of Witchcraft: Politics and the Occult in Postcolonial Africa*. Charlottesville: University of Virginia Press, 1997.

Gibson, Marian. "Thinking Witchcraft: Language, Literature and Intellectual History." In *Palgrave Advances in Witchcraft Historiography*, by Jonathan Barry and Owen Davies. New York: Palgrave Macmillan, 2007.

Giddens, Anthony. *The Constitution of Society: Outline of the Theory of Structuration*. Berkeley: University of California Press, 1984.

Giddens, Anthony, and Christopher Pierson. *Conversation with Anthony Giddens: Making Sense of Modernity*. Cambridge: Polity Press, 1998.

Gladwell, Malcolm. *The Tipping Point: How Little Things Can Make a Big Difference*. New York: Little, Brown, 2006.

Gleick, James. *Chaos Making a New Science*. New York: Viking Penguin, 1987.

Gluckman, Max. *Custom and Conflict in Africa*. London: Blackwell, 1955.

Gödelmann, Johann Georg. *Von Zäuberern, Hexen und Unholden, warhafftiger und wolgegründter Bericht*. Frankfurt am Main: Bassaeus, 1592.

Godfrey-Smith, Peter. *Philosophy of Biology*. Princeton Foundations of Contemporary Philosophy. Princeton: Princeton University Press, 2013.

Goertz, Hans-Jürgen. *Deutschland 1500–1648: eine zertrennte Welt*. Paderborn: Schöningh, 2004.

Goodare, Julian. *The European Witch-Hunt*. London: Routledge, 2016.

Goodheart, Eugene. *Darwinian Misadventures in the Humanities.* New Brunswick, N.J.: Transaction Publishers, 2007.
Gould, Stephen Jay. *Full House: The Spread of Excellence from Plato to Darwin.* New York: Harmony Books, 1996.
———. "Sociobiology: The Art of Storytelling." *New Scientist*, November 16, 1978.
———. *The Structure of Evolutionary Theory.* Cambridge, MA; London: Belknap Press of Harvard University Press, 2002.
———. *Wonderful Life: The Burgess Shale and the Nature of History.* New York: Norton, 1989.
Gouldner, Alvin Ward. *The Coming Crisis of Western Sociology.* London: Heinemann, 1977.
Gow, Andrew Colin, Robert B. Desjardins, and François V. Pageau. "Introduction: The Arras Witch Treatises in Context." In *The Arras Witch Treatises*, 1–18. University Park, Pennsylvania: Penn State University Press, 2016.
Grasmück, Gisela, Sönke Lorenz, and Jürgen Michael Schmidt. *Hexen Und Hexenverfolgung Im Deutschen Südwesten.* Volkskundliche Veröffentlichungen Des Badischen Landesmuseums Karlsruhe, 2: 1–2. Ostfildern: Cantz, 1994.
Grawlich, Tanja. "Der Hexenkommisar Heinrich von Schultheiß und die Hexenverfolgungen im Herzogtum Westfalen." In *Das Herzogtum Westfalen: Das kurkölnische Herzogtum Westfalen von den Anfängen der kölnischen Herrschaft im südlichen Westfalen bis zur Säkularisation 1803*, I:279–321. Münster: Aschendorff Verlag, 2009.
Greenblatt, Stephen Jay. *The Swerve: How the World Became Modern.* New York: Norton, 2011.
Gribbin, John. *In Search of the Multiverse.* London: Penguin Books, 2010.
Grimm, Jacob. *Deutsche mythologie.* Vol. I. Göttingen: Dieterische Büchhandlung, 1854.
Groß, Barbara. *Hexerei in Minden: Zur sozialen Logik von Hexereiverdächtigungen und Hexenprozessen.* Münster: Aschendorff Verlag, 2009.
Gross, Paul R., and Norman Levitt. *Higher Superstition: The Academic Left and Its Quarrels with Science.* Baltimore, MD.: Johns Hopkins University Press, 1994.
Guldi, Jo, and David Armitage. *The History Manifesto.* Cambridge: Cambridge University Press, 2017. https://doi.org/10.1017/9781139923880.
Gwynn, David M. *Christianity in the Later Roman Empire: A Sourcebook.* London: Bloomsbury Publishing, 2014.
Haas, Alexandra. *Hexen und Herrschaftspolitik: Die Reichsgrafen von Oettingen und ihr Umgang mit den Hexenprozessen im Vergleich.* Bielefeld: Verlag für Regionalgeschichte, 2018.
Haeckel, Ernst. *Die Welträtsel: Gemeinverständliche Studien über Monistische Philosophie.* Leipzig: Alfred Kröner Verlag, 1908.
Haggard P. "Human Volition: Towards a Neuroscience of Will." *Nature Reviews. Neuroscience* 9, no. 12 (2008): 934–46.

Haig, David. *From Darwin to Derrida: Selfish Genes, Social Selves, and the Meanings of Life*. 1st edition. Cambridge, Massachusetts: The MIT Press, 2020.

Hamilton, Malcolm. "Rational Choice Theory: A Critique." In *The Oxford Handbook of the Sociology of Religion*, by Peter B. Clarke, 116–33. Oxford: Oxford University Press, 2009.

———. *The Sociology of Religion*. Vol. IV. London: Routledge, 2007.

———. *The Sociology of Religion: Theoretical and Comparative Perspectives*. London: Routledge, 1995.

Hamilton, William D. "Altruism and Related Phenomena, Mainly in Social Insects." *Annual Review of Ecology and Systematics* 3, no. 1 (1972): 193–232.

Hanke, David. "Teleology: The Explanation That Bedevils Biology." In *Explanations: Styles of Explanation in Science*, by John Cornwell, 143–56. Oxford: Oxford University Press, 2004.

Hansen, Joseph. *Zauberwahn, Inquisition und Hexenproceß im Mittelalter und die Entstehung der großen Hexenverfolgung*. München: R. Oldenbourg, 1900.

Harari, Yuval Noah. *Sapiens: A Brief History of Humankind*. London: Vintage Books, 2014.

Haring, Bas. *For A Successful Life*. London: Beautiful Books, 2008.

Harper, Kyle. "The Sentimental Family: A Biohistorical Perspective." *The American Historical Review* 119, no. 5 (2014): 1547–62.

Harris, Marvin. *Cannibals and Kings: The Origins of Cultures*. New York: Vintage Books, 1991.

———. *Cows, Pigs, Wars, & Witches: The Riddles of Culture*. New York: Vintage Books, 1989.

———. *Cultural Materialism: The Struggle for a Science of Culture*. New York: Random House, 1979.

———. *The Rise of Anthropological Theory: A History of Theories of Culture*. Walnut Creek, CA: AltaMira Press, 2001.

Hatfield, Elaine, Cacioppo Hatfield, John T. Cacioppo, and Richard L. Rapson. *Emotional Contagion*. Cambridge: Cambridge University Press, 1994.

Hauser, Walter. *Der Justizmord an Anna Göldi: neue Recherchen zum letzten Hexenprozess in Europa*. Zürich: Limmat Verlag, 2007.

Haustein, Jörg. "Martin Luther als Gegner des Hexenwahns." In *Vom Unfug des Hexen-Processes: gegner der Hexenverfolgungen von Johann Weyer bis riedrich Spee*, by Hartmut Lehmann and Otto Ulbricht, 35–52. Wiesbaden: Wolfenbütteler Forschungen, 1992.

Hawkins, Mike. *Social Darwinism in European and American Thought 1860–1945: Nature as Model and Nature as Threat*. Cambridge: Cambridge University Press, 1997.

Hayek, Friedrich A. *The Constitution of Liberty*. London: Routlegde & Keegan Paul, 1960.

Heerikhuizen, Bart van. "Sociaal evolutionisme in de vormende jaren van de sociologie, in het bijzonder in Nederland." *Amsterdams Sociologisch Tijdschrift* 27, no. 1–2 (2000): 82–95.

Heinsohn, Gunnar, and Otto Steiger. *Die Vernichtung der weisen Frauen: Beiträge zur Theorie und Geschichte von Bevölkerung und Kindheit.* München: Heyne, 1985.

Henningsen, Gustav. *The Salazar Documents: Inquisitor Alonso de Salazar Frías and Others on the Basque Witch Persecution.* Leiden: Brill, 2004.

———. *The Witches' Advocate: Basque Witchcraft and the Spanish Inquisition, (1609–1614).* Basque Series. Reno, Nevada: University of Nevada Press, 1980.

———. "The Witches' Flying and the Spanish Inquisitors, or How to Explain (Away) the Impossible." *Folklore* 120, no. 1 (April 1, 2009): 57–74.

Henrich, Joseph. *The Secret of Our Success: How Culture Is Driving Human Evolution, Domesticating Our Species, and Making Us Smarter.* Princeton: Princeton University Press, 2016.

———. *The WEIRDest People in the World: How the West Became Psychologically Peculiar and Particularly Prosperous.* Illustrated edition. New York: Farrar, Straus and Giroux, 2020.

Herzig, Tamar. "Flies, Heretics, and the Gendering of Witchcraft." *Magic, Ritual, and Witchcraft* 5, no. 1 (June 5, 2010): 51–80.

———. "The Bestselling Demonologist: Heinrich Institutoris's 'Malleus Malificarum.'" In *The Science of Demons: Early Modern Authors Facing Witchcraft and the Devil*, by Jan Machielsen, 53–67. London: Routlegde, 2020.

———. "Witchcraft Prosecutions in Italy." In *The Oxford Handbook of Witchcraft in Early Modern Europe and Colonial America*, by Brian P. Levack, 249–67. Oxford: Oxford University Press, 2013.

Hocker, Jodocus. *Der Teufel selbst, Das ist Warhafftiger bestendiger und wolgegründter bericht von den Teufeln.* Henricus, 1568.

Hodgson, Geoffrey M., and Thorbjørn Knudsen. *Darwin's Conjecture: The Search for General Principles of Social and Economic Evolution.* Chicago: The University of Chicago Press, 2012.

Hofstadter, Richard. *Social Darwinism in American Thought, 1860–1915.* Philadelphia: University of Pennsylvania Press, 1944.

Holland, John H. *Complexity: A Very Short Introduction.* Oxford: Oxford University Press, 2014.

Homans, George Caspar. *Bringing Men Back In.* Indianapolis: Bobbs-Merrill, 1964.

Homza, Lu Ann. "An Expert Lawyer and Reluctant Demonologist: Alonso de Salazar Frías, Spanish Inquisitor." In *The Science of Demons: Early Modern Authors Facing Witchcraft and the Devil*, by Jan Machielsen, 299–312. London: Routlegde, 2020.

Honegger, Claudia. *Die Hexen der Neuzeit.* Frankfurt am Main: Suhrkamp, 1979.

Howell, Martha C, and Walter Prevenier. *From Reliable Sources: An Introduction to Historical Methods.* Ithaca, New York.; London: Cornell University Press, 2005.

Hume, David. *An Enquiry Concerning Human Understanding.* New York: Cosimo, 2008.

———. *Dialogues Concerning Natural Religion.* London, 1779.

Hunley, Keith. "The 10,000 Year Explosion: How Civilization Accelerated Human Evolution. Gregory Cochran, Henry Harpending." *Journal of Anthropological Research Journal of Anthropological Research* 65, no. 4 (2009): 643–44.

Hunt, Lynn. *History: Why It Matters.* Cambridge: Polity Press, 2018.

Hutton, Ronald. *The Witch: A History of Fear, from Ancient Times to the Present.* New Haven: Yale University Press, 2018.

Iannaccone, Laurence R. "Voodoo Economics? Reviewing the Rational Choice Approach to Religion." *Journal for the Scientific Study of Religion* 34, no. 1 (1995): 76–88.

Iggers, Georg G. *Historiography in the Twentieth Century: From Scientific Objectivity to the Postmodern Challenge.* Middletown, CT: Wesleyan University Press, 2005.

Ingold, Tim. "The Trouble with 'Evolutionary Biology.'" *Anthropology Today* 23, no. 2 (2007): 13–17.

Institoris, Heinrich. *The Hammer of Witches: A Complete Translation of the Malleus Maleficarum.* Translated by Christopher S Mackay. Cambridge: Cambridge University Press, 2017.

Irigaray, Luce. *Speculum: de l'autre femme.* Critique. Paris: Éd. de Minuit, 1979.

Jablonka, Eva, and Marion C. Lamb. *Evolution in Four Dimensions: Genetic, Epigenetic, Behavioral, and Symbolic Variation in the History of Life.* Cambridge, Mass.: MIT Press, 2014.

Jablonka, Eva, Marion J Lamb, and Anna Zeligowski. *Evolution in Four Dimensions: Genetic, Epigenetic, Behavioral, and Symbolic Variation in the History of Life.* Cambridge (MA) [etc.: MIT Press, 2014.

Jaeger, Friedrich, and Jörn Rüsen. *Geschichte des Historismus: eine Einführung.* München: Beck, 1992.

Jenkins, Richard. *Pierre Bourdieu.* Key Sociologists. London: Routledge, 1992.

Jerouscheck, Günther. *Die Hexen und ihr Prozess: Die Hexenverfolgung in der Reichsstadt Esslingen.* Esslinger Studien, Schriftenreihe; Bd. 11. Esslingen: Selbstverlag des Stadtarchivs Esslingen, 1992.

Johnson, Monte R. *Aristotle on Teleology.* Oxford: Clarendon Press, 2005.

Joseph Bulbulia, Armin W. Geertz, Quentin D. Atkinson, Emma Cohen, Nicholas Evans, Pieter François, Herbert Gintis, et al. "The Cultural Evolution of Religion." by Peter J. Richerson and Morten H. Christiansen, 381–404. Cambridge, Mass.: MIT Press, 2013.

Jung, Norbert. "Die Bamberger Hexenverfolgungen in Theologiegeschichtlicher Perspektive." In *"Hexenprozesse Und Hexenverfolgung" Im Hochstift Bamberg: Eine Vorlaüfige Bilanz*, by Ulrieke Siebenhaar, 62–91. Bamberg: Verlags-UG, 2013.

Kamen, Henry. *Early Modern European Society*. London: Routledge, 2000.

———. *The Iron Century: Social Change in Europe, 1550–1660*. London: Cardinal, 1976.

Kant, Immanuel. *Critique of Judgment*. Translated by Werner S Pluhar. Indianapolis: Hackett Publishing Company, 1987.

———. *Kritik Der Urteilskraft*. Frankfurt: Suhrkamp, 1989.

Kaplan, Benjamin J. *Divided by Faith Religious Conflict and the Practice of Toleration in Early Modern Europe*. Cambridge: Belknap Press of Harvard University Press, 2007.

Kay, Thomas, Laurent Lehmann, and Laurent Keller. "Kin Selection and Altruism." *Current Biology* 29, no. 11 (2019): R438–42.

Keesing, Roger M, and Andrew Strathern. *Cultural Anthropology: A Contemporary Perspective*. Belmont, CA: Wadsworth/Thompson Learning, 1998.

Keitt, Andrew. "Rethinking with Demons: The Campaign against Superstition in Late Medieval and Early Modern Europe from a Cognitive Perspective." *Preternature: Critical and Historical Studies on the Preternatural* 6, no. 2 (2017): 236–77.

Keller, Albert Galloway. *Societal Evolution; a Study of the Evolutionary Basis of the Science of Society,*. New York: The Macmillan Company, 1915.

Keller, Laurent, Michel Chapuisat, and Jonathan B. Losos. "Evolution of Eusociality." In *The Princeton Guide to Evolution*, 697–702. Princeton NJ: Princeton University Press, 2014.

Kieckhefer, Richard. *Hazards of the Dark Arts: Advice for Medieval Princes on Witchcraft and Magic*. Illustrated edition. University Park, Pennsylvania: Penn State University Press, 2017.

———. "Mythologies of Witchcraft in the Fifteenth Century." *Magic, Ritual, and Witchcraft* 1, no. 1 (2006): 79–108.

Kincaid, Harold. "Functional Explanation and Evolutionary Social Science," 2007a.

———. "Functional Explanation and Evolutionary Social Science." In *Philosophy of Anthropology and Sociology*, edited by Stephen P. Turner and Mark W. Risjord, 213–47. Handbook of the Philosophy of Science. Amsterdam: North-Holland, 2007b.

———. *Philosophical Foundations of the Social Sciences: Analyzing Controversies in Social Research*. Cambridge: Cambridge University Press, 1996.

King, Charles. *The Reinvention of Humanity: How a Circle of Renegade Anthropologists Remade Race, Sex and Gender*. Vintage, 2020.

King James I, and Donald Tyson. *The Demonology of King James I: Includes the Original Text of Daemonologie and News from Scotland*. Woodbury: Llewellyn Worldwide, 2012.

Kinsman, Robert S. *The Darker Vision of the Renaissance: Beyond the Fields of Reason*. University of California Press, 1974.

Kirby, Simon. "Language Is an Adaptive System: The Role of Cultural Evolution in the Origin of Structure." In *The Oxford Handbook of Language Evolution*, by Magie Tallerman and Kathleen R. Gibson, 589–604. Oxford: Oxford University Press, 2012.

Kleinberg, Ethan. "Just the Facts: The Fantasy of a Historical Science." *History of the Present*, January 1, 2016.

Kluckhohn, Clyde. *Navaho Witchcraft*. Boston: Beacon Press, 1944.

Knauft, Bruce M. *Good Company and Violence: Sorcery and Social Action in a Lowland New Guinea Society*. Studies in Melanesian Anthropology 3. Berkeley: University of California Press, 1985.

Knefelkamp, Ulrich. "Von der Ketzerei zur Hexerei, Ursachen und Hintergründe der grossen Hexenverfolgungen ca. 1570–1630." In *"Hexenprozesse und Hexenverfolgung" im Hochstift Bamberg: eine vorlaüfige Bilanz*, by Ulrieke Siebenhaar, 9–33. Verlags-UG, 2013.

Kofman, Ava. "Bruno Latour, the Post-Truth Philosopher, Mounts a Defense of Science." *The New York Times*, October 25, 2018, sec. Magazine. https://www.nytimes.com/2018/10/25/magazine/bruno-latour-post-truth-philosopher-science.html.

Koning, Niek. "Witchcraft Beliefs and Witch Hunts." *Human Nature* 24, no. 2 (June 1, 2013): 158–81.

Koopmans, Ruud. "Het mysterie van de naastenliefde: een evolutionair-sociologische verklaring" 2, no. 2 (2006): 114–38.

Korrodi-Aebli, Elisabeth. "Auf den Spuren der 'letzten Hexe'. Anna Göldi—Der Fall—Die Presseberichte." http://www.historicum.net, 1996. https://www.historicum.net/no_cache/persistent/artikel/auf-den-spuren/.

Krause, Johannes, and Thomas Trappe. *Die Reise unserer Gene: Eine Geschichte über uns und unsere Vorfahren*. Propyläen, Berlin.

Krause, Virginia. "The Will to Know and the Unknowable: Jean Bodin's 'De La Démonomanie.'" In *The Science of Demons: Early Modern Authors Facing Witchcraft and the Devil*, by Jan Machielsen, 119–33. London: Routledge, 2020.

———. *Witchcraft, Demonology, and Confession in Early Modern France*. New York: Cambridge University Press, 2015.

Krause, Virginia, Christian Martin, and Eric MacPhail. "Introduction." In *De la démonomanie des sorciers*, 9–56. Genève: Librairie Marcel Didier, 2016.

Kroeber, Alfred L. *The Nature of Culture*. Chicago: University of Chicago Press, 1952.

Kroeber, Alfred L., and Clyde Kluckhohn. *Culture: A Critical Review of Concepts and Definitions.* New York: Random House, 1950.

Kronfeldner, Maria. *Darwinian Creativity and Memetics.* London: Taylor & Francis Group, 2011.

Kunda, Z. "The Case for Motivated Reasoning." *Psychological Bulletin*, 1990.

Kunin, Victor, Leon Goldovsky, Nikos Darzentas, and Christos A. Ouzounis. "The Net of Life: Reconstructing the Microbial Phylogenetic Network." *Genome Research* 15, no. 7 (July 2005): 954–59.

Kuntzen, Hans. *Newe Zeitung von einer Erschrecklichen That, welche zu Dillingen, von einem Jhesuwider, und einer Hexen geschehen ist.* Niklaus Heinrich, 1580.

Kuper, Adam. "If Memes Are the Answer, What Is the Question?" In *Darwinizing Culture: The Status of Memetics as a Science*, by Robert Aunger, 175–88. Oxford: Oxford University Press, 2001.

Kurlander, Eric. *Hitler's Monsters: A Supernatural History of the Third Reich*, 2017.

Labouvie, Eva. *Zauberei und Hexenwerk: ländlicher Hexenglaube in der frühen Neuzeit.* Fischer Taschenbücher: 10493. Geschichte. Frankfurt am Main: Fischer Taschenbuch Verlag, 1991.

Laland, Kevin N. *Darwin's Unfinished Symphony: How Culture Made the Human Mind.* Princeton: Princeton University Press, 2017.

Laland, Kevin N., and Gillian R. Brown. *Sense and Nonsense: Evolutionary Perspectives on Human Behaviour.* Oxford: Oxford University Press, 2011.

Laland, Kevin N., Tobias Uller, Marcus W. Feldman, Kim Sterelny, Gerd B. Müller, Armin Moczek, Eva Jablonka, and John Odling-Smee. "The Extended Evolutionary Synthesis: Its Structure, Assumptions and Predictions." *Proceedings of the Royal Society B: Biological Sciences* 282, no. 1813 (August 22, 2015): 20151019.

Lamarck, Jean-Baptiste de Monet de. *Philosophie zoologique, ou Exposition des considérations relatives à l'histoire naturelle des animaux.* Vol. 1, 1809.

Lancre, Pierre de. *On the Inconstancy of Witches: Pierre's De Lancre's Tableau de l'inconstance Des Mauvais Anges et Demons (1612).* Edited by Gerhild Scholz Williams. Tempe: Arizona Center for Medieval and Renaissance Studies, 2006.

Lang, Berel. *Genocide: The Act as Idea.* Philadelphia: University of Pennsylvania Press, 2017.

Lapouge, Georges Vacher de. *Les sélections sociales.* Paris: A. Fontemoing, 1896.

Latour, Bruno. *Face à Gaïa: Huit conférances sur le Nouveau Régime Climatique.* Paris: La Découverte, 2015.

———. *Politics of Nature: How to Bring the Sciences into Democracy.* Translated by Catherine Porter. Cambridge, Mass: Harvard University Press, 2004.

———. "Ramsès II est-il mort de la tuberculose ? Jusqu'où faut-il mener des découvertes scientifiques." *La Recherche.* 29, no. 307 (1998): 84.

———. *Reassembling the Social: An Introduction to Actor-Network-Theory.* Clarendon Lectures in Management Studies. Oxford: Oxford University Press, 2005.

———. *Science in Action: How to Follow Scientists and Engineers through Society.* Stony Stratford: Open University Press, 1987.

———. *The Pasteurization of France.* Cambridge, Mass.: Harvard University Press, 1988.

Latour, Bruno, and Steve Woolgar. *Laboratory Life: The Social Construction of the Scientific Facts.* Sage Library of Social Research 80. Beverly Hills, CA: Sage Publications, 1979.

Law, John. "STS as Method." In *The Handbook of Science and Technology Studies*, edited by Ulrike Felt, Rayvon Fouché, Clark A. Miller, and Laurel Smith-Doerr, 4th ed., 31–58. Cambridge, Mass.: MIT Press, 2017.

Leeson, Peter T, and Jacob W Russ. "Witch Trials." *The Economic Journal*, no. 128 (2017): 2066–2105.

Leezenberg, Michiel, and Gerard de Vries. *History and Philosophy of the Humanities: An Introduction.* Amsterdam: Amsterdam University Press, 2018.

Lehmann, Hartmut. "Hexenverfolgungen und Hexenprozesse im alten Reich zwischen Reformation und Aufklärung." *Jahrbuch des Instituts für Deutsche Geschichte* 7 (1978): 13–70.

Lehmann, Hartmut, and Otto Ulbricht. "Motive und Argumente von Gegnern der Hexenverfolgung von Weyer bis Spee." In *Vom Unfug des Hexen-Processes: Gegner der Hexenverfolgung von Johann Weyer bis Friedrich Spee*, by Hartmut Lehmann and Otto Ulbricht, 1–14. Wiesbaden: Wolfenbütteler Forschungen, 1992a.

———, eds. *Vom Unfug des Hexen-Processes: Gegner der Hexenverfolgung von Johann Weyer bis Friedrich Spee.* Wolfenbütteler Forschungen, Bd. 55. Wiesbaden: Harrassowitz, 1992b.

Leroi, Armand Marie. *The Lagoon: How Aristotle Invented Science.* London: Bloomsbury, 2015.

Leunissen, Mariska. *Explanation and Teleology in Aristotle's Science of Nature.* Cambridge: Cambridge University Press, 2010.

Levack, Brian P. "The Decline and End of Witchcraft Prosecutions." In *The Oxford Handbook of Witchcraft in Early Modern Europe and Colonial America*, by Brian P. Levack, 429–88. Oxford: Oxford University Press, 2013.

———. *The Witchcraft Sourcebook.* New York: Routledge, 2004.

———. *The Witch-Hunt in Early Modern Europe.* New York: Routledge, 2016.

Levine, Philippa. *Eugenics. A Very Short Introduction.* Oxford: Oxford University Press, 2017.

Lewens, Tim. *Cultural Evolution: Conceptual Challenges.* Oxford: Oxford University Press, 2015a.

———. *Cultural Evolution: Conceptual Challenges.* OUP Oxford, 2015b.

———. "Cultural Evolution: Integration and Skepticism." In *The Oxford Handbook of Philosophy of Social Science*, by Harold Kincaid, 458–80. Oxford: Oxford University Press, 2012.

———. *Darwin*. London: Routledge, 2006.

Lewis, Herbert S. "Boas, Darwin, Science, and Anthropology." *Current Anthropology*. 42, no. 3 (2001): 381–406.

Lewontin, Richard C. "The Wars Over Evolution." *The New York Review of Books*, October 20, 2005. http://www.nybooks.com/articles/2005/10/20/the-wars-over-evolution/.

Lewontin, Richard C., Steven Rose, and Leon J. Kamin. *Not in Our Genes: Biology, Ideology, and Human Nature*. New York: Pantheon Books, 1984.

Loftin, John D. *Religion and Hopi Life in the Twentieth Century*. Bloomington: Indiana University Press, 2003.

Lorenz, Chris. *De constructie van het verleden: een inleiding in de theorie van de geschiedenis*. Amsterdam: Boom, 1998.

Lowie, Robert H. *Culture and Ethnology*. New York: Basic Books, 1966.

Luckhurst, Roger. "Bruno Latour's Scientifiction: Networks, Assemblages, and Tangled Objects." *Science Fiction Studies* 33, no. 1 (2006): 4–17.

Lumsden, Charles J., and Edward O. Wilson. *Genes, Mind, and Culture: The Coevolutionary Process*. Cambridge, Mass.: Harvard University Press, 1981.

MacCulloch, Diarmaid. *Reformation: Europe's House Divided, 1490–1700*. London: Penguin Books, 2004.

Macfarlane, Alan. *Witchcraft in Tudor and Stuart England: A Regional and Comparative Study*. New York: Harper & Row, 1970.

Machielsen, Jan. *Martin Delrio: Demonology and Scholarship in the Counter-Reformation*. Oxford: Oxford University Press, 2015.

———. "Thinking with Montaigne: Evidence, Skepticism and Meaning in Early Modern Demonology." *French History* 25, no. 4 (2011): 427–52.

Mackay, Charles. *Memoirs of Extraordinary Popular Delusions and the Madness of Crowds*. London: Office of the National Illustrated Library, 1852.

Maderspacher, Florian. "Epigenetics." In *The Princeton Guide to Evolution*, by Jonathan B. Losos, 420–27. Princeton NJ: Princeton University Press, 2014.

Mährle, Wolfgang. "Ellwangen—Hexenverfolgungen." In *Lexikon Zur Geschichte Der Hexenverfolgung*, edited by Gudrun Gersmann, Katrin Moeller, and Jürgen Michael Schmidt. Historicum.Net, 2006. https://www.historicum.net/purl/jfzqr/.

———. "Fürstpropstei Ellwangen." In *Wider alle Hexerei und Teufelswerk: Die Europäische Hexenverfolgung und ihre Auswirkungen auf Südwestdeutschland*, by Sönke Lorenz and Jürgen Michael Schmidt, 377–86. Ostfildern: Jan Thorbeckeverlag, 2004.

———. "'O wehe der armen seelen': Hexenverfolgungen in der Fürstpropstei Ellwangen (1588–1694)." In *Zum Feuer verdammt: Die Hexenverfolgungen in*

der Grafschaft Hohenberg, der Reichstadt Reutlingen und der Fürstpropstei Ellwangen, by Johannes Dillinger, Thomas Fritz, and Wolfgang Mährle, 325–500. Stuttgart: Franz Steiner Verlag, 1998.
Malebranche, Nicolas. *Recherche de la vérité*. Vol. I. Paris: Ernest Flammarion, n.d.
Malinowski, Bronislaw. *A Scientific Theory of Culture, and Other Essays*. Chapel Hill: University of North Carolina Press, 1944.
———. *Coral Gardens and Their Magic. A Study of the Methods of Tilling the Soil and of Agricultural Rites in the Trobriand Islands*. London, 1935.
———. *Magic, Science and Religion and Other Essays*. New York: Doubleday Anchor Books, 1954.
———. "The Group and the Individual in Functional Analysis." *American Journal of Sociology* 44, no. 6 (1939): 938–64.
Malm, Andreas. *The Progress of This Storm: Nature and Society in a Warming World*. London: Verso, 2018.
Martin, Raymond. "The Essential Difference between History and Science." *History and Theory* 36, no. 1 (1997): 1–14.
Marwick, Max. *Sorcery in Its Social Setting: A Study of the Northern Rhodisian Cewa*. Manchester: Manchester University Press, 1965.
Marx, Karl. *Das Kapital: Kritik der politischen Oekonomie*. 2nd ed. Vol. I. Hamburg: Verlag von Otto Meissner, 1872.
———. *Der junge Marx: philosophische Schriften*. Edited by Stefan Kraft and Karl Reitter. Vienna: Promedia Verlag, 2007.
Matthews, John A., and Keith R. Briffa. "The 'Little Ice Age': Re-Evalaution of an Evolving Concept." *Geografiska Annaler*, Physical Geography, 87, no. 1 (2005): 17–36.
Mayr, Ernst. *What Evolution Is*. New York: Basic Books, 2001.
McGee, R. Jon, and Richard L. Warms. *Anthropological Theory: An Introductory History*. New York: McGraw-Hill, 2012.
McNeill, William H. "History and the Scientific Worldview." *History and Theory* 37, no. 1 (1998): 1–13.
———. *Plagues and Peoples*. Oxford: Blackwell, 1976.
Mead, Margaret. *Coming of Age in Samoa: A Psychological Study of Primitive Youth for Western Civilisation*. New York: Morrow, 1928.
Meder, David. *Acht Hexenpredigten*. Leipzig: Jacob Apels, 1605.
Meier, Christine. "Die Anfänge der Hexenprozesse in Lemgo." In *Hexenverfolgung und Regionalgeschichte*, by Gisela Wilbertz, Gerd Schwerhoff, and Jürgen Scheffler, 83–107. Bielefeld: Verlag für Regionalgeschichte, 1994.
Mercier, Franck. *La vauderie d'Arras: Une chasse aux sorcières à l'automne du Moyen Âge*. Rennes: Presses universitaires de Rennes, 2015.
Mercier, Hugo. *Not Born Yesterday: The Science of Who We Trust and What We Believe*. Princeton: Princeton University Press, 2020.

Merton, Robert King. *Social Theory and Social Structure.* New York: Simon and Schuster, 1968.

Mesoudi, Alex. "Cultural Evolution: A Review of Theory, Findings and Controversies." *Evolutionary Biology* 43, no. 4 (2016): 481–97.

———. *Cultural Evolution: How Darwinian Theory Can Explain Human Culture and Synthesize the Social Sciences.* Chicago: University of Chicago Press, 2011.

———. "Pursuing Darwin's Curious Parallel: Prospects for a Science of Cultural Evolution." *Proceedings of the National Academy of Sciences of the United States of America*, 2017.

Meyfart, Johannes. *Christliche Erinnerung, An Gewaltige Regenten und Gewissenhafte Praedicanten, wie das abschewliche Laster der Hexerey mit Ernst auszurotten, aber in Verfolgung desselbigen auff Cantzeln und in Gerichtshaeusern sehr bescheidentlich zu handeln sey.* Schleusingen: Peter Schmiden, 1635.

Midelfort, H. C. Erik. "Witch Craze?: Beyond the Legends of Panic." *Magic, Ritual, and Witchcraft* 6, no. 1 (June 2, 2011): 11–33.

———. *Witch Hunting in Southwestern Germany: 1562–1684: The Social and Intellectual Foundations.* Stanford California: Stanford University Press, 1972.

Midgley, Mary. "Gene-Juggling." *Philosophy* 54, no. 210 (1979): 439–58.

———. "Why Memes." In *Alas, Poor Darwin: Arguments against Evolutionary Psychology*, by Hilary Rose and Steven Rose, 67–84. London: Random House, 2000.

Millikan, Ruth Garrett. *Varieties of Meaning: The 2002 Jean Nicod Lectures.* S.L.: Bradford Books, 2006.

Mirandola, Giovanni Francesco Pico della. *La sorcière: dialogue en trois livres sur la tromperie des démons.* Translated by Alfredo Perifano. Turnhout: Brepols, 2007.

Modrow, Susanne. *Viren: Grundlagen, Krankheiten, Therapien.* München: Beck, 2001.

Moeller, Katrin. *Dass Willkür über Recht ginge: Hexenverfolgung in Mecklenburg im 16. und 17. Jahrhundert.* Bielefeld: Verlag für Regionalgeschichte, 2007.

Mokyr, Joel. *A Culture of Growth: The Origins of the Modern Economy.* Princeton: Princeton University Press, 2017.

Montaigne, Michel de. *Les Essais.* Paris: Gallimard, 2007.

Monter, William E. "Witch Trials in Continental Europe 1560–1660." In *Witchcraft and Magic in Europe: The Period of the Witch Trials*, by Bengt Ankarloo and Stuart Clark, 1–52. Athlone History of Magic and Witchcraft in Europe. London: The Athlone Press, 2002.

Moore, Jerry D. *Visions of Culture: An Introduction to Anthropological Theories and Theorists.* Lanham: AltaMira Press, 2012.

Moore, R. I. *The Formation of a Persecuting Society: Authority and Deviance in Western Europe, 950–1250.* Malden, MA: Blackwell Publishing, 2008.

"Most Cited Authors of Books in the Humanities, 2007 | Times Higher Education (THE)." Accessed March 29, 2018. https://www.timeshighereducation.com/news/most-cited-authors-of-books-in-the-humanities-2007/405956.article.

Muchembled, Robert. *Sorcières: justice et société aux 16e et 17e siècles*. Paris: Editions Imago, 1987.

Müller, Andreas. "Elaborated Concepts of Witchcraft? Applying the 'Elaborated Concept of Witchcraft' in a Comparative Study on the Witchcraft Trials of Rostock (1584) and Hainburg (1617–18)." *E-Rhizome* 1, no. 1 (2019): 1–18.

Murray, Margaret Alice. *The Witch-Cult in Western Europe*. Lexington, KY: Forgotten Books, 2008.

Nelson, Richard R., and Sidney G. Winter. *An Evolutionary Theory of Economic Change*. Cambridge, Mass.: Harvard University Press, 1982.

Newer Tractat von der Verführten Kinder Zauberey. Cologne: Brachel, 1629.

Nider, Johannes. "Formicarius." In *L'imaginaire du sabbat: Edition critique des textes les plus anciens (1430 c.–1440 c.)*, by Martine Ostorero, Agostino Paravicini Bagliani, and Kathrin Utz Tremp, 122–200. translated by Katherine Chène. Lausanne: Cahier Lausannois d'histoire médiévale, 1999.

Norenzayan, Ara. *Big Gods: How Religion Transformed Cooperation and Conflict*. Princeton: Princeton University Press, 2013.

Norman, Andy. *Mental Immunity: Infectious Ideas, Mind-Parasites, and the Search for a Better Way to Think*. New York: Harper Wave, 2021.

Oberste, Jörg. *Ketzerei Und Inquisition Im Mittelalter*. Darmstadt: Wissenschaftliche Buchgesellschaft, 2012.

Odling-Smee, John. "Niche Construction." In *Evolution, the Extended Synthesis*, by Massimo Pigliucci and Gerd B. Müller, 175–208. Cambridge, Mass.: MIT Press, 2010.

Okasha, Samir. "Units and Levels of Selection." In *The Princeton Guide to Evolution*, by Jonathan B. Losos, 200–205. Princeton NJ: Princeton University Press, 2014.

Opitz-Belakhal, Claudia. *Böse Weiber: Wissen und Geschlecht in der Dämonologie der Frühen Neuzeit*. Sulzbach: Ulrike Helmer Verlag, 2017.

———. *Das Universum des Jean Bodin: Staatsbildung, Macht und Geschlecht im 16. Jahrhundert*. Frankfurt: Campus Verlag, debating. 2006

———. "Der Magistrat als Hexenjäger: Hexenverfolgung und staatliche Ordnung bei Jean Bodin." In *Hexenprozess und Staatsbildung*, by Johannes Dillinger, Jürgen Michael Schmidt, and Dieter R Bauer, 41–58. Bielefeld: Verlag für Regionalgeschichte, 2008.

Ostorero, Martine. *Le diable au sabbat: Littérature démonologique et sorcellerie (1440–1460)*. Florence: Sismel—Edizione del Galluzzo, 2011.

Ostorero, Martine, Agostino Paravicini Bagliani, and Kathrin Utz Tremp. *L'imaginaire du sabbat: Edition critique des textes es plus anciens (1430 c.–1440 c.)*. Lausanne: Cahier Lausannois d'histoire médiévale, 1999.

Otten, Johan. *Duivelskwartier: 1595: heksen, heren en de dood in het vuur.* Nijmegen: Vantilt, 2015.

Otterspeer, Willem. *Orde en trouw: over Johan Huizinga.* Amsterdam: De Bezige Bij, 2006.

Pagel, Mark. *Wired for Culture: The Natural History of Human Cooperation.* London: Penguin Books, 2013.

Paley, William. *Natural Theology: Or, Evidences of the Existence and Attributes of the Deity, Collected from the Appearances of Nature.* Boston: Gould and Lincoln, 1854.

Parker, Geoffrey. *Global Crisis: War, Climate Change and Catastrophe in the Seventeenth Century.* Padstow, Cornwall: Yale University Press, 2013.

Parren, Nora. "The (Possible) Cognitive Naturalness of Witchcraft Beliefs: An Exploration of the Existing Literature." *Journal of Cognition and Culture* 17, no. 5 (2017): 396–418.

Parsons, Talcott. *The Social System.* New York: Free Press of Glencoe, 1951.

Paulson, Steve. "The Critical Zone of Science and Politics: An Interview with Bruno Latour." *Los Angeles Review of Books*, February 23, 2018. https://lareviewofbooks.org/article/the-critical-zone-of-science-and-politics-an-interview-with-bruno-latour/.

Perlman, Mark. "The Modern Philosophical Resurrection of Teleology." *The Monist* 87, no. 1 (2004): 3–51.

Perry, George, and Ruth Mace. "The Lack of Acceptance of Evolutionary Approaches to Human Behaviour." *Journal of Evolutionary Psychology Journal of Evolutionary Psychology* 8, no. 2 (2010): 105–25.

Pettit, Philip. "Functional Explanation and Virtual Selection." *The British Journal for the Philosophy of Science.* 47, no. 2 (1996): 291.

Pfeifer, Hans. *Ellwangen: Kunst und Geschichte aus 1250 Jahren.* Ulm: Süddeutsche Verlagsgesellschaft Ulm, 2000.

Pichot, André. *La société pure: de Darwin à Hitler.* Paris: Flammarion, 2009.

Pietro Corsi. "Jean-Baptiste Lamarck: From Myth to History." In *Transformations of Lamarckism: From Suble Fluids to Molecular Biology*, by Snait B Gissis and Eva Jablonka, 9–20. Cambridge, Mass.: The MIT Press, 2011.

Pigliucci, Massimo, and Gerd B. Müller. *Evolution, the Extended Synthesis.* Cambridge, Mass.: MIT Press, 2010.

Pinker, Steven. *The Better Angels of Our Nature: Why Violence Has Declined.* New York: Penguin Books, 2012a.

———. *The Blank Slate: The Modern Denial of Human Nature.* London: Allen Lane, 2002.

———. "The False Allure of Group Selection." *Edge*, June 18, 2012b. https://www.edge.org/conversation/steven_pinker-the-false-allure-of-group-selection.

Pleier, Hermann. *Malleus Judicum, das ist: Gesetzhammer der unbarmhertzigen Hexenrichter, ausz dem besten Ertz Göttlicher, Natürlicher, und Weltlicher*

Rechaten, von etlichen Christlichen Meistern geschmiedet, und jetzo durch einen barmhertzigen Christen auffs newe bestielet, 1630.

Pluckrose, Helen, and James A. Lindsay. *Cynical Theories: How Activist Scholarship Made Everything about Race, Gender, and Identity—and Why This Harms Everybody*. Durham, North Carolina: Pitchstone Publishing, 2020.

Pohl, Herbert. *Zauberglaube und Hexenangst im Kurfürstentum Mainz: ein Beitrag zur Hexenfrage im 16. und beginnenden 17. Jahrhundert*. 2nd ed. Hexenforschung 3. Stuttgart: Steiner, 1998.

Prinz, Jesse J. *Beyond Human Nature How Culture and Experience Shape the Human Mind*. New York: W.W. Norton, 2012.

Purkiss, Diane. *The Witch in History: Early Modern and Twentieth-Century Representations*. London: Routledge, 1996.

Queller, David C., and Joan E. Strassmann. "Kin Selection and Inclusive Fitness." In *The Princeton Guide to Evolution*, by Jonathan B. Losos, 213–20. Princeton NJ: Princeton University Press, 2014.

Rabinow, Paul. *The Foucault Reader*. New York: Pantheon Books, 1984.

Radcliffe-Brown, Alfred Reginald. *Structure and Function in Primitive Society*. London: Cohen & West, 1952.

Rampton, Martha. *European Magic and Witchcraft: A Reader*. Toronto: University of Toronto Press, 2018.

Ranke, Leopold von. *The Theory and Practice of History*. The European Historiography Series. Indianapolis: The Bobbs-Merrill Co, 1973.

Ratzel, Friedrich. *Der Lebensraum: eine biogeographische Studie*. Tübingen: H. Laupp, 1901.

Rees, Martin. "Explaining the Universe." In *Explanations: Styles of Explanation in Science*, by John Cornwell, 39–66. Oxford: Oxford University Press, 2004.

Reich, David. *Who We Are and How We Got Here: Ancient DNA and the New Science of the Human Past*. Oxford: Oxford University Press, 2018.

Rémy, Nicolas. *Demonolatry*. Translated by E.A. Ashwin. Kessinger Publishing, 2010.

Renczes, Andrea. *Wie löscht man eine Familie aus?: eine Analyse Bamberger Hexenprozesse*. Pfaffenweiler: Centaurus-Verlagsgesellschaft, 1992.

Repko, Allen F. *Interdisciplinary Research: Process and Theory*. Thousand Oaks: Sage Publications, 2008.

Richard Kieckhefer. "The First Wave of Trials for Diabolical Witchcraft." In *The Oxford Handbook of Witchcraft in Early Modern Europe and Colonial America*, by Brian P. Levack, 159–79. Oxford: Oxford University Press, 2013.

Richards, Robert J. *Was Hitler a Darwinian?: Disputed Questions in the History of Evolutionary Theory*. Chicago: University of Chicago Press, 2013.

Richerson, Peter J., and Robert Boyd. *Not By Genes Alone: How Culture Transformed Human Evolution*. Chicago: University of Chicago Press, 2006.

Rickert, Heinrich. *Kulturwissenschaft und Naturwissenschaft.* 2nd ed. Tübingen: Mohr, 1910.

Ricœur, Paul. "Le retour de l'Événement." *Mélanges de l'École française de Rome. Italie et Méditerranée* 104, no. 1 (1992): 29–35.

Ridley, Mark. *Evolution.* Malden, Mass.: Blackwell, 2004.

Ridley, Matt. *Nature via Nurture: Genes, Experience and What Makes Us Human.* London: HarperCollins Publishers, 2003.

———. *The Evolution of Everything: How Small Changes Transform Our World.* London: HarperCollins Publishers, 2015.

Rio, Martin del, and Peter G. Maxwell-Stuart. *Investigations into Magic.* Manchester: Manchester University Press, 2000.

Rippon, Gina. *Gender and Our Brains: How New Neuroscience Explodes the Myths of the Male and Female Minds.* New York: Penguin Random House, 2019.

Ritchie, David G. *Darwin and Hegel and Other Philosophical Studies.* Edited by Peter P. Nicholson. Vol. 2. Collected Works of D.G. Ritchie. Bristol: Bristol Thoemmes, 1998.

———. "Social Evolution." *International Journal of Ethics* 6, no. 2 (1896): 165–81.

Ritzer, George. *Contemporary Sociological Theory and Its Classical Roots: The Basics.* 2nd ed. New York: McGraw-Hill, 2007.

———. "Sociology: A Multi Paradigm Science." *The American Sociologist* 10, no. 3 (August 1975): 156–67.

Robbins, Rossell Hope. *The Encyclopedia of Witchcraft and Demonology.* New York: Crown, 1959.

Robert Aunger. *Darwinizing Culture: The Status of Memetics as a Science.* Oxford: Oxford University Press, 2001.

Robisheaux, Thomas. "The German Witch Trials." In *Oxford Handbook of Witchcraft in Early Modern Europe and North America*, by Brian P. Levack, 179–98. Oxford: Oxford University Press, 2013.

Roper, Lyndal. *Witch Craze: Terror and Fantasy in Baroque Germany.* New Haven: Yale University Press, 2006.

Rorty, Richard. *Objectivity, Relativism, and Truth: Philosophical Papers.* Cambridge: Cambridge University Press, 1990.

———. *Philosophy and Social Hope.* London: Penguin Books, 1999.

Rose, Nikolas. "The Human Sciences in a Biological Age." *Theory, Culture & Society* 30, no. 1 (2013): 3–34.

Rose, Steven. Richard Dawkins interviews Steven Rose. Youtube, November 25, 2012. https://www.youtube.com/watch?v=QceGqKZMqIM.

———. "The Biology of the Future and the Future of Biology." In *Explanations: Styles of Explanation in Science*, by John Cornwell, 125–43. Oxford: Oxford University Press, 2004.

Rose, Steven P. R. *Lifelines: Biology, Freedom, Determinism.* London: Vintage, 2005.

Rosenberg, Karen R., and Wenda Trevathan. *Costly and Cute Helpless Infants and Human Evolution*. Santa Fe: University of New Mexico Press, 2016.

Rowlands, Alison. "Witchcraft and Gender in Early Modern Europe." In *The Oxford Handbook of Witchcraft in Early Modern Europe and Colonial America*, by Levack, 449–67. Oxford: Oxford University Press, 2013.

———. *Witchcraft Narratives in Germany: Rothenburg, 1561–1652*. Studies in Early Modern European History. Manchester etc.: Manchester University Press, 2003.

———. "Würzburg, Prince-Bishopric Of." In *Encyclopedia of Witchcraft: The Western Tradition*, IV:1230–31. Santa Barbara: ABC-CLIO, 2006.

Rublack, Ulinka. *Reformation Europe*. New York: Cambridge University Press, 2005.

———. *The Astronomer & the Witch: Johannes Kepler's Fight for His Mother*. First edition. Oxford: Oxford University Press, 2015.

Rügge, Nicolas. *Die Hexenverfolgung in der Stadt Osnabrück: Überblick—Deutungen—Quellen*. Osnabrück: Verein für Geschichte und Landeskunde von Osnabrück, 2015.

Rummel, Walter. *Bauern, Herren und Hexen: Studien zur Sozialgeschichte sponheimischer und kurtrierischer Hexenprozesse, 1574–1664*. Göttingen: Vandenhoeck & Ruprecht, 1991.

———. "Das 'ungestüme Umherlaufen der Untertanen'. Zum Verhältnis von religiöser Ideologie, sozialem Interesse und Staatsräson in den Hexenverfolgungen im Rheinland." *Rheinische Vierteljahrsblätter: Mitteilungen des Instituts für Geschichtliche Landeskunde der Rheinlande an der Universität Bonn*, 2003, 121–61.

———. "Phasen Und Träger Kurtrierischer Und Sponheimischer Hexenvervolgungen." In *Hexenglaube Und Hexenprozesse Im Raum Rhein-Mosel-Saar*, by Günther Franz and Franz Irsigler, 255–54. Trier: Spee Buchverlag, 1995.

———. "So mögte auch eine darzu kommen, so mich belädiget: Zur sozialen Motivation und Nutzung von Hexereianklagen." In *Hexenverfolgung und Herrschaftspraxis*, edited by Rita Voltmer, 205–28. Trier: Paulinus Verlag, 2006.

———. "Vom Umgang mit Hexen und Hexerei: Das Wirken des Alltags in Hexenprozessen und die alltägliche Bedeutung des Hexenthemas." In *Methoden und Konzepte der historischen Hexenforschung*, by Günther Franz and Franz Irsigler, 79–108. Trier: Spee, 1998.

Rummel, Walter, and Rita Voltmer. *Hexen und Hexenverfolgung in der Frühen Neuzeit*. Darmstadt: Wissenschaftliche Buchgesellschaft, 2008.

Runciman, W. G. "Was Max Weber a Selectionist in Spite of Himself?" *Journal of Classical Sociology* 1, no. 1 (March 1, 2001): 13–32.

Rutting, Lucas, Ger Post, Machiel Keestra, Mieke de Roo, Sylvia Blad, and Linda de Greef. *An Introduction to Interdisciplinary Research: Theory and Practice.*

Edited by Machiel Keestra and Steph Menken. Amsterdam: Amsterdam University Press, 2016.
Saad, Gad. *The Parasitic Mind: How Infectious Ideas Are Killing Common Sense.* Washington, D.C.: Regnery Publishing, 2020.
Sahlins, Marshall. *Culture and Practical Reason.* University of Chicago Press, 1976.
———. "Culture as Protein and Profit." *The New York Review of Books*, November 23, 1978. https://www.nybooks.com/articles/1978/11/23/culture-as-protein-and-profit/.
Saini, Angela. *Superior: The Return of Race Science.* London: Beacon Press, 2019.
Salganik, Matthew J., Peter Sheridan Dodds, and Duncan J. Watts. "Experimental Study of Inequality and Unpredictability in an Artificial Cultural Market." *Science* 311, no. 5762 (February 10, 2006): 854–56.
Sanderson, Stephen K. *The Evolution of Human Sociality: A Darwinian Conflict Perspective.* Lanham: Rowman & Littlefield, 2001.
Saplakoglu, Yasemin. "Here's Why the New Coronavirus Is so Good at Infecting Human Cells." livescience.com, April 3, 2020. https://www.livescience.com/why-coronavirus-attaches-stronger-human-cells.html.
Sauer, Hanno. *Moral: Die Erfindung von Gut und Böse.* Munich: Piper, 2023.
Scarre, Geoffrey. *Witchcraft and Magic in Sixteenth and Seventeenth Century Europe.* Studies in European History. Basingstoke: Macmillan Education, 1987.
Scheidel, Walter. "Evolutionary Psychology and the Historian." *The American Historical Review* 119, no. 5 (2014): 1563–75.
Schilling, Heinz. *Aufbruch und Krise: Deutschland 1517–1648.* Siedler Deutsche Geschichte. Berlin: Goldmann Verlag, 1998a.
———. *Die neue Zeit: Vom Christenheitseuropa zum Europa der Staaten.* Siedler Geschichte Europas. Berlin: Siedler, 1998b.
Schmidt, Jürgen Michael. "Ein Politisches Ausrottungsprogramm? Kurpfalz, Kurmainz, St. Alban Und Die Große Hexenverfolgung in Bodenheim 1612–1615." In *Hexenverfolgung Und Herrschaftspraxis*, by Rita Voltmer, 147–82. Trier: Paulinus Verlag, 2005.
———. *Glaube und Skepsis: Die Kurpfalz und die abendländische Hexenverfolgung 1446–1685.* Bielefeld: Verlag für Regionalgeschichte, 2000.
Schmitt, D. P., and J. J. Pilcher. "Evaluating Evidence of Psychological Adaptation: How Do We Know One When We See One?" *Psychological Science* 15, no. 10 (2004): 643–49.
Schormann, Gerhard Albert. *Der Krieg gegen die Hexen: das Ausrottungsprogramm des Kurfürsten von Köln.* Sammlung Vandenhoeck. Göttingen: Vandenhoeck & Ruprecht, 1991.
———. *Hexenprozesse in Deutschland.* Göttingen: Vandenhoeck und Ruprecht, 1981.
Schreier, Herb, Miriam Rosenthal, Reed Pyeritz, Larry Miller, Chuck Madansky, Richard C. Lewontin, Anthony Leeds, et al. "Against 'Sociobiology.'" *The*

New York Review of Books, November 13, 1975. https://www.nybooks.com/articles/1975/11/13/against-sociobiology/.

Schultheis, Hinrichen. *Eine Außführliche Instruction Wie in Inquisition Sachen des grewlichen Lasters der Zauberey gegen Die Zaubere der Göttlichen Majestät und der Christenheit Feinde ohn gefahr der Unschuldigen zu procediren*. Cologne: Hinrich Berchem, 1634.

Schwerhoff, Gerd. "Hexenverfolgungen in der frühneizeitlichen Germania Sacra." In *Weltliche Herrschaft in Geistlicher Hand: Die Germania Sacra im 17. und 18. Jahrhundert*, by Hedwig Röckelein and Dietmar Schiersner, 297–324, De Gruyter Akademie Forschung. Berlin, 2018.

———. "Hexerei, Geslecht und Regionalgeschichte: Überlegungen zur Erklärung des scheinbar selbstverständlichen." In *Hexenverfolgung und Regionalgeschichte: Die Grafschaft Lippe im Vergleich*, by Gisela Wilbertz, Gerd Schwerhoff, and Jürgen Scheffler, 325–55. Bielefeld: Verlag für Regionalgeschichte, 1994.

———. "Rationalität im Wahn: zum gelehrten Diskurs über die Hexen in der frühen Neuzeit." *Saeculum* 37, no. 1 (1986): 45–82.

Schwillus, Harald. *Kleriker Im Hexenprozess: Geistliche Als Opfer Der Hexenprozesse Des 16. Und 17. Jahrhunderts in Deutschland*. Würzburg: Echter, 1992.

Scott, Joan W. "Gender: A Useful Category of Historical Analysis." *The American Historical Review* 91, no. 5 (1986): 1053–75.

Searle, John R. "Rationality and Realism, What Is at Stake?" *Daedalus* 122, no. 4 (1993): 55–83.

Segerstråle, Ullica. *Defenders of the Truth: The Sociobiology Debate*. Oxford: Oxford University Press, 2002.

Seitz, Jonathan. *Witchcraft and Inquisition in Early Modern Venice*. Cambridge: Cambridge University Press, 2011.

Sharpe, James. *Witchcraft in Early Modern England*. Harlow: Longman, 2001.

Shoshani, Jeheskel. "Understanding Proboscidean Evolution: A Formidable Task." *Trends in Ecology & Evolution* 13, no. 12 (December 1, 1998): 480–87.

Sigwart, Johann Georg. *Ein Predigt Vom Reiffen und Gefröst*. Tübingen: Erhardo Cellio, 1602.

Sil, Rudra, and Peter J. Katzenstein. "Analytic Eclecticism in the Study of World Politics: Reconfiguring Problems and Mechanisms across Research Traditions." *Perspectives on Politics* 8, no. 2 (June 2010): 411–31.

Singh, Manvir. "Magic, Explanations, and Evil: On the Origins and Design of Witches and Sorcerers," August 6, 2018.

Slingerland, Edward. *What Science Offers the Humanities: Integrating Body and Culture*. Cambridge: Cambridge University Press, 2008.

Slingerland, Edward, and Brenton Sullivan. "Durkheim with Data: The Database of Religious History." *Journal of the American Academy of Religion* 85, no. 2 (2017): 312–47.

Smail, Daniel Lord. *On Deep History and the Brain*. Chicago: University of California Press, 2008.

Smith, John Maynard, and Eors Szathmary. *The Major Transitions in Evolution*. Oxford: Oxford University Press, 1997.

Snowden, Frank M. *Epidemics and Society: From the Black Death to the Present*. 1st edition. New Haven, CT: Yale University Press, 2019.

Sobo, Elisa Janine, and Sandra Bell. *Celibacy, Culture, and Society: The Anthropology of Sexual Abstinence*. Madison: University of Wisconsin Press, 2001.

Sokal, Alan D. *Beyond the Hoax: Science, Philosophy and Culture*. Oxford: Oxford University Press, 2010.

———. "Pourquoi j'ai écrit ma parodie." *Le Monde*, January 31, 1997. https://www.lemonde.fr/archives/article/1997/01/31/pourquoi-j-ai-ecrit-ma-parodie_3738447_1819218.html.

Sokal, Alan D., and Jean Bricmont. *Intellectual Impostures: Postmodern Philosophers' Abuse of Science*. London: Profile Books, 1998.

Soldan, Wilhelm Gottlieb, and Heinrich Heppe. *Geschichte der Hexenprozesse*. Stuttgart: Cotta, 1880.

Sontag, Susan. *Illness as Metaphor*. Harmondsworth: Penguin Books, 1977.

Sosis, Richard, and Anders K. Petersen et al. "Why Cultural Evolutionary Models of Religion Need a Systemic Approach." In *Evolution, Cognition, and the History of Religion: A New Synthesis*, 45–61. Leiden: Brill, 2018.

Spee, Friedrich von. *Cautio criminalis oder rechtliches Bedenken wegen der Hexenprozesse*. Deutscher Taschenbuch Verlag, 2000.

———. *Cautio Criminalis. Seu De Processibus Contra Sagas Liber. Das ist/ Peinliche Warschawung von Anstell: und Führung deß Processes gegen die angegebene Zauberer/ Hexen und Unholden*. Frankfurt am Main: Weiss, 1649.

Spee von Langenfeld, Friedrich. *Cautio criminalis*. Deutsche Barock-Literatur, Bd. 3. Tübingen: Francke, 1992.

Spier, Fred. *Big History and the Future of Humanity*. Chichester; Malden: Wiley-Blackwell, 2010.

Spinks, Jennifer, and Charles Zika. *Disaster, Death and the Emotions in the Shadow of the Apocalypse, 1400–1700*. London: Palgrave Macmillan, 2016.

Stamenkovic, Philippe. "The Contradictions and Dangers of Bruno Latour's Conception of Climate Science." *Disputatio. Philosophical Research Bulletin* 9, no. 13 (November 24, 2019).

Stark, Rodney. "Micro Foundations of Religion: A Revised Theory*:" *Sociological Theory* 17, no. 3 (June 23, 1999): 264–89.

Stein, Rebecca L., and Philip L. Stein. *The Anthropology of Religion, Magic, and Witchcraft*. 3rd ed. London; New York: Routledge, 2011.

Stephens, Walter. *Demon Lovers: Witchcraft, Sex, and the Crisis of Belief*. Chicago: The University of Chicago Press, 2002.

———. "The Skeptical Tradition." In *The Oxford Handbook of Witchcraft in Early Modern Europe and Colonial America*, by Brian P. Levack, 101–21. Oxford: Oxford University Press, 2013.

Sterelny, Kim, and Paul E. Griffiths. *Sex and Death: An Introduction to Philosophy of Biology*. Chicago: University of Chicago Press, 2012.

Stewart-Williams, Steve. *The Ape That Understood the Universe: How the Mind and Culture Evolve*. Cambridge: Cambridge University Press, 2018.

Stickler, Andrea. *Eine Stadt im Hexenfieber: aus dem Tagebuch des Zeiler Burgermeisters Johann Langhans (1611–1628)*. Pfaffenweiler: Centaurus, 1994.

Stokes, Laura Patricia. *Demons of Urban Reform: Early European Witch Trials and Criminal Justice, 1430–1530*. 1st ed. 2011 edition. Basingstoke: Palgrave Macmillan, 2011.

Stone, Lawrence. "The Revival of Narrative: Reflections on a New Old History." *Past & Present* 85, no. 1 (November 1, 1979): 3–24.

Strassmann, Joan E., and David C. Queller. "Cooperation and Conflict: Microbes to Humans." In *The Princeton Huide to Evolution*, by Jonathan B. Losos, 671–76. Princeton NJ: Princeton University Press, 2014.

Tanesini, Alessandra. "Epistemic Relativism." In *Routledge Encyclopedia of Philosophy*, 1st ed. London: Routledge, 2016. https://doi.org/10.4324/0123456789-P016-2.

Tehrani J.J. "The Phylogeny of Little Red Riding Hood." *PLOS ONE* 8, no. 11 (2013).

Theunissen, Bert. "Sociaal-darwinisme. Enkele kanttekeningen bij een meerduidig begrip." *Amsterdams Sociologisch Tijdschrift* 27, no. 1–2 (July 1, 2000): 96–110.

Thomas, Keith. *Religion and the Decline of Magic: Studies in Popular Beliefs in Sixteenth- and Seventeenth-Century England*. Harmondsworth: Penguin Books, 1971.

Thompson, Willie. *Postmodernism and History*. Basingstoke: Palgrave Macmillan, 2004.

Thurston, Robert W. *The Witch Hunts: A History of the Witch Persecutions in Europe and North America*. Harlow, England; New York: Pearson Longman, 2007.

Tollebeek, Jo. "Het Duitse debat: geschiedenis rond 1900." In *Geschiedschrijving in de twintigste eeuw: discussie zonder eind*, by Herman Beliën and Gert Jan van Setten, 15–40. Amsterdam: Agon BV, 1991.

Tooby, John, and Leda Cosmides. "The Psychological Foundations of Culture." In *The Adapted Mind: Evolutionary Psychology and the Generation of Culture*, by Jerome H. Barkow, Leda Cosmides, and John Tooby, 19–136. Oxford: Oxford University Press, 1992.

Trethowan, W. H. "The Demonopathology of Impotence." *The British Journal of Psychiatry* 109, no. 460 (May 1, 1963): 341–47.

Trevor-Roper, Hugh Redwald. *The European Witch-Craze of the 16th and 17th Centuries.* Harmondsworth, Middlesex: Penguin Books, 1969.

Tschaikner, Manfred. "Nutzung oder Instrumentalisierung? Hexenverfolgung und Herrschaftspraxis in Vorarlberg, Liechtenstein und der Stadt St. Gallen." In *Hexenverfolgung und Herrschaftspraxis*, by Rita Voltmer, 95–112. Trier: Paulinus Verlag, 2005.

Tuczay, Christa. "The Nineteenth Century: Medievalism and Witchcraft." In *Palgrave Advances in Witchcraft Historiography*, by Jonathan Barry and Owen Davies, 52–68. New York: Palgrave Macmillan, 2007.

Turchin, Peter. *Ultrasociety: How 10,000 Years of War Made Humans the Greatest Cooperators on Earth.* Chaplin, Connecticut: Beresta Books, 2015.

Turner, Jonathan H. "Conflict Theory." In *Encyclopedia of Social Theory*, by George Ritzer, 134–39. London: Sage Publications, 2005.

Tylor, Edward Burnett. *Primitive Culture 1. 1.* London: Murray, 1871.

Valente, Michaela. "'Against the Devil, the Subtle and Cunning Enemy': Johann Wier's 'De Praestigiis Daemonum.'" In *The Science of Demons: Early Modern Authors Facing Witchcraft and the Devil*, by Jan Machielsen, 103–18. London: Routlegde, 2020.

Vernon, Richard. "Unintended Consequences." *Political Theory* 7, no. 1 (1979): 57–73.

Verschuer, Nynke van. "'Hoop is nu een laffe uitvlucht.'" NRC, November 24, 2017. https://www.nrc.nl/nieuws/2017/11/24/hoop-is-nu-een-laffe-uitvl ucht-14191677-a1582405.

Voltaire. *Oeuvres completes de Voltaire.* De L'Imprimerie de la Société Littéraire-Typographique, 1784.

Voltmer, Rita. "Claudius Musiel oder die Karriere eines Hexenrichters. Auch ein Beitrag zur Trierer Sozialgeschichte des späten 16. Jahrhunderts." In *Methoden und Konzepten der historischen Hexenforschung*, by Günther Franz and Franz Irsigler, 211–54. Trier: Spee Buchverlag, 1998.

———. "Debating the Devil's Clergy. Demonology and the Media in Dialogue with Trials (14th to 17th Century)." *Religions* 10, no. 12 (December 2019a): 648. https://doi.org/manvir.

———. "Demonology and Anti-Demonology: Binsfeld's 'De Confessionibus' and Loos's 'De Vera et False Magia.'" In *The Science of Demons: Early Modern Authors Acing Witchcraft and the Devil*, by Jan Machielsen, 149–64. London: Routledge, 2020a.

———. "Die Hexenverfolgungen im Raum des Erzbistums Trier (15.-17. Jahrhundert)—Strukturen und Deutungen." In *Kirchenreform und Konfessionsstaat 1500–1801*, edited by Bernhard Schneider, 709–49. Geschichte des Bistums Trier 3. Trier: Paulinus Verlag GmbH, 2010a.

———. "Die politischen Funktionen der frühneuzeitlichen Hexenverfolgungen: Machtdemonstration, Kontrolle und Herrschatsverdichtung im Rhein-Maas-

Raum." In *Chasses aux sorcières et démonologie: entre discours et pratiques (XIVe-XVIIe siècles)*, by Martine Ostorero, Georg Modestin, and Kathrin Utz Tremp, 89–116. Florence: SISMEL—Edizione del Galluzzo, 2010b.

———. "'Germany's First "Superhunt"?'. Rezeption Und Konstruktion Der so Genannten Trierer Verfolgungen (16.-21. Jahrhundert." In *Realität Und Mythos: Hexenverfolgung Und Rezeptionsgeschichte*, by Katrin Moeller and Burghart Schmidt, 225–59. Hamburg: Dobu Verlag, 2003.

———. "'Gott ist tot, und der Teufel jetzt Meister!' Hexenverfolgungen und dörfliche Krisen im Trierer Land des 16. und 17. Jahrhunderts." *Kurtrierisches Jahrbuch*, no. 39 (1999): 175–223.

———. "Herren und Hexen. Adlige Hexenpolitik in der Nordeifel und in angrenzenden Gebieten." In *Herren und Hexen in der Nordeifel: Darstellung—Edition—Vergleiche*, 3–154. Geschichte im Kreis Euskirchen 30. Weilerwist: Verlag Ralf Liebe, 2018.

———. *Hexen. Wissen was stimmt.* Freiburg im Breisgau: Verlag Herder, 2008.

———. "Hexenbrenner Und Hexenbischöfe—Zur (De)Konstruktion Eines Verfolgungsparadigma (16.-21. Jahrhundert)." In *Ideal Und Praxis—Bischöfe Und Bischofsamt Im Heiligen Römischen Reich 1570–1620*, by Peter Walter, 165–253. Münster: Aschendorff Verlag, 2020b.

———. "Hexenverfolgung und Herrschaftspraxis. Einführung und Ergebniss." In *Hexenverfolgung und Herrschaftspraxis*, by Rita Voltmer, 1–22. Trierer Hexenprozesse: Quellen und Darstellungen, Band 7. Trier: Paulinus Verlag GmbH, 2005.

———. "'Hört an neu schrecklich abentheuer / von den unholden ungeheuer'—Zur multimedialen Vermittlung des Fahndungsbildes 'Hexerei' im Kontext konfessioneller Polemik." In *Repräsentationen von Kriminalität und öffentlicher Sicherheit*, by Michael Stolleis and Karl Härter, 89–165. Frankfurt am Main: Vittorio Klostermann, 2010c.

———. "Jesuiten und Kinderhexen: Thesen zur Entstehung, Rezeption und Verbreitung eines Verfolgungsmusters." In *Hexenkinder—Kinderbanden—Straßenkinder*, by Wolfgang Behringer and Claudia Opitz-Belakhal, 201–32. Bielefeld: Verlag für Regionalgeschichte, 2016.

———. "'Krieg, uffrohr und teufelgespenst'- Das erzbistum Trier und seine Bevölkerung während der Frühen Neuzeit." In *Kirchenreform und Konfessionsstaat 1500–1801*, edited by Bernhard Schneider, 20–37. Geschichte des Bistums Trier 3. Trier: Paulinus Verlag GmbH, 2010d.

———. "Preaching on Witchcraft? The Sermons of Johannes Geiler of Kaysersberg." In *Contesting Orthodoxy in Medieval and Early Modern Europe: Heresy, Magic and Witchcraft*, by Lousie Nyholm Kallestrup and Raisa Maria Toivo, 193–215. New York: Palgrave Macmillan, 2017.

———. "Schultheiss, Heinrich von (ca 1580-ca 1646)." In *Encyclopedia of Witchcraft: The Western Tradition*, IV:1012–13. Santa Barbara: ABC-CLIO, 2006.

———. "Von der besonderen Alchimie, aus Menschenblut Gold zu machen, oder von den Möglichkeiten Hexereiverdacht und Hexenprozesse zu instrumentalisieren." In *Hexenwahn: Ängste der Neuzeit*, by Rosmarie Beier-de Haan, Rita Voltmer, and Franz Irsigler, 130–53. Berlin: Deutsches Historisches Museum, 2002.

———. "Witchcraft in the City. Patterns of Persecution in the Holy Roman Empire." In *La Sorcellerie et La Ville*, by Antoine Follain and Maryse Simon, 149–74. Sciences de l'histoire. Strasbourg: Presses universitaires de Strasbourg, 2019b.

———. "Zwischen Herrschaftskrise, Wirtschaftsdepression und Jesuitenpropaganda: Hexenverfolgungen in der Stadt Trier (15.-17. Jahrhundert)." *Jahrbuch für westdeutsche Landesgeschichte* 27 (2001): 37–107.

Voltmer, Rita, and Karl Weisenstein. *Das Hexenregister des Claudius Musiel: ein Verzeichnis von hingerichteten und besagten Personen aus dem Trierer Land (1586–1594)*. Trier: Spee Buchverlag, 1996.

Vries, Gerard de. *Bruno Latour*. Key Contemporary Thinkers. Cambridge, UK; Malden, MA: Polity Press, 2016.

Vries, Jan de. "Changing the Narrative: The New History That Was and Is to Come." *The Journal of Interdisciplinary History* 48, no. 3 (November 1, 2017): 313–34.

Vrieze, Jop de. "Bruno Latour, a Veteran of the 'Science Wars,' Has a New Mission." Science | AAAS, October 6, 2017. http://www.sciencemag.org/news/2017/10/bruno-latour-veteran-science-wars-has-new-mission.

Waardt, Hans de. *Toverij en samenleving: Holland, 1500–1800*. Den Haag: Stichting Hollandse Historische Reeks, 1991.

Wade, Nicholas. *A Troublesome Inheritance*. Penguin Group USA, 2015.

Wagner, Andreas. *Arrival of the Fittest: Solving Evolution's Greatest Puzzle*. New York: Penguin, 2014.

Waite, Gary K. *Eradicating the Devil's Minions: Anabaptists and Witches in Reformation Europe, 1525–1600*. Toronto: University of Toronto Press, 2012.

———. "Sixteenth-Century Religious Reform and the Witch-Hunts." In *The Oxford Handbook of Witchcraft in Early Modern Europe and Colonial America*, by Brian P. Levack, 485–506. Oxford: Oxford University Press, 2013.

Walinski-Kiehl, Robert. "Males, 'Masculine Honor', and Witch Hunting in Seventeenth-Century Germany." *Men and Masculinities* 6, no. 3 (2004): 254–71.

———. "The Devil's Children: Child Witch-Trials in Early Modern Germany." *Continuity and Change* 11, no. 2 (August 1996): 171–89.

———. "Witch-Hunting and State Building in the Bishoprics of Bamberg and Würzburg: 1570–1630." In *Hexenprozess Und Staatsbilding*, by Johannes Dillinger, Jürgen Michael Schmidt, and Dieter R Bauer, 245–64. Hexenforschung, Band 12. Bielefeld: Verlag für Regionalgeschichte, 2008.

Wallace, Ruth A., and Alison Wolf. *Contemporary Sociological Theory: Expanding the Classical Tradition*. Upper Saddle River, NJ: Pearson: Prentice Hall, 2005.

Walz, Rainer. *Hexenglaube und magische Kommunikation im Dorf der frühen Neuzeit: die Verfolgungen in der Grafschaft Lippe*. Paderborn: Schöningh, 1993.

———. "Kinder in Hexenprozessen: Die Grafschaft lippe 1654–1663." In *Hexenverfolgung und Regionalgeschichte: Die Grafschaft Lippe im Vergleich*, by Gisela Wilbertz, Gerd Schwerhoff, and Jürgen Scheffler, 211–32. Bielefeld: Verlag für Regionalgeschichte, 1994.

Warachtighe ende verschrickelycke beschryuinge van vele toouenaers ende toouenerssen oft vandoysen, hoe ende waerom men die herwaerts ende ghentswaerts verbrandt heeft in dit teghenwoordich jaar 1589. Antwerpen: Jan van Ghelen, 1589.

Ward, Peter D, and Joseph L Kirschvink. *A New History of Life: The Radical New Discoveries about the Origins and Evolution of Life on Earth*. New York: Bloomsbury, 2015.

Warren, Zachary J, and Séamus A Power. "It's Contagious: Rethinking a Metaphor Dialogically." *Culture & Psychology* 21, no. 3 (September 1, 2015): 359–79.

Watson, Robert. *Cultural Evolution and Its Discontents: Cognitive Overload, Parasitic Cultures, and the Humanistic Cure*. 1st edition. London: Routledge, 2018.

Weber, Max. *Wirtschaft Und Gesellschaft: Grundriss Der Verstehenden Soziologie*. Tübingen: Mohr Siebeck, 2002.

Weikart, Richard. *Hitler's Ethic: The Nazi Pursuit of Evolutionary Progress*. Basingstoke: Palgrave Macmillan, 2009.

Weiner, Irving B. *Handbook of Psychology, History of Psychology*. Hoboken, New Jersey: John Wiley & Sons, 2003.

Wekhrlin, Wilhelm Ludwig. *Chronologen. Ein periodisches Werk: Zehnter Band*. In der Felßeckecischen Buchhandlung, 1782.

Weyer, Johann. *Witches, Devils, and Doctors in the Renaissance*. Translated by George Mora and Benjamin G Kohl. Tempe, Arizona: Medieval & Renaissance Texts & Studies, 1998.

White, Andrew Dickson. *A History of the Warfare of Science with Theology in Christendom*. New York: D. Appleton & Company, 1896.

White, Hayden. "Afterword." In *Beyond the Cultural Turn: New Directions in the Study of Society and Culture*, edited by Victoria E. Bonnel, Lynn Hunt, and Richard Biernacki, 315–24. Berkeley: University of California Press, 1999.

———. *Metahistory: The Historical Imagination in Nineteenth-Century Europe*. Baltimore: The John Hopkins University Press, 1973.

White, Leslie A. *The Evolution of Culture: The Development of Civilization to the Fall of Rome*. New York: McGraw-Hill, 1959.
Whitehouse, Harvey. "Dying for the Group: Towards a General Theory of Extreme Self-Sacrifice." *Behavioral and Brain Sciences* 41 (ed 2018a).
———. "Why Is a Science of the Sociocultural so Difficult?" In *Evolution, Cognition, and the History of Religion: A New Synthesis*, by Anders K. Petersen et al., 413–26. Leiden: Brill, 2018b.
Whitney, Elspeth. "The Witch 'She'/the Historian 'He': Gender and the Historiography of the European Witch-Hunts." *Journal of Women's History* 7, no. 3 (1995): 77–101.
Wick, Johann Jakob. *Die Wickiana: Johann Jakob Wicks Nachrichtensammlung aus dem 16. Jahrhundert*. Edited by Matthias Senn. Zurich: Raggi-Verlag, 1975.
Wiersing, Erhard. *Geschichte des historischen Denkens: zugleich eine Einführung in die Theorie der Geschichte*. Paderborn: Schöningh, 2007.
Wiesner-Hanks, Merry E. *Women and Gender in Early Modern Europe*. 3rd ed. New Approaches to European History. Cambridge: Cambridge University Press, 2008.
Wilson, David Sloan. *Darwin's Cathedral: Evolution, Religion, and the Nature of Society*. Chicago: University of Chicago Press, 2002.
———. *Darwin's Cathedral: Evolution, Religion, and the Nature of Society*. University of Chicago Press, 2010.
———. *Evolution for Everyone: How Darwin's Theory of Evolution Can Change the Way We Think about Our Lives*. New York: Delta Science, 2008.
———. *This View of Life: Completing the Darwinian Revolution*. New York: Pantheon Books, 2019.
Wilson, Edward O. *Consilience: The Unity of Knowledge*. London: Little, Brown and Company, 1998.
———. "For Sociobiology." *The New York Review of Books*, December 11, 1975a. https://www.nybooks.com/articles/1975/12/11/for-sociobiology/.
———. *Naturalist*. Washington, D.C.: Island Press/Shearwater Books, 2006.
———. *On Human Nature*. Cambridge, Mass.: Harvard University Press, 1978.
———. *Sociobiology: The New Synthesis*. Cambridge, Mass.: Harvard University Press, 1975b.
———. *The Diversity of Life*. New York: W.W. Norton, 1993.
———. *The Social Conquest of Earth*. New York; London: Liveright, 2012.
Wilson, Liz. "Starvation and Self-Mutilation in Religious Traditions." In *The Oxford Handbook of Religion and Violence*, by Mark Juergensmeyer, Margo Kitts, and Micheal Jerryson, 242–49. Oxford: Oxford University Press, 2013.
Wilson, Stephen. *The Magical Universe: Everyday Ritual and Magic in Pre-Modern Europe*. London: Hambledon and London, 2000.

Wilterdink, Nico. "Darwinisme in de Sociale Wetenschappen." *De Gids*, no. 172 (2009): 444–54.

———. *Omstreden Wetenschap: Goede En Slechte Sociologie*. Amsterdam: Vossiuspers UvA, 2011. https://dare.uva.nl/search?identifier=9233ab5a-1526-4115-8e5e-4d6dc03dbfb7.

———. "Te veel darwinisme, te weinig sociologie." *Sociologie* 2, no. 3 (2006): 302–10.

———. "The Concept of Social Evolution: Its Meanings and Uses." In *Soziale Evolution: Die Evolutionstheorie Und Die Sozialwissenschaften*, by Tamas Meleghy and Heinz-Jürgen Niedenzu, 53–73. Wiesbaden: Springer Fachmedien, 2003.

———. "The Sociogenesis of Postmodernism." *European Journal of Sociology / Archives Européennes de Sociologie / Europäisches Archiv Für Soziologie* 43, no. 2 (2002): 190–216.

Windelband, Wilhelm. *Geschichte Und Naturwissenschaft: Rede Zum Antritt Des Rectorats Der Kaiser-Wilhelm-Universität Strassburg*. Heitz und Mundel, 1894.

Witekind, Hermann. *Christlich Bedencken und Erinnerung von Zauberey*, 1586.

Woolfson, Jonathan, ed. *Palgrave Advances in Renaissance Historiography*. Houndmills, Basingstoke: Palgrave Macmillan, 2005.

Wrangham, Richard W. *Catching Fire: How Cooking Made Us Human*. New York: Basic Books, 2010.

Wunder, Heide. *"Er ist die Sonn', sie ist der Mond": Frauen in der Frühen Neuzeit*. München: C.H. Beck, 1992.

Wynn, Melissa L., Christofer Clemente, Ami Fadhillah Amir Abdul Nasir, and Robbie S. Wilson. "Running Faster Causes Disaster: Trade-Offs between Speed, Manoeuvrability and Motor Control When Running around Corners in Northern Quolls (Dasyurus Hallucatus)." *Journal of Experimental Biology* 218, no. 3 (February 1, 2015): 433–39.

Yong, Ed. *I Contain Multitudes: The Microbes within Us and a Grander View of Life*. London: The Bodley Head, 2016.

Zeck, Mario. *"Im Rauch gehn Himmel geschüggt": Hexenverfolgungen in der Reichstadt Rottweil*. Stuttgart: Ibidem-Verlag, 2000.

Zenz, Emil. *Die Taten der Trierer: Von Johann VII. bis zum Tode Karl Kaspers von der Leyen 1581–1676*. Vol. VII. Trier: Paulinus Verlag, 1964.

———. *Ein Opfer des Hexenwahns: das Schicksal des Doctor Dietrich Flade aus Trier*. Trier: Spee-Verlag, 1977.

Zika, Charles. *The Appearance of Witchcraft: Print and Visual Culture in Sixteenth-Century Europe*. London: Routledge, 2007.

Zimmer, Carl. *A Planet of Viruses*. Chicago: The University of Chicago Press, 2011.

———. *Evolution: The Triumph of an Idea*. London: Arrow, 2003.

Zwo erschreckliche und unerhörte Geschicht, welches in diesem XCVI Jar geschehen ist auff dem Brockersberg. Cologne: Betram Bückholtz, 1596.

Zwo Hexenzeitung, Die Erste, Von Dreyen Hexen Pfaffen, Und Einem Organisten Zu Ellwang [...] Die Ander: Von Einer Unholdin Oder Hexen [...]. Nürnberg, 1615.

Zwo Warhafftige, und doch Männiglich zuvor bekante Newe Zeitung. Die Erste, Von dem grossen Jammer, welcher sich begeben in der Marggraffschafft Baden. Die Ander, Aus der Elsasserland. Moltzen, 1626.

Index[1]

[1] Note: Page numbers followed by 'n' refer to notes.

S. Hofhuis, *The Evolutionary History of Witch-hunting*, https://doi.org/10.1007/978-3-031-66936-1

GPSR Compliance

The European Union's (EU) General Product Safety Regulation (GPSR) is a set of rules that requires consumer products to be safe and our obligations to ensure this.

If you have any concerns about our products, you can contact us on ProductSafety@springernature.com

In case Publisher is established outside the EU, the EU authorized representative is:

Springer Nature Customer Service Center GmbH
Europaplatz 3
69115 Heidelberg, Germany

The manufacturer's authorised representative in the EU is Springer Nature Customer Service Centre GmbH, Europaplatz 3, 69115 Heidelberg, Germany. If you have any concerns regarding our products, please contact ProductSafety@springernature.com

Printed and bound by CPI Group (UK) Ltd, Croydon, CR0 4YY

01/10/2026

02225903-0015